T0191891

Fundamentals of IP and SoC Security

Swarup Bhunia · Sandip Ray
Susmita Sur-Kolay
Editors

Fundamentals of IP and SoC Security

Design, Verification, and Debug

Springer

Editors
Swarup Bhunia
Department of Electrical and Computer Engineering
University of Florida
Gainesville, FL
USA

Susmita Sur-Kolay
Advanced Computing and Microelectronics Unit
Indian Statistical Institute
Kolkata
India

Sandip Ray
NXP Semiconductors
Austin, TX
USA

ISBN 978-3-319-84308-7 ISBN 978-3-319-50057-7 (eBook)
DOI 10.1007/978-3-319-50057-7

Softcover reprint of the hardcover 1st edition 2017

Printed on acid-free paper

This Springer imprint is published by Springer Nature
The registered company is Springer International Publishing AG
The registered company address is: Gewerbestrasse 11, 6330 Cham, Switzerland

Contents

Chapter 1
The Landscape of SoC and IP Security

Sandip Ray, Susmita Sur-Kolay and Swarup Bhunia

1.1 Introduction

It has been almost a decade since the number of smart, connected computing devices has exceeded the human population, ushering in the regime of the Internet of things [1]. Today, we live in an environment containing tens of billions of computing devices of wide variety and form factors, performing a range of applications often including some of our most private and intimate data. These devices include smartphones, tablets, consumer items (e.g., refrigerators, light bulbs, and thermostats), wearables, etc. The trend is toward this proliferation to increase exponentially in the coming decades, with estimates going to trillions of devices as early as by 2030, signifying the fastest growth by a large measure across any industrial sector in the history of the human civilization.

Security and trustworthiness of computing systems constitute a critical and gating factor to the realization of this new regime. With computing devices being employed for a large number of highly personalized activities (e.g., shopping, banking, fitness tracking, providing driving directions, etc.), these devices have access to a large amount of sensitive, personal information which must be protected from unauthorized or malicious access. On the other hand, communication of this information to other peer devices, gateways, and datacenters is in fact crucial to providing the kind of adaptive, "smart" behavior that the user expects from the device. For example,

S. Ray (✉)
Strategic CAD Labs, Intel Corporation, Hillsboro, OR 97124, USA
e-mail: sandip.ray@intel.com

S. Sur-Kolay
Advanced Computing and Microelectronics Unit, Indian Statistical Institute, Kolkata 700108, India
e-mail: ssk@isical.ernet.in

S. Bhunia
Department of ECE, University of Florida, Gainesville, FL 32611, USA
e-mail: swarup@ece.ufl.edu

S. Bhunia et al. (eds.), *Fundamentals of IP and SoC Security*,
DOI 10.1007/978-3-319-50057-7_1

a smart fitness tracker must detect from its sensory data (e.g., pulse rate, location, speed, etc.) the kind of activity being performed, the terrain on which the activity is performed, and even the motivation for the activity in order to provide anticipated feedback and response to the user; this requires a high degree of data processing and analysis much of which is performed by datacenters or even gateways with higher computing power than the tracker device itself. The communication and processing of one's intimate personal information by the network and the cloud exposes the risk that it may be compromised by some malicious agent along the way. In addition to personalized information, computing devices contain highly confidential collateral from architecture, design, and manufacturing, such as cryptographic and digital rights management (DRM) keys, programmable fuses, on-chip debug instrumentation, defeature bits, etc. Malicious or unauthorized access to secure assets in a computing device can result in identity thefts, leakage of company trade secrets, even loss of human life. Consequently, a crucial component of a modern computing system architecture includes authentication mechanisms to protect these assets.

1.2 SoC Design Supply Chain and Security Assets

Most computing systems are developed today using the system-on-chip (SoC) design architecture. An SoC design is architected by a composition of a number of pre-designed hardware and software blocks, often referred to as *design intellectual properties* or design IPs (IPs for short). Figure 1.1 shows a simple toy SoC design, including some "obvious" IPs, e.g., CPU, memory controller, DRAM, various controllers for peripherals, etc. In general, an IP can refer to any design unit that can be viewed as a standalone sub-component of a complete system. An SoC design architecture then entails connecting these IPs together to implement the overall system functionality. To achieve this connection among IPs, an SoC design includes a network-on-chip (NoC) that provides a standardized message infrastructure for the IPs to coordinate and cooperate to define the complete system functionality. In industrial practice today, an SoC design is realized by procuring many third-party IPs. These IPs are then integrated and connected by the SoC design integration house which is responsible for the final system design. The design includes both hardware components (written in a hardware description language such as Verilog of VHDL language) as well as software and firmware components. The hardware design is sent to a foundry or fabrication house to create the silicon implementation. The fabricated design is transferred to platform developers or Original Equipment Manufacturers (OEMs), who create computing platforms such as a smartphone, tablet, or wearable devices, which are shipped to the end customer.

The description above already points to a key aspect of complexity in SoC design fabrication, e.g., a complex supply chain and stake holders. This includes various IP providers, the SoC integration house, foundry, and the OEMs. Furthermore, with increasing globalization, this supply chain is typically long and globally distributed. Chapter 2 discusses some ramifications of this infrastructure, e.g., the possibility

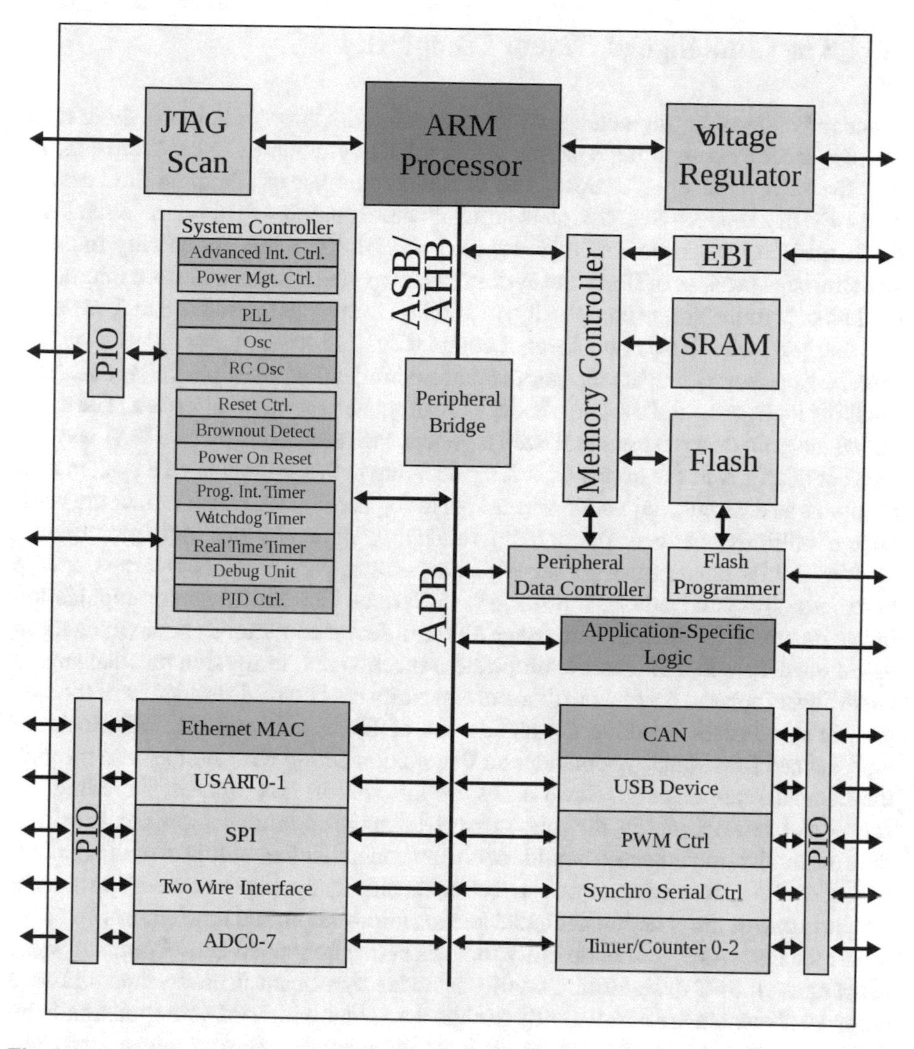

Fig. 1.1 A representative SoC design. SoC designs are created by putting together intellectual property (IP) blocks of well-defined functionality

of any component of the supply chain incorporating malicious or inadvertent vulnerability into the design or the manufacturing process. Malicious activities can include insertion of specific design alterations or Trojans by IP providers, leaking of a security asset by the SoC integration house, overproduction or counterfeiting by a malicious foundry, and even overlooked or apparently benign design errors or features that can be exploited on-field. Security architectures and assurance techniques and methodologies must be robust enough to address challenges arising from this plethora of sources, arising from different points of the system design life cycle.

1.3 The Challenge of Design Complexity

A second dimension of challenges with the secure SoC design is in the sheer complexity. Modern computing systems are inordinately complex. Note from Fig. 1.1 that the CPU represents "merely" one of a large number of IPs in an SoC design. The CPU in a modern SoC design is arguably more complex than many of the high-performance microprocessors of a decade back. Multiply this complexity increase with the large number of IPs in the system (many of which include custom microcontrollers of commensurate complexity, in addition to custom hardware and firmware), and one gets some sense of the level of complexity. Add some other cross-design features, e.g., power management, performance optimization, multiple voltage islands, clocking logic, etc., and the complexity perhaps goes beyond imagination. The number of different design states that such a system can reach exceeds by a long way the number of atoms in the universe. It is challenging to ensure that such a system ever functions as desired even under normal operating conditions, much less in the presence of millions of adversaries looking to identify vulnerabilities for exploitation.

Why is this complexity a bottleneck for security in particular? For starters, secure assets are sprinkled across the design, in various IPs and their communication infrastructure. It is difficult to envisage all the different conditions under which these assets are accessed and insert appropriate protection and mitigation mechanisms to ensure unauthorized access. Furthermore, security cross-cuts different IPs of the system, in some cases breaking the abstraction of IPs as coherent, distinct blocks of well-defined functionality. Consider an IP communicating with another one through the communication fabric. Several IPs are involved in this process, including the source and destination IPs, the routers involved in the communication, etc. Ensuring the communication is secure would require an understanding of this overall architecture, identifying trusted and untrusted components, analyzing the consequences of a Trojan in one of the constituent blocks leaking information, and much more. To exacerbate the issue, design functionality today is hardly contained entirely in hardware. Most modern SoC design functionality includes significant firmware and software components which are concurrently designed together with hardware (potentially by different players across the supply chain). Consequently, security design and validation become a complex hardware/software co-design and co-validation problem distributed across multiple players with potentially untrusted participants. Finally, the security requirements themselves vary depending on how an IP or even the SoC design is used in a specific product. For example, the same IP when used in a wearable device will have a different security requirement from when it is used as a gaming system. The security requirements also vary depending on the stage of the life cycle of the product, e.g., when it is with a manufacturer, OEM, or end customer. This makes it hard to compositionally design security features without a global view.

1.4 State of Security Design and Validation: Research and Practice

There has been significant research in recent years to address the challenges outlined above. There have been techniques to define security requirements [2, 3], architectures to facilitate such implementation [4–7], testing technologies to define and emulate security attacks [8], and tools to validate diverse protection and mitigation strategies [9–12]. There have been cross-cutting research too, on understanding trade-offs between security and functionality, energy requirements, validation, and architectural constraints [13, 14].

In spite of these advances, the state of the industrial practice is still quite primitive. We still depend on security architects, designers, and validators painstakingly mapping out various security requirements, architecting and designing various tailored and customized protection mechanisms, and coming up with attack scenarios to break the system by way of validation. There is a severe lack of disciplined methodology for developing security, in the same scale as there is methodology for defining and refining architectures and micro-architectures for system functionality or performance. Unsurprisingly, security vulnerabilities are abound in modern SoC designs, as evidenced by the frequency and ease in which activities like identity theft, DRM override, device jailbreaking, etc. are performed.

1.5 The Book

This book is an attempt to bridge across the research and practice in SoC security. It is conceived as an authoritative reference on all aspects of security issues in SoC designs. It discusses research issues and progresses in topics ranging from security requirements in SoC designs, definition of architectures and design choices to enforce and validate security policies, and trade-offs and conflicts involving security, functionality, and debug requirements, as well as experience reports from the trenches in design, implementation, and validation of security-critical embedded systems.

In addition to providing an extensive reference to the current state-of-the-art, the book is anticipated to serve as a conduit for communication between different stake holders of security in SoC designs. Security is one of the unique areas of SoC designs, which cross-cuts a variety of concerns, including architecture, design, implementation, validation, and software/hardware interfaces, in many cases with conflicting requirements from each domain. With a unified material documenting the various concerns side-by-side, we hope this book will help each stake holder better understand and appreciate the others points of view and ultimately foster an overall understanding of the trade-offs necessary to achieve truly secure systems.

The book includes eleven -chapters focusing on diverse aspects of system-level security in modern SoC designs. The book is intended for researchers, students, and practitioners interested in understanding the spectrum of challenges in architecture, validation, and debug of IP and SoC securities. It can serve as a material for a research course on the topic, or as a source of advanced research materials in a graduate class.

Chapter 2 focuses on SoC security validation. Validation is a critical and highly important component in SoC security. This chapter introduces the different security activities performed at different stages of the design life cycle, and the complex dependencies involved. It draws from examples in industrial practice, identifies the holes in current state of the practice, and discusses emergent research in industry and academia to plug these holes.

Chapter 3 discusses interoperability challenges to security. For a system to be usable, it is not sufficient for it to be secure but also to provide meaningful functionality. The trade-offs between security and functionality is an old one, stemming from the notion of availability itself as a security requirement in addition to confidentiality and integrity. Interoperability between security and functionality is a cornerstone for design and validation of modern computing systems. The twist in this tale is that validation itself introduces trade-offs and conflicts with security. In particular, post-silicon validation imposes observability and controllability requirements which can conflict with the security restrictions of the system. This chapter discusses the various ramifications of this conflict, and discusses the current state of the practice in its resolution.

Chapter 4 introduces the challenge of IP trust validation and assurance. Given the high proliferation of third-party IPs that are included in a modern SoC design as well as the complex global supply chain alluded to above, trustworthiness of the IP is a matter of crucial concern. This chapter enumerates the challenges with the current IP trust assurance, discusses protection technologies, and describes certification methods to validate trustworthiness of acquired third-party IPs.

Chapter 5 discusses challenges in developing trustworthy cryptographic implementations for modern SoC designs. Cryptographic protocols are crucial to various protecting a large number of system assets. Virtually, all SoC designs include a number of IP cores implementing such protocols, which form a critical component of the root-of-trust for the system. It is obviously crucial that the cryptographic implementations themselves are robust against adversarial attacks. Attacks on cryptographic implementations include both attacks on the protocols themselves as well as inference of implementation behavior and secure data (e.g., key) through side channel analysis. This chapter provides an overview of these attacks and discusses countermeasures.

Chapter 6 discusses a promising technology for secure authentication, namely Physical unclonable function (PUF). PUF is one of the key security primitives that provides source of randomness for cryptographic implementations (among others) in an SoC design. The idea is to exploit variations in the structural and electrical characteristics as a source of entropy unique to each individual integrated circuit. PUFs have become a topic of significant research over the last decade, with various novel PUF-based authentication protocols emerging in recent years. This chapter discusses this exciting and rapidly evolving area, compares PUF-based authentication with other standard approaches, and identifies several open research problems.

Chapter 7 discusses security with IP and SoC designs based on field-programmable gate arrays (FPGA). FPGAs have been the focus of attention because they permit dynamic reconfigurability while still providing the energy efficiency and performance compatible with a custom hardware implementation for many applications. Unfortunately, FPGA-based IP implementations induce a number of significant security challenges of their own. An FPGA-based IP is essentially a design implementation in a low-level hardware description language (also referred to as an FPGA bitstream) which is loaded on a generic FPGA architecture. To ensure authentication and prevent unauthorized access, the bitstream needs to be encrypted, and must thereafter be decrypted on-the-fly during load or update. However, bitstreams are often updated on field and the encryption may be attacked through side channel or other means. If the entire SoC is implemented in FPGA, IP management and coordination may become even more challenging. This chapter discusses the various facets of security techniques for IP and SOC FPGA, open problems, and areas of research.

Chapter 8 discusses PUFs and IP protection techniques. IP protection techniques are techniques to ensure robustness of IPs against various threats, including supply chain challenges, Trojan, and counterfeiting. The chapter provides a broad overview of the use of PUFs and IP protection techniques in modern SoC designs, and various conflicts, cooperations, and trade-offs involved.

Chapter 9 discusses SoC design techniques that are resistant to fault injection attacks. Fault injection attack is a complex, powerful, and versatile approach to subvert SoC design protections, particularly cryptographic implementations. The chapter provides an overview of fault injection attacks and describes broad class of design techniques to develop systems that are robust against such attacks.

Chapter 10 looks closely into one of the core problems of SoC security, e.g., hardware Trojans. With increasing globalization of the SoC design supply chain there has been increasing threat of such Trojans, i.e., hardware circuitry that may perform intentionally malicious activity including subverting communications, leaking secrets, etc. The chapter looks closely at various facets of this problem from IP security perspective, the countermeasures taken in the current state of practice, their deficiencies, and directions for research in this area.

Chapter 11 discusses logic obfuscation techniques, in particular for FPGA designs based on nonvolatile technologies. Logic obfuscation is an important technique to provide robustness of an IP against a large class of adversaries. It is particularly critical for FPGA-based designs which need to be updated on field. The chapter proposes a scheme for loading obfuscated configurations into nonvolatile memories to protect design data from physical attacks.

Chapter 12 presents a discussion on security standards in embedded SoCs used in diverse applications, including automotive systems. It notes that security is a critical consideration in the design cycle of every embedded SoC. But the level of security and resulting cost–benefit trade-off depend on a target application. This chapter provides a valuable industry perspective to this critical problem. It describes two layers of a hierarchical security model that a system designer typically uses to achieve application-specific security needs: (1) foundation level security targeted at basic security services, and (2) security protocols like TLS or SSL.

It has been a pleasure and honor for the editors to edit this material, and we hope the broad coverage of system-level security challenges provided here will bridge a key gap in our understanding of the current and emergent security challenges. We believe the content of the book will provide a valuable reference for SoC security issues and solutions to a diverse readership including students, researchers, and industry practitioners. Of course, it is impossible for any book on the topic to be exhaustive on this topic: it is too broad, too detailed, and touches too many areas of computer science and engineering. Nevertheless, we hope that the book will provide a flavor of the nature of the needed and current research in this area, and cross-cutting challenges across different areas that need to be done to achieve the goal of trustworthy computing systems.

References

1. Evans, D.: The Internet of Things—How the Next Evolution of the Internet is Changing Everything. White Paper, Cisco Internet Business Solutions Group (IBSG) (2011)
2. Li, X., Oberg, J.V.K., Tiwari, M., Rajarathinam, V., Kastner, R., Sherwood, T., Hardekopf, B., Chong, F.T.: Sapper: a language for hardware-level security policy enforcement. In: International Conference on Architectural Support for Programming Languages and Operating Systems (2014)
3. Srivatanakul, J., Clark, J.A., Polac, F.: Effective security requirements analysis: HAZOPs and use cases. In: 7th International Conference on Information Security, pp. 416–427 (2004)
4. ARM: Building a secure system using trustzone technology. ARM Limited (2009)
5. Basak, A., Bhunia, S., Ray, S.: A flexible architecture for systematic implementation of SoC security policies. In: Proceedings of the 34th International Conference on Computer-Aided Design (2015)
6. Intel: Intel® Software Guard Extensions Programming Reference. https://software.intel.com/sites/default/files/managed/48/88/329298-002.pdf
7. Samsung: Samsung KNOX. www.samsungknox.com
8. Microsoft Threat Modeling & Analysis Tool version 3.0 (2009)
9. JasperGold Security Path verification App. https://www.cadence.com/tools/system-design-and-verification/formal-and-static-verification/jasper-gold-verification-platform/security-path-verification-app.html
10. Bazhaniuk, O., Loucaides, J., Rosenbaum, L., Tuttle, M.R., Zimmer, V.: Excite: symbolic execution for BIOS security. In: Workshop on Offensive Technologies (2015)
11. Kannavara, R., Havlicek, C.J., Chen, B., Tuttle, M.R., Cong, K., Ray, S., Xie, F.: Challenges and opportunities with concolic testing. In: NAECON 2015 (2015)
12. Takanen, A., DeMott, J.D., Mille, C.: Fuzzing for software security testing and quality assurance. Artech House (2008)
13. Ray, S., Hoque, T., Basak, A., Bhunia, S.: The power play: trade-offs between energy and security in IoT. In: ICCD (2016)
14. Ray, S., Yang, J., Basak, A., Bhunia, S.: Correctness and security at odds: post-silicon validation of modern SoC designs. In: Proceedings of the 52nd Annual Design Automation Conference (2015)

Chapter 2
Security Validation in Modern SoC Designs

Sandip Ray, Swarup Bhunia and Prabhat Mishra

2.1 Security Needs in Modern SoC Designs

System-on-Chip (SoC) architecture pervades modern computing devices, being the prevalent design approach for devices in embedded, mobile, wearable, and Internet-of-Things (IoT) applications. Many of these devices have access to highly sensitive information or data (often collectively called "assets"), that must be protected against unauthorized or malicious access. The goal of SoC security *architecture* is to develop mechanisms to ensure this protection. The goal of SoC security *validation* is to ensure that such mechanisms indeed provide the protection needed. Clearly the two activities are closely inter-related in typical SoC security assurance methodologies. This chapter is about the security validation component, but we touch upon architectural issues as necessary.

To motivate the critical role of security validation activities, it is necessary to clarify (1) what kind of assets is being protected, and (2) what kind of attacks we are protecting against. One can get some flavor of the kind (and diversity) of assets by looking at the kind of activities we perform on a typical mobile system. Figure 2.1 tabulates some obvious end user usages of a standard smartphone and the kind of end user information accessed during these usages. Note that it includes such intimate information as our sleep pattern, health information, location, and finances. In addition to private end user information, there are other assets in a smartphone that may have been put by the manufacturers and OEMs, which *they* do not want to be leaked

S. Ray (✉)
Strategic CAD Labs, Intel Corporation, Hillsboro, OR 97124, USA
e-mail: sandip.ray@intel.com

S. Bhunia · P. Mishra
Department of ECE, University of Florida, Gainesville, FL 32611, USA
e-mail: swarup@ece.ufl.edu

P. Mishra
e-mail: prabhat@cise.ufl.edu

S. Bhunia et al. (eds.), *Fundamentals of IP and SoC Security*,
DOI 10.1007/978-3-319-50057-7_2

Usages	Assets Exposed
Browsing	Browsing history
Fitness tracking	Health information,sleep pattern
GPS	Location
Phone call	Contacts
Banking,Stock trading	Finances

Fig. 2.1 Some typical smartphone applications and corresponding private end user information

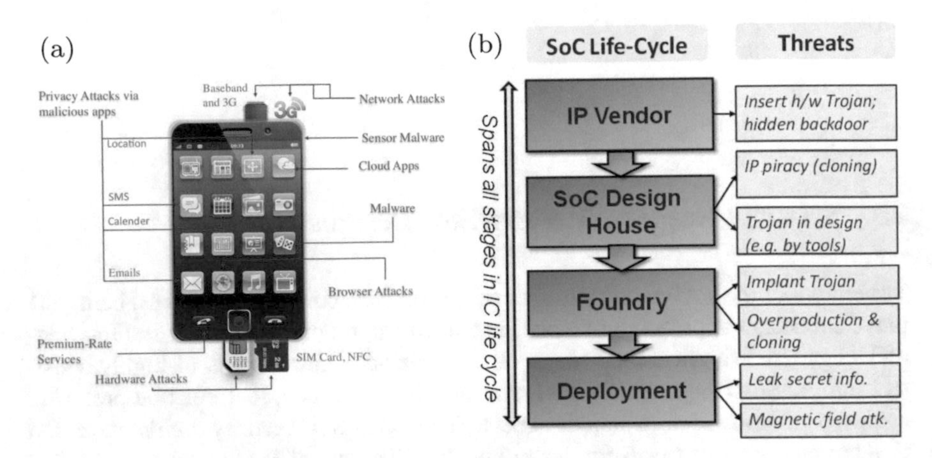

Fig. 2.2 Some potential attacks on a modern SoC design. **a** Potential attack areas for a smartphone after production and deployment. **b** Potential threats from untrusted supply chain during the design life cycle of an SoC design

out to unauthorized sources. This includes cryptographic and DRM keys, premium content locks, firmware execution flows, debug modes, etc. Note that the notion of "unauthorized source" changes based on what asset we are talking about: end user may be an unauthorized source for DRM keys while manufacturer/OEM may be an unauthorized source for end user private information.

In addition to criticality of the assets involved, another factor that makes SoC security both critical and challenging is the high diversity of attacks possible. Figure 2.2 provides a flavor of potential attacks on a modern SoC design. Of particular concern are the following two observations:

- Because of the untrusted nature of the supply chain, there are security threats at most stages of the design development, even before deployment and production.
- A deployed SoC design inside a computing device (e.g., smartphone) in the hand of the end user is prone to a large number of potential attacker entry points, including applications, software, and network, browser, and sensors. Security assurance must permit protection against this large attack surface.

We discuss security validation for the continuum of attacks from design to deployment. Given that the attacks are diverse, protection mechanisms are also varied, and each induces a significantly different validation challenge. However, validation *technology* is still quite limited. For most of the security requirements, we still very much depend on the perspicuity, talent, and experience of the human validators to identify potential vulnerabilities.

2.2 Supply Chain Security Threats

The life cycle of a SoC from concept to deployment involves number of security threats at all stages involving various parties. Figure 2.2b shows the SoC life cycle and the security threats that span the entire life cycle. These threats are increasing with the rapid globalization of the SoC design, fabrication, validation, and distribution steps, driven by the global economic trend.

This growing reliance on reusable pre-verified hardware IPs during SoC design, often gathered from untrusted third-party vendors, severely affects the security and trustworthiness of SoC computing platforms. Statistics show that the global market for third-party semiconductor IPs grew by more than 10 % to reach more than 2.1 billion in late 2012 [1]. The design, fabrication, and supply chain for these IP cores is generally distributed across the globe involving USA, Europe, and Asia. Figure 2.3 illustrates the scenario for an example SoC that includes processor, memory controllers, security, graphics, and analog core. Due to growing complexity of the IPs as well as the SoC integration process, SoC designers increasingly tend to treat these IPs as black box and rely on the IP vendors on the structural/functional integrity of these IPs. However, such design practices greatly increase the number of untrusted components in a SoC design and make the overall system security a pressing concern.

Hardware IPs acquired from untrusted third-party vendors can have diverse security and integrity issues. An adversary inside an IP design house involved in the IP design process can deliberately insert a malicious implant or design modification to incorporate hidden/undesired functionality. In addition, since many of the IP providers are small vendors working under highly aggressive schedules, it is difficult to ensure a stringent IP validation requirement in this ecosystem. Design features may also introduce unintentional vulnerabilities, e.g., intentional information leakage through hidden test/debug interfaces or side-channels through power/performance profiles. Similarly, IPs can have uncharacterized parametric behavior (e.g., power/thermal) which can be exploited by an attacker to cause irrecoverable damage to an electronic system. There are documented instances of such attacks. For example, in 2012, a study by a group of researchers in Cambridge revealed an undocumented silicon level backdoor in a highly secure military-grade ProAsic3 FPGA device from MicroSemi (formerly Actel) [2], which was later described as a vulnerability induced unintentionally by on-chip debug infrastructure. In a recent report, researchers have demonstrated such an attack where a malicious upgrade of a firmware destroys the processor it is controlling by affecting the power management system [3]. It manifests a new attack mode for IPs, where

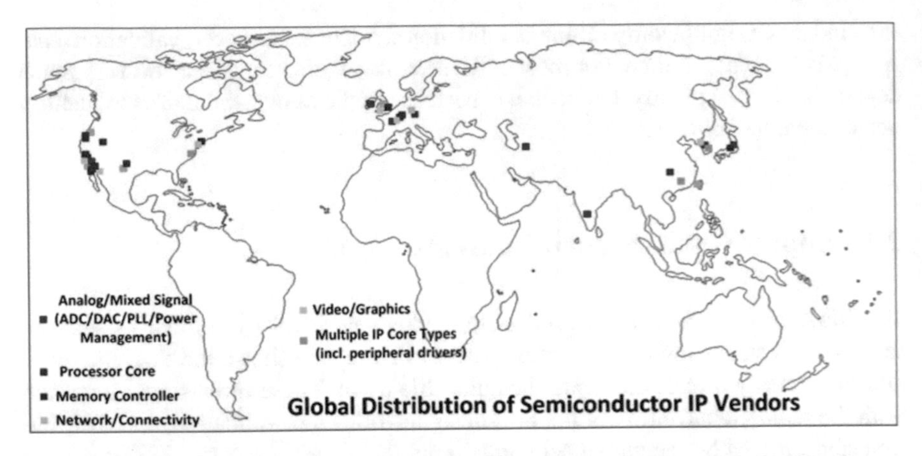

Fig. 2.3 An SoC would often contain hardware IP blocks obtained from entities distributed across the globe

firmware/software update can maliciously affect the power/performance/temperature profile of a chip to either destroy a system or reveal secret information through appropriate side-channel attack, e.g., a fault or timing attack.

Trusted and untrusted CAD tools pose similar trust issues to the SoC designers. Such tools are designed to optimize a design for power, performance, and area. Security optimization is not an option in today's tools, hence sometimes during the optimization new vulnerabilities are introduced [4]. Rogue designers in an untrusted design facility, e.g., in case of a design outsourced to a facility for Design-for-Test (DFT) or Design-for-Debug (DFD) insertion, can compromise the integrity of a SoC design through insertion of stealthy hardware Trojan. These Trojans can act as backdoor or compromise the functional/parametric properties of a SoC in various ways.

Finally, many SoC manufacturers today are fabless and hence must rely upon external untrusted foundries for fabrication service. An untrusted foundry would have access to the entire SoC design and thus brings in several serious security concerns, which include reverse-engineering and piracy of the entire SoC design or the IP blocks as well as tampering in the form of malicious design alterations or Trojan attacks. During distribution of fabricated SoCs through a typically long globally distributed supply chain, consisting of multiple layers of distributors, wholesalers, and retailers, the threat of counterfeits is a growing one. These counterfeits can be low-quality clones, overproduced chips in untrusted foundry, or recycled ones [5]. Even after deployment, the systems are vulnerable to physical attacks, e.g., side-channel attacks which target information leakage, and magnetic field attacks that aim at corrupting memory content to cause denial-of-service (DoS) attacks.

2.3 Security Policies: Requirements from Design

In addition to supply-chain threats, the design itself may have exploitable vulnerabilities. Vulnerabilities in system design, in fact, forms the quintessential objective of security study, and has been the focus of research for over three decades. At a high level, the definition of security requirement for assets in a SoC design follows the well-known "CIA" paradigm, developed as part of information security research [6]. In this paradigm, accesses and updates to secure assets are subject to the following three requirements:

- **Confidentiality**: An asset cannot be accessed by an agent unless authorized to do so.
- **Integrity**: An asset can be mutated (e.g., the data in a secure memory location can be modified) only by an agent authorized to do so.
- **Availability**: An asset must be accessible to an agent that requires such access as part of correct system functionality.

Of course, mapping these high-level requirements to constraints on individual assets in a system is nontrivial. This is achieved by defining a collection of security policies that specify which agent can access a specific asset and under what conditions. Following are two examples of representative security policies. Note that while illustrative, these examples are made up and do not represent security policy of a specific company or system.

Example 1 During boot time, data transmitted by the cryptographic engine cannot be observed by any IP in the SoC other than its intended target.

Example 2 A programmable fuse containing a secure key can be updated during manufacturing but not after production.

Example 1 is an instance of confidentiality, while Example 2 is an instance of integrity policy; however, the policies are at a lower level of abstraction since they are intended to be translated to "actionable" information, e.g., architectural or design features. The above examples, albeit hypothetical, illustrate an important characteristic of security policies: the same agent may or may not be authorized access (or update) of the same security asset depending on (1) the phase of the execution (i.e., boot or normal), or (2) the phase of the design life cycle (i.e., manufacturing or production). These factors make security policies difficult to implement. Exacerbating the problem is the fact that there is typically no central documentation for security policies; documentation of policies can range from microarchitectural and system integration documents to informal presentations and conversations among architects, designers, and implementors. Finally, the implementation of a policy is an exercise in concurrency, with different components of the policy implemented in different IPs (in hardware, software, or firmware), that coordinate together to ensure adherence to the policy.

Unfortunately, security policies in a modern SoC design are themselves significantly complex, and developed in ad hoc manner based on customer requirements and product needs. Following are some representative policy classes. They are not complete, but illustrate the diversity of policies employed.

Access Control. This is the most common class of policies, and specifies how different agents in an SoC can access an asset at different points of the execution. Here an "agent" can be a hardware or software component in any IP of the SoC. Examples 1 and 2 above represent such policy. Furthermore, access control forms the basis of many other policies, including information flow, integrity, and secure boot.

Information Flow. Values of secure assets can sometimes be inferred without direct access, through indirect observation or "snooping" of intermediate computation or communications of IPs. Information flow policies restrict such indirect inference. An example of information flow policy might be the following.

- *Key Obliviousness*: A low-security IP cannot infer the cryptographic keys by snooping only the data from crypto engine on a low-security communication fabric.

Information flow policies are difficult to analyze. They often require highly sophisticated protection mechanisms and advanced mathematical arguments for correctness, typically involving hardness or complexity results from information security. Consequently they are employed only on critical assets with very high confidentiality requirements.

Liveness. These policies ensure that the system performs its functionality without "stagnation" throughout its execution. A typical liveness policy is that a request for a resource by an IP is followed by an eventual response or grant. Deviation from such a policy can result in system deadlock or livelock, consequently compromising system availability requirements.

Time-of-Check Versus Time of Use (TOCTOU). This refers to the requirement that any agent accessing a resource requiring authorization is indeed the agent that has been authorized. A critical example of TOCTOU requirement is in firmware update; the policy requires firmware eventually installed on update is the same firmware that has been authenticated as legitimate by the security or crypto engine.

Secure Boot. Booting a system entails communication of significant security assets, e.g., fuse configurations, access control priorities, cryptographic keys, firmware updates, debug and post-silicon observability information, etc. Consequently, boot imposes more stringent security requirements on IP internals and communications than normal execution. Individual policies during boot can be access control, information flow, and TOCTOU requirements; however, it is often convenient to coalesce them into a unified set of boot policies.

2.4 Adversaries in SoC Security

To discuss security validation, one of the first steps is to identify how a security policy can be subverted. Doing so is tantamount to identifying potential adversaries and charactertizing the power of the adversaries. Indeed, effectiveness of virtually all security mechanisms in SoC designs today are critically dependent on how realistic the model of the adversary is, against which the protection schemes are considered. Conversely, most security attacks rely on breaking some of the assumptions made regarding constraints on the adversary while defining protection mechanisms. When discussing adversary and threat models, it is worth noting that the notion of adversary can vary depending on the asset being considered: in the context of protecting DRM keys, the end user would be considered an adversary, while the content provider (and even the system manufacturer) may be included among adversaries in the context of protecting private information of the end user. Consequently, rather than focusing on a specific class of users as adversaries, it is more convenient to model adversaries corresponding to each policy and define protection and mitigation strategies with respect to that model.

Defining and classifying the potential adversary is a highly creative process. It needs considerations such as whether the adversary has physical access to the system, which components they can observe, control, modify, or reverse-engineer, etc. Recently, there have been some attempts at developing a disciplined, clean categorization of adversarial powers. One potential categorization, based on the interfaces through which the adversary can gain access to the system assets, can be used to classify them into the following six broad categories (in order of increasing sophistication). Note that there has been significant research into specific attacks in different categories, and a comprehensive treatment of different attacks is beyond the scope of this chapter; the interested reader is encouraged to look up some of the references for a thorough description of specific details.

Unprivileged Software Adversary: This form of adversary models the most common type of attack on SoC designs. Here the adversary is assumed to not have access to any privileged information about the design or architecture beyond what is available for the end user, but is assumed to be smart enough to identify or "reverse-engineer" possible hardware and software bugs from observed anomalies. The underlying hardware is also assumed to be trustworthy, and the user is assumed to have no physical access to the underlying IPs. The importance of this naïve adversarial model is that any attack possible by such an adversary can be potentially executed by any user, and can therefore be easily and quickly replicated on-field on a large number of system instances. For these types of attacks, the common "entry point" of the attack is assumed to be user-level application software, which can be installed or run on the system without additional privileges. The attacks then rely on design errors (both in hardware and software) to bypass protection mechanisms and typically get a higher privilege access to the system. Examples of these attacks include buffer overflow, code injection, BIOS infection, return-oriented programming attacks, etc. [7, 8].

System Software Adversary: This provides the next level of sophistication to the adversarial model. Here we assume that in addition to the applications, potentially the operating system itself may be malicious. Note that the difference between the system software adversary and unprivileged software adversary can be blurred, in the presence of bugs in the operating system implementation leading to security vulnerabilities: such vulnerabilities can be seen as unprivileged software adversaries exploiting an operating system bug, or a malicious operating system itself. Nevertheless, the distinction facilitates defining the root of trust for protecting system assets. If the operating system is assumed untrusted, then protection and mitigation mechanisms must rely on lower level (typically hardware) primitives to ensure policy adherence. Note that system software adversary model can have a highly subtle and complex impact on how a policy can be implemented, e.g., recall from the masquerade prevention example above that it can affect the definition of communication fabric architecture, communication protocol among IPs, etc.

Software Covert-Channel Adversary: In this model, in addition to system and application software, a side-channel or covert-channel adversary is assumed to have access to nonfunctional characteristics of the system, e.g., power consumption, wall-clock time taken to service a specific user request, processor performance counters, etc., which can be used in subtle ways to identify how assets are stored, accessed, and communicated by IPs (and consequently subvert protection mechanisms) [9, 10].

Naïve Hardware Adversary: Naive hardware adversary refers to the attackers who may gain the access to the hardware devices. While the attackers may not have advanced reverse-engineering tools, they may be equipped with basic testing tools. Common targets for these types of attacks include exposed debug interfaces and glitching of control or data lines [11]. Embedded systems are often equipped with multiple debugging ports for quick prototype validation and these ports often lack proper protection mechanisms, mainly because of the limited on-board resources. These ports are often left on purpose to facilitate the firmware patching or bug-fixing for errors and malfunctions detected on-field. Consequently, these ports also provide potential weakness which can be exploited for violating security policies. Indeed, some of the "celebrated" attacks in recent times make use of available hardware interfaces including the XBOX 360 Hack [12], Nest Thermostat Hack [13], and several smartphone jailbreaking techniques.

Hardware Reverse-Engineering Adversary: In this model, the adversary is assumed to be able to reverse-engineer the silicon implementation for on-chip secrets identification. In practice, such reverse-engineering may depend on sniffing interfaces as discussed for naïve hardware adversaries. In addition, they can depend on advanced techniques such as laser-assisted device alteration [14] and advanced chip-probing techniques [15]. Hardware reverse engineering can be further divided into two categories: (1) chip-level and (2) IP core functionality reconstruction. Both attack vectors bring security threats into the hardware systems, and permit extraction of secret information (e.g., cryptographic and DRM keys coded into hardware), which cannot be otherwise accessed through software or debugging interfaces.

Malicious Hardware Intrusion Adversary: A hardware intrusion adversary (or hardware Trojan adversary) is a malicious piece of hardware inside the SoC design. It is different from a hardware reverse-engineering adversary in that instead of "passively" observing and reverse-engineering functionality of the rest of the design components, it has the ability to communicate with them (and "fool" them into violating requisite policies). Note that as with the difference between system software and unprivileged software adversaries above, many attacks possible by an intrusion adversary can, in principle, be implemented by a reverse-engineering adversary in the presence of hardware bugs. Nevertheless, the root of trust and protection mechanisms required are different. Furthermore, in practice, hardware Trojan attacks have become a matter of concern specifically in the context of SoC designs that include untrusted third-party IPs as well as those integrated in an untrusted design house. Protection policies against such adversaries are complex, since it is unclear a priori which IPs or communication fabric to trust under this model. The typical approach taken for security in the presence of intrusion adversaries (and in some cases, reverse-engineering adversaries) is to ensure that a rogue IP $\mathscr{A}$ cannot subvert a non-rogue IP $\mathscr{B}$ into deviating from a policy.

2.5 IP-Level Trust Validation

One may wonder, why is it not possible to reuse traditional functional verification techniques to this problem? This is due to the fact that IP trust validation focuses on identifying malicious modifications such as hardware Trojans. Hardware Trojans typically require two parts: (1) a trigger, and (2) a payload. The trigger is a set of conditions that their activation deviates the desired functionality from the specification and their effects are propagated through the payload. An adversary designs trigger conditions such that they are satisfied in very rare situations and usually after long hours of operation [16]. Consequently, it is extremely hard for a naïve functional validation technique to activate the trigger condition. Below we discuss a few approaches based on simulation-based validation as well as formal methods. A detailed description of various IP trust validation techniques is available in [17, 18].

Simulation-Based Validation: There are significant research efforts on hardware Trojan detection using random and constrained-random test vectors. The goal of logic testing is to generate efficient tests to activate a Trojan and to propagate its effects to the primary output. These approaches are beneficial in detecting the presence of a Trojan. Recent approaches based on structural/functional analysis [19–21] are useful to identify/localize the malicious logic. Unused Circuit Identification (UCI) [19] approaches look for unused portions in the circuit and flag them as malicious. The FANCI approach [21] was proposed to flag suspicious nodes based on the concept of control values. Oya et al. [20] utilized well-crafted templates to identify Trojans in TrustHUB benchmarks [22]. These methods assume that the attacker uses rarely occurring events as Trojan triggers. Using "less-rare" events as trigger will void these approaches. This was demonstrated in [23], where Hardware Trojans were designed to defeat UCI [19].

Side-Channel Analysis: Based on the fact that a trigger condition usually has extremely low probability, the traditional ATPG-based method for functional testing cannot fulfill the task of Trojan activation and detection. Bhunia et al. [16] proposed the multiple excitation of rare occurrence (MERO) approach to generate more effective tests to increase the probability to trigger the Trojan. A more recent work by Saha et al. [24] can improve MERO to get higher detection coverage by identifying possible payload nodes. Side-channel analysis focuses on the side channel signatures (e.g., delay, transient, and leakage power) of the circuit [25], which avoids the limitations (low trigger probability and propagation of payload) of logic testing. Narasimhan et al. [26] proposed the Temporal Self-Referencing approach on large sequential circuits, which compares the current signature of a chip at two different time windows. This approach can completely eliminate the effect of process noise, and it takes optimized logic test sets to maximize the activity of the Trojan.

Equivalence Checking: In order to trust an IP block, it is necessary to make sure that the IP is performing the expected functionality—nothing more and nothing less. From security point of view, verification of correct functionality is not enough. The verification engineer has to confirm that there are no other activities besides the desired functionality. Equivalence checking ensures that the specification and implementation are equivalent. Traditional equivalence checking techniques can lead to state space explosion when large IP blocks are involved with significantly different specification and implementation. One promising direction is to use Gröbner basis theory to verify arithmetic circuits [27]. Similar to [28], the reduction of specification polynomial with respect to Gröbner basis polynomials is performed by Gaussian elimination to reduce verification time. In all of these methods, when the remainder is nonzero, it shows that the specification is not exactly equivalent with the implementation. Thus, the nonzero remainder can be analyzed to identify the hidden malfunctions or Trojans in the system.

Model Checking: Model checking is the process of analyzing a design for the validity of properties stated in temporal logic. A model checker takes the Register Transfer Level (RTL) (e.g., Verilog) code along with the property written as a Verilog assertion and derives a Boolean satisfiability (SAT) formulation for validating/invalidating the property. This SAT formulation is fed to a SAT engine, which then searches for an input assignment that violates the property [29]. In practice, designers know the bounds on the number of steps (clock cycles) within which a property should hold. In Bounded Model Checking (BMC), a property is determined to hold for at least a finite sequence of state transitions. The Boolean formula for validating/ invalidating the target property is given to a SAT engine, and if a satisfying assignment is observed within specific clock cycles, that assignment is a witness against the target property [30]. The properties can be developed to detect Trojans that corrupt critical data and verify the target design for satisfaction of these properties using a bounded model checker.

Theorem Proving: Theorem provers are used to prove or disprove properties of systems expressed as logical statements. However, verifying large and complex systems using theorem provers require excessive effort and time. Despite these limitations, theorem provers have currently drawn a lot of interest in verification of security properties on hardware. In [31–33], the Proof-Carrying Hardware (PCH) framework was used to verify security properties on soft IP cores. Supported by the Coq proof assistant [34], formal security properties can be formalized and proved to ensure the trustworthiness of IP cores. The PCH method is inspired from the proof-carrying code (PCC), which was proposed by Necula [35]. The central idea is that untrusted developers/vendors certify their IP. During the certification process, the vendor develops *safety proof* for the safety policies provided by IP customers. The vendor then provides the user with the IP design, which includes the formal proof of the safety properties. The customer becomes assured of the safety of the IP by validating the design using a proof checker. A recent approach presented a scalable trust validation framework using a combination of theorem proving and model checking [36].

2.6 Security Along SoC Design Life Cycle

We now turn to the problem of system-level security validation for the SoC designs. This process takes place in the SoC design house and continues across the system design life cycle. When performing system-level validation, the constituent IPs are assumed to have undergone a level of standalone trust validation before integration.

Figure 2.4 provides a high-level overview of the SoC design life cycle. Each component of the life cycle, of course, involves a large number of design, development, and validation activities. Here, we summarize the key activities involved along the life cycle, that pertain to security. Subsequent sections will elaborate on the individual activities.

Risk Assessment. Security requirements definition is a key part of product planning, and happens concurrently with (and in close collaboration with) the definition of architectural features of the product. This process involves identifying the security assets in the system, their ownership, and protection requirements, collectively defined as *security policies* (see below). The result of this process is typically the generation of a set of documents, often referred to as *product security specification* (PSS), which provides the requirements for downstream architecture, design, and validation activities.

Security Architecture. The goal of a security architecture is to design mechanisms for protection of system assets as specified by the PSS. It includes several components, as follows: (1) identifying and classifying potential adversary for each asset; (1) determining attacker entry points, also referred to as threat modeling; and (3) developing protection and mitigation strategies. The process can identify additional security policies—typically at a lower level than those identified during risk assessment (see below)—which are added to the PSS. The security definition typi-

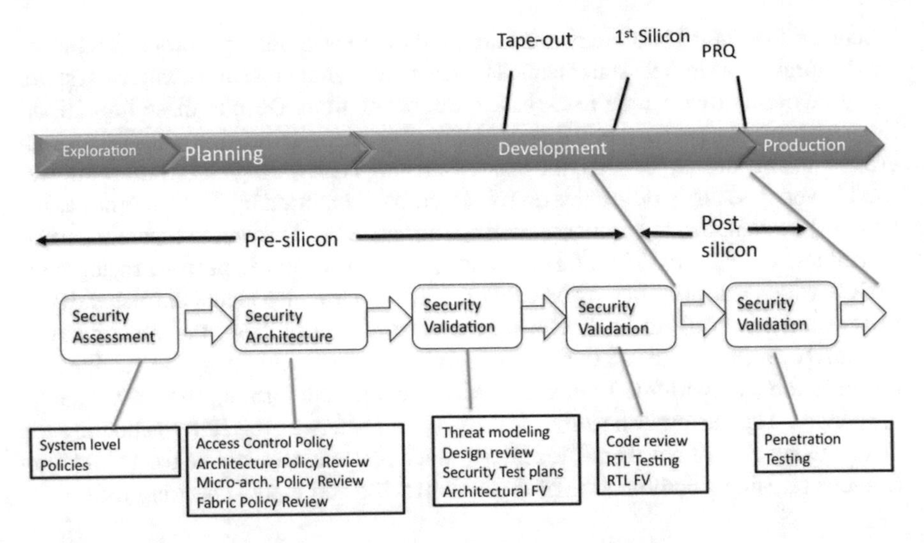

Fig. 2.4 A typical SoC life cycle from exploration to production

cally proceeds in collaboration with architecture and design of other system features, including speed, power management, thermal characteristics, etc., with each component potentially influencing the others.

Security Validation. Security validation represents one of the longest and most critical part of security assurance for industrial SoC designs, spanning the architecture, design, and post-silicon components of the system life cycle. The actual validation target and properties validated at any phase, of course, depends on the collateral available in that phase. For example, we target, respectively, architecture, design, implementation, and silicon artifacts as the system development matures. Below we will discuss some of the key validation activities and associated technologies. One key component of security validation is to develop techniques to subvert the advertised security requirements of the system, and identify mitigation measures. Mitigation measures for early-stage validation targeting architecture and early system design often include significant refinement of the security architecture itself. At later stages of the system life cycle, when architectural changes are no longer feasible due to product maturity, mitigation measures can include software or firmware patches, product defeature, etc.

2.7 Security Validation Activities

Unfortunately, the role of security validation is different from most other kinds of validation (such as functional or power-performance or timing) since the requirements are typically less precise. In particular, the goal of security validation is to "validate conditions related to security and privacy of the system that are not covered by other validation activities." The requirement that security validation focuses on targets not covered by other validation is important given the strict time-to-market constraints, which preclude duplication of resources for the same (or similar) validation tasks; however, it puts onus on the security validation organization to understand activities performed across the spectrum of the SoC design validation and identify holes that pertain to security. To exacerbate the problem, a significant amount of security objectives are not clearly specified, making it difficult to (1) identify validation tasks to be performed, and (2) develop clear coverage/success criteria for the validation. Consequently, the validation plan includes a large number of diverse activities that range from the science to the art and sometimes even "black magic."

At a high level, security validation activities can be divided roughly among the following four categories.

Functional Validation of Security-sensitive Design Features. This is essentially extension to functional validation, but pertain to design elements involved in critical security feature implementations. An example is the cryptographic engine IP. A critical functional requirement for the crypographic engine is that it encrypts and decrypts data correctly for all modes. As with any other design block, the cryptographic engine is also a target of functional validation. However, given that it is a critical component of a number of security-critical design features, security validation planning may determine that correctness of cryptographic functionality to be crucial enough to justify further validation beyond the coverage provided by vanilla functional validation activities. Consequently, such an IP may undergo more rigorous testing, or even formal analysis in some cases. Other such critical IPs may include IPs involved in secure boot, on-field firmware patching, etc.

Validation of Deterministic Security Requirements. Deterministic security requirements are validation objectives that can be directly derived from security policies. Such objectives typically encompass access control restrictions, address translations, etc. Consider an access control restriction that specifies a certain range of memory to be protected from Direct Memory Access (DMA) access; this may be done to ensure protection against code-injection attacks, or protect a key that is stored in such location, etc. An obvious derived validation objective is to ensure that all DMA calls for access to a memory whose address translates to an address in the protected range must be aborted. Note that validation of such properties may not be included as part of functional validation, since DMA access requests for DMA-protected addresses are unlikely to arise for "normal" test cases or usage scenarios.

Negative Testing. Negative testing looks beyond the functional specification of designs to identify if security objectives can be subverted or are underspecified. Continuing with the DMA-protection example above, negative testing may extend the deterministic security requirement (i.e., abortion of DMA access for protected memory ranges) to identify if there are any other paths to protected memory in addition to address translation activated by a DMA access request, and if so, potential input stimulus to activate such paths.

Hackathons. Hackathons, also referred to as *white-box hacking* fall in the "black magic" end of the security validation spectrum. The idea is for expert hackers to perform goal-oriented attempts at breaking security objectives. This activity depends primarily on human creativity, although some guidelines exist on how to approach them (see discussion on penetration testing in the next section). Because of their cost and the need for high human expertise, they are performed for attacking complex security objectives, typically at hardware/firmware/software interfaces or at the chip boundary.

2.8 Validation Technologies

Recall from above that focused functional validation of security-critical design components form a key constituent of security validation. From that perspective, security validation includes and supersedes all functional validation tools, flows, and methodologies. Functional validation of SoC designs is a mature and established area, with a number of comprehensive surveys covering different aspects [37, 38]. In this section, we instead consider validation technologies to support other validation activities, e.g., negative testing, white-box hacking, etc. As discussed above, these activities inherently depend on human creativity; tools, methodologies, and infrastructures around them primarily act as assistants, filling in gaps in human reasoning and providing recommendations.

Security validation today primarily uses three key technologies: fuzzing, penetration testing, and formal or static analysis. Here we provide a brief description of these technologies. Note that fuzzing and static analysis are very generic techniques with applications beyond security validation; our description will be confined to their applications only on security.

Fuzzing. Fuzzing, or fuzz testing [39], is a testing technique for hardware or software that involves providing invalid, unexpected, or random inputs and monitoring the result for exceptions such as crashes, or failing built-in code assertions or memory leaks. Figure 2.5 demonstrates a standard fuzzing framework. It was developed as a software testing approach, and has since been adapted to hardware/software systems. It is currently a common practice in industry for system-level validation. In the context of security, it is effective for exposing a number of potential attacker entry points, including through buffer or integer overflows, unhandled exceptions, race conditions, access violations, and denial of service. Traditionally, fuzzing uses either random inputs or random mutations of valid inputs. A key attraction to this approach is its high automation compared to other validation technologies such as penetration testing and formal analysis. Nevertheless, since it relies on randomness, fuzzing may miss security violations that rely on unique corner-case scenarios. To address that deficiency, there has been recent work on "smart" input generation for fuzzing, based on domain-specific knowledge of the target system. Smart fuzzing

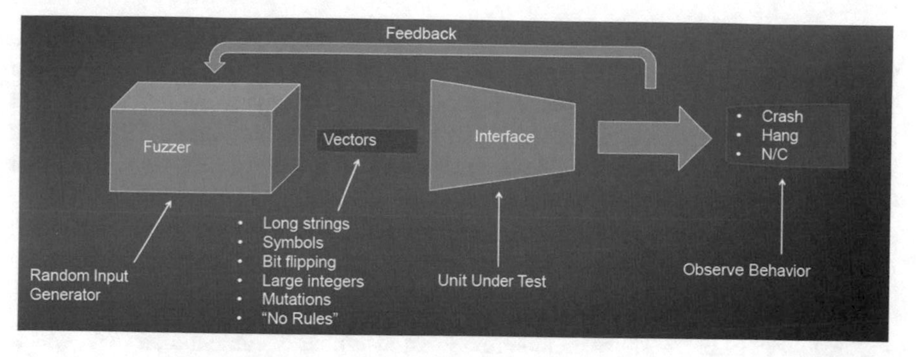

Fig. 2.5 A pictorial representation of fuzzing framework used in post-silicon SoC security validation

may provide a greater coverage of security attack entry points, at the cost of more up front investment in design understanding.

Penetration Testing. A penetration test is an attack on a computer system with the intention to find security weakness, potentially gaining access to it, its functionality, and data. It is typically performed by expert hackers often with deep knowledge of system architecture, design, and implementation characteristics. Note that while there are commonalities between penetration testing and testing done on functional validation, there are several important differences. In particular, roughly, penetration testing involves iterative application of the following three phases:

1. **Attack Surface Enumeration**. The first task is to identify the features or aspects of the system that are vulnerable to attack. This is typically a creative process involving a smorgasbord of activities, including documentation review, network service scanning, and even fuzzing or random testing (see below).
2. **Vulnerability Exploitation**. Once the potential attacker entry points are discovered, applicable attacks and exploits are attempted against target areas. This may require research into known vulnerabilities, looking up applicable vulnerability class attacks, engaging in vulnerability research specific to the target, and writing/creating the necessary exploits.
3. **Result Analysis**. If the attack is successful, then in this phase the resulting state of the target is compared against security objectives and policy definitions to determine if the system was indeed compromised. Note that even if a security objective is not directly compromised, a successful attack may identify additional attack surface which must then be accounted for with further penetration testing.

Note that while there are commonalities between penetration testing and testing done functional validation, there are several important differences. In particular, the goal of functional testing is to simulate benign user behavior and (perhaps) accidental failures under normal environmental conditions of operation of the design as defined by its specification. Penetration testing goes outside the specification to the limits set by the security objective, and simulates deliberate attacker behavior.

Clearly, the efficacy of penetration testing critically depends on the ability to identify the attack surface in the first phase above. Unfortunately, rigorous methodologies for achieving this are lacking. Following are some of the typical activities in current industrial practice to identify attacks and vulnerabilities. We classify them below as "easy," "medium," and "hard" depending on the creativity necessary. Note that there are tools to assist the human in many of the activities below [40, 41]. However, determining the relevancy of the activity, identifying the degree to which each activity should be explored, and inferring a potential attack from the result of the activity involve significant creativity.

- **Easy Approaches**. These include review of available documentation (e.g., specification, architectural materials, etc.), known vulnerabilities or misconfigurations of IPs, software, or integration tools, missing patches, use of obsolete or out-of-date software versions, etc.
- **Medium Approaches**. These include inferring potential vulnerabilities in the target of interest from information about misconfigurations, vulnerabilities, and attacks in related or analogous products, e.g., a competitor product, a previous software version, etc. Other activities of similar complexity involve executing relevant public security tools or published attack scenarios against the target.
- **Hard Approaches**. This includes full security evaluation of any utilized third-party components, integration testing of the whole platform, and identification of vulnerabilities involving communications among multiple IPs or design components. Finally, *vulnerability research* involves identifying new classes of vulnerabilities for the target which have never been seen before. The latter is particularly relevant for new IPs or SoC designs for completely new market segments.

Static or Formal Reasoning. This involves making use of mathematical logic to either derive a security assurance requirement formally, or identifying flaws in the target system (architecture, design, or implementation). Application of formal methods typically involve significant effort, either in the manual exercise of performing deductive reasoning or in developing abstractions of the security objective which are amenable to analysis by automated formal tools [38, 42]. In spite of the cost, however, the effort is justified for highly critical security objectives, e.g., cryptographic algorithm implementation. Furthermore, for some critical properties, automated formal methods can be used in a light-weight manner as effective state exploration tools. For example, TOCTOU property violations often involve scenarios of overlapping execution of different instances of the same protocol, which are effectively exposed by formal methods tools [43]. Finally, formal proofs have also been used as certification mechanisms for third party IP vendors to convey security assurance to SoC system integration teams [33].

2.9 Summary

We have provided a tutorial overview of the industrial practices in security assurance and validation of modern SoC designs. The goal has been to give the reader an overall big picture, provide an understanding of the current state of the practice, and describe the different pieces of a highly complex ecosystem that must interact and cooperate to ensure trustworthiness of our computing devices. The picture of the current practice is scary. On the one hand, the complexity involved is staggering and increasing at an alarming rate. On the other hand, the state of the art in current practice is to depend on human creativity and experience to identify innovative attacks within a small time window before the system goes on field (and is exposed to attacks from the "bad guys")—an approach that we know cannot scale over the complexity we are encountering. While there are promising emergent approaches, we are very far from solving the problem of creating trustworthy computing devices. The need is to develop a disciplined approach to security assurance, from the ground up. Perhaps more importantly, it may require a highly cooperative research initiative involving the different participants, viz., architects, designers, validators, and even cross-cutting stake-holders such as power/performance architects, physical design engineers, etc. Our objective for this chapter has been to serve as the starting point for researchers to understand the overall complexity and contribute to development of trustworthy and secure systems.

Although we covered a broad spectrum of activities on security, we only scratched the surface. There are more complexities involved, including trade-offs with power management, physical design, testing, etc., as well as complex supply chain issues, which we only touched peripherally. The readers interested in deeper exploration are encouraged to explore into some of the references, which include challenges and surveys of specific components, and use the discussions in this paper as a glue for connecting the different pieces.

References

1. Ramamoorthy, G.: Market share analysis: semiconductor design intellectual property, worldwide (2012). https://www.gartner.com/doc/2403015/market-share-analysis-semiconductor-design
2. Skorobogatov, S., Woods, C.: Breakthrough silicon scanning discovers backdoor in military chip. In: CHES, pp. 23–40 (2012)
3. Messmer, E.: RSA security attack demo deep-fries Apple Mac components (2014). http://www.networkworld.com/news/2014/022614-rsa-apple-attack-279212.html
4. Nahiyan, A., Xiao, K., Forte, D., Jin, Y., Tehranipoor, M.: AVFSM: a framework for identifying and mitigating vulnerabilities in FSMs. In: Design Automation Conference (DAC) (2016)
5. Tehranipoor, M., Guin, U., Forte, D.: Counterfeit Integrated Circuits: Detection and Avoidance. Springer (2014)
6. Greenwald, S.J.: Discussion topic: what is the old security paradigm. In: Workshop on New Security Paradigms, pp. 107–118 (1998)

7. Davi, L., Sadeghi, A.R., Winandy, M.: Dynamic integrity measurement and attestation: towards defense against return-oriented programming attacks. In: Proceedings of the 2009 ACM workshop on Scalable trusted computing, STC'09 (2009)
8. Schuster, F., Tendyck, T., Liebchen, C., Davi, L., Sadeghi, A.R., Holz, T.: Counterfeit object-oriented programming: On the difficulty of preventing code reuse attacks in C++ applications. In: Proceedings of the 36th IEEE Symposium on Security and Privacy (2015)
9. Kocher, P.C.: Timing attacks on implementations of Diffie-Hellman, RSA, DSS, and other systems. In: 16th Annual International Cryptology Conference, pp. 104–113 (1996)
10. Kocher, P.C., Jaffe, J., Jun, B.: Differential power analysis. In: 19th Annual International Cryptology Conference, pp. 398–412 (1999)
11. Ray, S., Yang, J., Basak, A., Bhunia, S.: Correctness and security at odds: post-silicon validation of modern SoC designs. In: Proceedings of the 52nd Annual Design Automation Conference (2015)
12. Homebrew Development Wiki: JTAG-Hack. http://dev360.wikia.com/wiki/JTAG-Hack
13. Hernandez, G., Arias, O., Buentello, D., Jin, Y.: Smart nest thermostat: a smart spy in your home. In: Black Hat USA (2014)
14. Rowlette, R., Eiles, T.: Critical timing analysis in microprocessors using near-IR laser assisted device alteration (LADA). In: IEEE International Test Conference, pp. 264–273 (2003)
15. http://www.chipworks.com/
16. Chakraborty, R.S., Wolff, F., Paul, S., Papachristou, C., Bhunia, S.: MERO: A statistical approach for hardware trojan detection. In: Workshop on Cryptographic Hardware and Embedded Systems (2009)
17. Mishra, P., Bhunia, S., Tehranipoor, M.: Hardware IP Security and Trust. Springer (2016)
18. Guo, X., Dutta, R.G., Jin, Y., Farahmandi, F., Mishra, P.: Pre-silicon security verification and validation: a formal perspective. In: ACM/IEEE Design Automation Conference (DAC) (2015)
19. Hicks, M., Finnicum, M., King, S., Martin, M., Smith, J.: Overcoming an untrusted computing base: detecting and removing malicious hardware automatically. In: IEEE Symposium on Security and Privacy (SP), pp. 159–172 (2010)
20. Oya, M., Shi, Y., Yanagisawa, M., Togawa, N.: A score-based classification method for identifying hardware-trojans at gate-level netlists. In: Design Automation and Test in Europe (DATE), pp. 465–470 (2015)
21. Waksman, A., Suozzo, M., Sethumadhavan, S.: Fanci: identification of stealthy malicious logic using boolean functional analysis. In: ACM SIGSAC Conference on Computer and Communications Security, pp. 697–708 (2013)
22. Trust-HUB. https://www.trust-hub.org/
23. Sturton, C., Hicks, M., Wagner, D., King, S.: Defeating UCI: building stealthy and malicious hardware. In: 2011 IEEE Symposium on Security and Privacy (SP), pp. 64–77 (2011)
24. Saha, S., Chakraborty, R., Nuthakki, S., Anshul, Mukhopadhyay, D.: Improved test pattern generation for hardware trojan detection using genetic algorithm and boolean satisfiability. In: Cryptographic Hardware and Embedded Systems (CHES), pp. 577–596 (2015)
25. Aarestad, J., Acharyya, D., Rad, R., Plusquellic, J.: Detecting trojans through leakage current analysis using multiple supply pad I_{ddq}s. In: IEEE Transactions on Information Forensics and Security, pp. 893–904 (2010)
26. Narasimhan, S., Wang, X., Du, D., Chakraborty, R., Bhunia, S.: Tesr: a robust temporal self-referencing approach for hardware trojan detection. In: Hardware-Oriented Security and Trust (HOST), pp. 71–74 (2011)
27. Farahmandi, F., Mishra, P.: Automated test generation for debugging arithmetic circuits. In: Design Automation and Test in Europe (DATE) (2016)
28. Lv, J., Kalla, P., Enescu, F.: Efficient groebner basis reductions for formal verification of galois field arithmetic circuits. IEEE Trans. CAD (TCAD) **32**, 1409–1420 (2013)
29. Cadence Berkeley Lab: The cadence SMV model checker. http://www.kenmcmil.com
30. Biere, A., Cimatti, A., Clarke, E., Zhu, Y.: Symbolic model checking without BDDs. In: Tools and Algorithms for the Construction and Analysis of Systems, p. 193207 (1999)

31. Jin, Y.: Design-for-security vs. design-for-testability: A case study on dft chain in cryptographic circuits. In: IEEE Computer Society Annual Symposium on VLSI (ISVLSI) (2014)
32. Jin, Y., Yang, B., Makris, Y.: Cycle-accurate information assurance by proof-carrying based signal sensitivity tracing. In: IEEE International Symposium on Hardware-Oriented Security and Trust (HOST), pp. 99–106 (2013)
33. Love, E., Jin, Y., Makris, Y.: Proof-carrying hardware intellectual property: a pathway to trusted module acquisition. IEEE Trans. Inf. Forensics Secur. **7**(1), 25–40 (2012)
34. INRIA: The coq proof assistant (2010). http://coq.inria.fr/
35. Necula, G.C.: Proof-carrying code. In: POPL '97: Proceedings of the 24th ACM SIGPLAN-SIGACT Symposium on Principles of Programming Languages, pp. 106–119 (1997)
36. Guo, X., Dutta, R., Mishra, P., Jin, Y.: Scalable SoC trust verification using integrated theorem proving and model checking. In: IEEE International Symposium on Hardware-Oriented Security and Trust (HOST) (2016)
37. Bhadra, J., Abadir, M.S., Wang, L., Ray, S.: A survey of hybrid technqiues for functional verification. IEEE Des. Test Comput. **24**(2), 112–122 (2007)
38. Gupta, A.: Formal hardware verification methods: a survey. Formal Methods Syst. Des. **2**(3), 151–238 (1992)
39. Takanen, A., DeMott, J.D., Mille, C.: Fuzzing for Software Security Testing and Quality Assurance. Artech House (2008)
40. Corporation, M.: Microsoft free security tools microsoft baseline security analyzer (2015). https://blogs.microsoft.com/cybertrust/2012/10/22/microsoft-free-security-tools-microsoft-baseline-security-analyzer/
41. Software, F.: (2012). http://secunia.com
42. Clarke, E.M., Grumberg, O., Peled, D.A.: Model-Checking. The MIT Press, Cambridge, MA (2000)
43. Krstic, S., Yang, J., Palmer, D.W., Osborne, R.B., Talmor, E.: Security of SoC firmware load protocol. In: IEEE HOST (2014)

Chapter 3
SoC Security and Debug

Wen Chen, Jayanta Bhadra and Li-C. Wang

3.1 Introduction

Post-silicon debug includes a diverse range of activities performed after chip manufacturing to diagnose issues on a chip. The debugging activities are performed at several post-silicon stages. One of such stages is post-silicon validation. Due to the increasing complexity of hardware implementation, SoC development nowadays often requires multiple tapeouts. Post-silicon validation has become a necessary step for validating the functionality and performance of an SoC. Post-silicon validation offers the benefits of running tests at the chip's working frequency compared with a typical frequency of thousands of Hz in pre-silicon simulation. It accelerates discovery of issues in both hardware and software and thus reduces validation time. Once an issue is found, validation engineers must be able to get access to the internal states and signals of the SoC in order to localize the bug and resolve the issue. Therefore, comprehensive support of post-silicon debug capabilities is mandatory for modern SoC development. Often, such debugging capabilities need to be extended after post-silicon validation and manufacturing test. For example, authorized application software developers need to diagnose why an application software crashes on a specific SoC. Moreover, when a chip fails in field and is sent back to the manufacturer for hardware evaluation, the analyst needs the debugging capabilities to find out the root cause of the failure. The debugging capabilities of an SoC can be needed during its

W. Chen (✉) · J. Bhadra
NXP Semiconductors N.V., Austin, USA
e-mail: wen.chen@nxp.com

J. Bhadra
e-mail: jayanta.bhadra@nxp.com

L.-C. Wang
University of California, Santa Barbara, USA
e-mail: licwang@ece.ucsb.edu

S. Bhunia et al. (eds.), *Fundamentals of IP and SoC Security*,
DOI 10.1007/978-3-319-50057-7_3

entire product life cycle, spanning from post-silicon validation and chip bring-up to platform software development and field return evaluation.

One of the most notable challenges in post-silicon debug is the reducedpg observability and controllability compared with that in pre-silicon debug. Therefore, a variety of Design-for-Debug (DfD) structures were developed to be instrumented on-chip for increasing observability and controllability of the internal states of an SoC. Such instrumentation circuitry enables getting access to the processor registers and memory from external test and debug interfaces. While the DfD circuitry facilitates post-silicon debug, it exhibits security risks of exposing the secrets or intellectual property (IP) stored on-chip under attack. It can be exploited as a backdoor by attackers to steal on-chip secrets or make unauthorized modification of the IP. The needs for observability and controllability for debugging purposes seemingly have inherent conflicts with the security requirements of an SoC. Deliberate considerations must be taken in SoC design to make balance between the requirements of post-silicon debug and system security. Ideally, the goal is to prevent access to confidential or critical information from unauthorized entities and yet to allow debugging functions from trusted entities. Toward this goal, many solutions have been proposed by academia and industry, and it is still an open research area.

In this chapter, we introduce the basics of SoC debug circuitry and discuss the security risks imposed by it. Countermeasures to address the security issues proposed by both academia and industry and their virtues and limitations are reviewed. The rest of the chapter is outlined as follows. Section 3.2 reviews the requirements of SoC post-silicon debug and major components of an SoC debug architecture. Section 3.3 discusses the security hazards induced by the DfD circuitry. Countermeasures protecting the SoC against security hazards are reviewed in Sect. 3.4. Section 3.5 summarizes the chapter.

3.2 SoC Debug Architecture

Post-silicon debug can be performed at several different stages in the product life cycle of an SoC: post-silicon validation, laboratory bring-up, application software debugging by authorized developers, and field return evaluation. It is aiming to uncover varieties of issues, including functional bugs, electrical errors and performance issues in hardware design, application software bugs, and defects that escaped manufacturing test. Compared with pre-silicon debug, the observability and controllability of SoC internal signals in a post-silicon debug environment is quite limited. Therefore, the ultimate goal of DfD techniques is to allow the observation and manipulation of internal circuit states via externally accessible interfaces. An SoC debug architecture is a system comprising protocols for such observation and manipulation and the supporting DfD circuitry. The architecture should be able to provide debug capabilities for different debugging scenarios at different stages of SoC life cycle. Some general requirements for such an SoC debug architecture are listed as below:

- Observability of system registers and processor states combined with the capability to modify their contents out of the program execution flow.
- Ability to halt and run the processors as per need.
- Ability to obtain information of multiple software threads running on an SoC so as to debug and tune the software for better performance. Provision for triggering the collection of such information upon occurrence of a particular run-time event.
- Mechanism of securing the SoC from unauthorized access using DfD circuitry.

In a typical debugging environment, a user connects a host computer to the SoC under debug. The debugger software running on the host sends debugging commands to the SoC via debug interfaces following a certain protocol. The commands trigger debugging events of the on-chip DfD instrumentation such as halting the processors. The information gathered from the debugging events can be sent back to the host as responses to the commands. On the SoC side, the components of the debug architecture include the debug interface and on-chip DfD instrumentation.

3.2.1 Debug Interface

The debug interface is the port on the SoC that is used to communicate with the external debugger. It consists of the physical interface (external pins of the SoC) and hardware implementation of the standard communication protocol for receiving debug commands and sending the required response. We introduce three commonly used debug interfaces as follows.

BDM

Background debug mode (BDM) has a single-wire bidirectional debug interface along with a Background Debug controller (BDC). It appears in many earlier Freescale Semiconductor Inc. products such as the HCS08 microcontroller family [19]. The external pin of BDM is a pseudo open-drain with an on-chip pull-up and the communication is asynchronous. The external debugger, acting as a BDM master, can issue commands with arguments to the target SoC. The BDM commands provide almost all debugging features (e.g., halt, run, memory read/write, and tracing), except boundary scan. BDM is ideal for small SoCs and microcontrollers with a limited pin count.

SWD

ARM proposes a reduced-pin-count debug interface similar to BDM, known as serial wire debug (SWD), where the external debugger communicates with the SoC via a two-wire interface using a packet-based protocol [4]. The protocol packet is split into header, response and data, with the data being skipped if the interface is not ready. SWD provides full access to the debug and trace functionality on an SoC. It provides the communication channel to get access to the internal debug bus in an ARM CoreSight compliant system. SWD also provides simple parity checking for error detection. SWD is present in most ARM-based SoCs.

JTAG

IEEE Std. 1149.1, Standard Test Access Port and Boundary-Scan Architecture [26], which came out from recommendations of the joint test action group (JTAG), was originally proposed to allow effective testing of the interconnections between chips on a board. Mostly referred to as JTAG, it defines a device architecture comprising the following components as illustrated in Fig. 3.1:

- A Test Access Port (TAP) that includes four mandatory pins-Test Data In (TDI), Test Data Out (TDO), Test Mode Select (TMS), and Test Clock (TCK)- and one optional asynchronous Test Reset (TRST) pin.
- A series of boundary-scan cells on the device primary input and primary output pins, connected internally to form a serial boundary-scan register (Boundary Scan).
- An n-bit ($n \geq 2$) instruction register (IR), holding the current instruction.

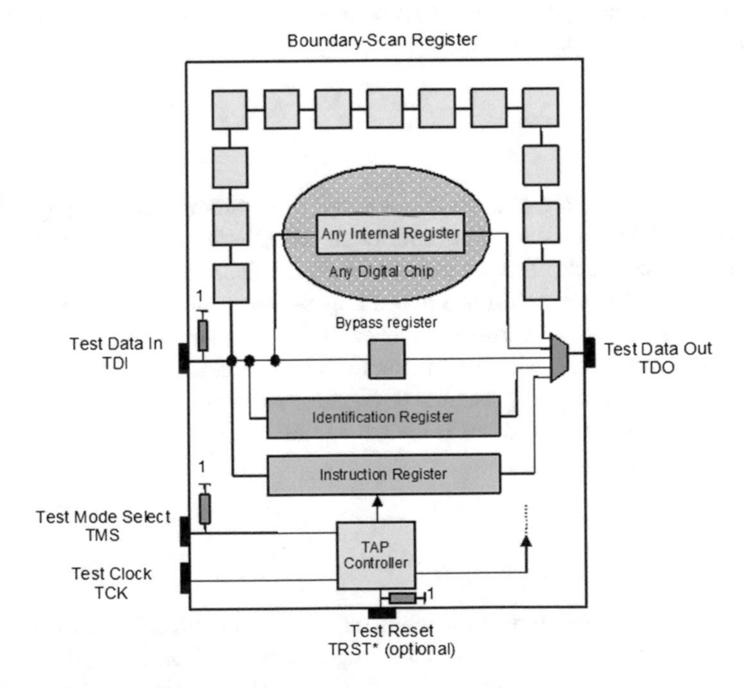

Fig. 3.1 JTAG chip architecture [6]

- A TAP controller, which allows instructions to be shifted into the IR and data to be shifted into the Boundary Scan (test data register). State transitions of the TAP controller are controlled by the value of TMS on the rising edge of TCK.
- A 1-bit bypass register (Bypass).
- An optional 32-bit identification register (Ident) that can be loaded with a permanent device identification code.

At any time, only one register can be connected from TDI to TDO (e.g., IR, Bypass, Boundary-scan, Ident, or even some appropriate register inside the core logic). The selected register is determined by the decoded output of the IR. IEEE Std. 1149.1 defines several mandatory instructions including BYPASS, EXTEST, and SAMPLE, and optional instructions such as RUNBIST and INTEST. It also allows adding custom instructions to the controller to aid in configuration, test, or debug.

Although JTAG was originally proposed for board test, it was being exploited for other purposes such as post-silicon debug. In early days, its utility was deployed to support access to chips for in-circuit emulation (ICE), albeit often with additional pins for proprietary signals [30]. Chip designers had been creative in leveraging the JTAG capabilities for debug. A few examples of JTAG debug capabilities are listed as follows [39]:

- Loading an internal counter used as a breakpoint
- Shadow capturing key registers (with a SAMPLE-like function)
- Masking or overwriting key registers (with an EXTEST-like function)
- Replacing data in key registers (with an UPDATE-like function)
- Selection of scan dump mode (enabling scan-out)

The use of the JTAG TAP as a debug interface was first standardized by NEXUS 5001 (although still requiring additional signaling for many cases) [30]. Today, thanks to its ubiquity and extensibility, the JTAG TAP is one of the most widely used debug interfaces. For example, ARM Coresight debug architecture supports JTAG as the physical interface to get access to its Debug Access Port (DAP).

3.2.2 *On-Chip DfD Instrumentation*

The principal purpose of on-chip DfD instrumentation is to fulfill all the post-silicon debug requirements mentioned earlier without noticeable performance impact on the SoC. Those requirements can be fulfilled by different types of instrumentation circuitry. For example, observation of system registers and modification of their contents can be realized by inserting scan chains, which was originally a design-for-test (DfT) technique. Scan chain insertion replaces the normal flip-flops with scan flip-flops at the SoC design phase. The scan flip-flops act like normal flip-flops in the functional mode and can be connected as a shift register chain (scan chain) in the test mode. By scan chain insertion targeting the key registers on the SoC, the important system states can be controlled and observed.

Another example is halting the processors using the hardware watchpoint/breakpoint. One of the most common methods of debugging is halting the processors or getting system states at a particular point of code execution. One approach to realize this is using software breakpoints, where an instruction is inserted in the code stored on RAM so that the processor will halt when it executes the inserted instruction. However, the software breakpoint cannot work when debugging the code from ROM, which does not allow modifications of the code. In this case, hardware watchpoint/breakpoint is essential for supporting the debugging functionality. Hardware watchpoint support is implemented in the form of watchpoint registers, which are programmed with values of address, control and data signals at which a watchpoint should occur. Comparison and mask circuitry compares the current values of the signals with that programmed in the watchpoint register and generates an output in case of a match, indicating the occurrence of a watchpoint. When a processor encounters a watchpoint, usually a trace message is emitted or the system state at that point gets reflected on the debug software. Watchpoints can be programmed to act as breakpoints, which halt the processor when the program counter reaches a certain address.

The two example features mentioned so far are primarily concerned with the observation and control of system state at a single point of time. For complex debugging scenarios such as those involving multiple threads, the user needs to obtain information on part of the system states in a contiguous period of time. Such information is referred to as traces and the collection of traces is realized by the tracing mechanism. Tracing instrumentation in the SoC captures and compresses the state data in real time upon triggering, to form traces. Then the traces can be made available to external debuggers either via a trace port or by being stored in an embedded trace memory that can be read offline.

The implementation of on-chip DfD instrumentation varies from one SoC design to another. There have been industrial efforts to establish standards for common DfD components that can be reused across different SoC implementations. These efforts result in several popular hardware debug architectures and ecosystems, such as Nexus and ARM Coresight, which standardize the DfD components and their communication protocols. We will take the ARM Coresight Debug Architecture as an example to illustrate DfD components commonly implemented in today's SoC debug architecture.

ARM Coresight Debug Architecture

CoreSight is an ARM standard of debug architecture created for ARM-based systems [3]. The following are the primary components used in this architecture, many of which are shown in Fig. 3.2.

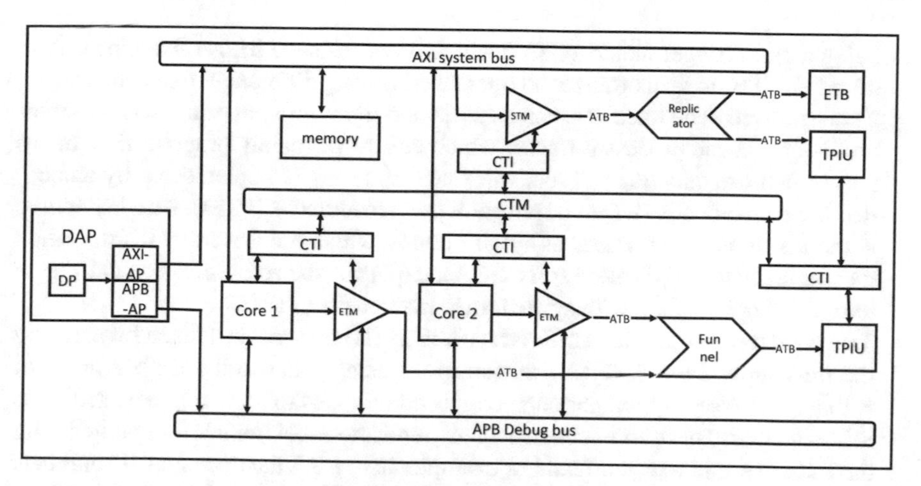

Fig. 3.2 Example of a system of Coresight components for dual-processor SoC

- **Debug Access Port (DAP)**: The debug access port (DAP) acts as a bridge between the external debug interface and multiple core domains and memory mapped peripherals on the SoC. Each DAP has a debug port (DP), which serves as a master device on the DAP bus. External debuggers send debug commands to DP via interfaces such as a full-fledged JTAG port or a reduced-pin-count SWD port. The debug commands are then translated as read or write transactions sent to access ports (AP), which act as slave devices on the DAP bus. As shown in Fig. 3.2, APs can be connected to system buses (e.g., AXI) or peripheral buses (e.g., Debug APB), acting as bus masters, thus providing memory-based access to SoC components. In addition, an AP can also be connected to an on-chip JTAG scan chain.
- **ROM Table**: The ROM table, as part of DAP, lists the memory mapped addresses of all CoreSight components present in an SoC. It is to be noted that one ROM table can point to another ROM table. The ROM table is used for discovery of on-chip Coresight debug components by the external debugger.
- **EmbeddedICE**: The processor debug and monitor features can vary on different processors. Watchpoints and breakpoints are among the most typical ones. EmbeddedICE is a Coresight macrocell containing watchpoint control and status registers to facilitate watchpoint functionality on ARM cores which can also act as breakpoints when debugging from ROM or Flash.
- **Cross Triggerring**: Cross triggering refers to triggering a particular operation in one debug component from a debug event that happened in another debug component. It is essential when debugging complex interactions among multiple cores [42]. In Coresight, wherever there are signals to sample or drive, a cross trigger interface (CTI) is used to control the selection of signals that are trigger sources or trigger targets. Most systems will implement a CTI per processor, and at least one CTI for system-level components. The CTIs in the system are interconnected

using a cross trigger matrix (CTM) which broadcasts the trigger from the CTI to all other CTIs, to synchronize the operations among different components.

- **Trace Sources** The trace data can be collected from different sources. One important source is the processor traces, which consist of mainly program flow traces and sometimes data traces. Processor trace capturing is implemented by embedded trace macrocells (ETM) or program trace macrocells (PTM). Another source is the instrumentation traces or system traces, which are driven by instrumented messaging in the application software, for capturing the active contexts in the system. It is implemented with system trace macrocells (STM).
- **Trace Interconnect**: The AMBA trace bus (ATB) protocol is defined for carrying the trace around the SoC. One advantage of using a standard trace bus protocol is that a small set of modular components can be used to form sophisticated trace infrastructure. These components include replicators and funnels for manipulating data streams, and trace buffers. For example, in Fig. 3.2 the trace funnel combines the trace data from ETM of two cores and the replicator replicates the trace from a single STM and sends them to two trace sinks (TPIU and ETB).
- **Trace Sinks**: A trace sink is the terminal CoreSight component in a trace interconnect. A system can have more than one trace sink, configured to collect overlapping or distinct sets of trace data. Trace streams are stored in on-chip trace buffers called embedded trace buffers (ETB), sent through an interface called trace port interface unit (TPIU) to be stored in an off-chip buffer to be read by external debuggers, or routed to shared system memory.

3.3 Security Concerns and Hazards

3.3.1 SoC Security Requirements

Ubiquitous computing devices now contain an increasing portion of credential and private information or Intellectual Property (IP) on-chip. For example, set-top boxes, which are used to allow only authorized users to get access to broadcast contents, rely on security keys stored on-chip to perform the decryption of the encrypted digital contents. The emergence of internet-of-things allows pervasive devices to be connected to each other. The security of such connections is a concern and thus imposes more stringent security requirements on each individual device. Regarding information security, there are two fundamental properties to preserve, on which nearly every higher level property can be based: confidentiality and integrity.

- **Confidentiality** requires that an asset that is confidential cannot be accessed by an unauthorized entity. This property is essential for assets such as passwords and cryptographic keys.
- **Integrity** requires that an asset that has its integrity assured is protected from unauthorized modifications. This property is essential for assets serving as the

root of system security, e.g., configuration fuses, chip unique ID, and the secure boot firmware.

In some circumstances, when integrity cannot be fully ensured, an alternative property called **authenticity** is a must, which requires detecting whether the integrity of an asset has been compromised. In this case the content of the asset can be altered by an attacker, but the defender will be able to detect the alteration before the asset is used and thus mitigate the risks of further loss.

3.3.2 *DfD Induced Security Hazards*

To satisfy the security requirements, chip architects and designers propose various security policies to defend against the possible risks of information leak or compromised integrity. However, the enforcement of security policies requires diverse components across the entire system to coordinate with each other in a seamless manner, which is very difficult to implement correctly. Moreover, security risks are aggravated by the existence of debug components mentioned in Sect. 3.2. The increased observability and controllability of the internal states of the SoC enabled by the debug circuitry inherently conflict with the requirements of preserving confidentiality and integrity. Attackers can exploit the debug access to obtain and affect the internal states of the SoC, which might reveal the secrets stored on-chip or compromise integrity of the IPs.

The security hazards induced by SoC debug circuitry fall into two categories. The first is related to the extensive use of external debug interfaces as a mechanism to transport configuration data to the SoC. This capability can be abused by a malicious entity to reprogram the firmware or the system configuration. This type of hazard mostly violates the integrity requirement. The most well-known example is JTAG, which can be used by attackers to upload corrupted firmware in flash memories. The second can be attributed to the enhanced controllability and observability of the on-chip DfD instrumentation, often accompanied by the communication channel provided by external debug interfaces. The selected signals to be controlled and observed by DfD instrumentation are usually related to important system states. Simply snooping those signals themselves sometimes reveals secret information on the SoC. Moreover, attackers can manipulate the control and data flow at their will to reach system states that might leak confidential information. This type of hazard usually violates the confidentiality requirement. A notable example is a category of attacks called scan-based attacks, which exploit the internal scan chain to derive secret keys used in cryptographic engines.

Of course, the two types of hazards can often be jointly exploited. For example, attackers might first try to hack the secret key for verifying signed firmware and then use the key to sign malicious code and overwrite the embedded flash with the malicious code. In this example, not only confidentiality and integrity but also the authenticity requirement is violated. Also note that exploits of on-chip DfD instrumentation do not always need to go through external debug interfaces. Attack vectors injected from the supply chain such as hardware Trojans and malicious third-party IPs can also leverage the on-chip instrumentation to aggravate their attacks.

In this chapter, we will review two exemplary security hazards induced by DfD circuitry: firmware hazards and scan hazards. These two are the most well-known in publications. The review by no means includes all the DfD induced security hazards but suffices to illustrate how the hazards would be incurred.

Firmware Hazards

IEEE Std. 1532, Boundary-Scan-Based In-System Configuration of Programmable Devices, has extended JTAG even further to support on-board programming [25]. It allows device programmers to transfer data to non-volatile memory like Flash, which often stores the firmware of the system. This capability is hazardous to the integrity of the IPs on-chip and can be exploited by attackers to upload unsigned firmware. If there are security holes in the firmware authentication scheme, the unsigned firmware will be running at the attacker's will. A well-known example of such an exploit is the first hack of XBox 360 gaming consoles [24]. Using JTAG to upload a version of the firmware that was hacked to bypass the authentication process, users can run code that was not initially allowed by Microsoft. This hack cannot be mitigated by releasing new versions of the firmware since the users can always overwrite the firmware to the known insecure version using JTAG.

The firmware update of set-top boxes used in pay-TV subscriptions also happens in most cases via the JTAG interface. An insecure JTAG access would allow reprogramming the firmware of set-top boxes, thus unlocking premium services that were supposed to be only available to paid users. On the other hand, it could be used to sniff configuration bits thus allowing retrieving the secret keys. There have been many practical attacks on secure devices such as set-top box decoders by leveraging the JTAG's ability to upload the firmware [17].

Other victims include SoCs in smartphones. ARM11 (Cortex) processor, which is used in latest smartphones, has extensive test and debug facilities through the JTAG port. This is a well-known backdoor that was used to jailbreak iPhones/iPads, or to unlock protected services in mobile phones [21].

Scan Hazards

Scan chain insertion is a widely used DfT technique for SoC testing as it provides full observability and controllability of state elements included in the scan chain and tremendously reduces the complexity of test generation for sequential circuits. Furthermore, the scan chains can be connected to the JTAG interface to provide on-chip debug capability in field [27]. Security hazards induced by the scan-based DfT technique fall into two categories, the observability and controllability ones [22].

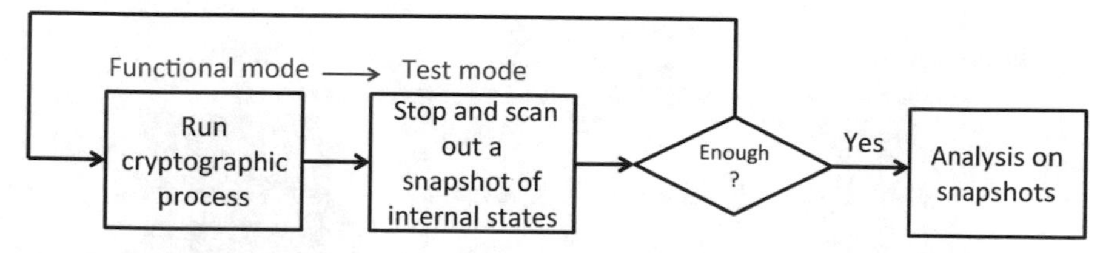

Fig. 3.3 Illustration of scan-based attack flow

The enhanced observability features of a scan-based design may be used for dumping out the values of flip-flops and thus can possibly be used to snoop the key retrieving during cryptographic processes. Though the cryptographic engines can be designed in a way that the keys are obfuscated so that a direct extraction from the flip-flops are not feasible, the internal states provide information for scan-based attacks. A scan-based attack can be performed by running a cryptographic process and then scanning out the internal states at different time points of the process. A conceptual flow of scan-based attacks is illustrated in Fig. 3.3. The attacker runs an encryption with a known plain-text in the functional mode. At a certain time point of the cryptographic process, the attacker switches the circuit to the test mode and thus scans out a snapshot of the internal state of the circuit. This process is iterated several times in order to gather enough information. Flip-flops storing intermediate results of cryptographic process are identified with knowledge of the cryptographic algorithm and the keys can be derived by analysis.

From the controllability point of view, a scan path can be used for deliberately inserting data into flip-flops deeply embedded in the circuit. The scan path can be deemed as a shortcut to fault injection [7]. Randomly injecting faults into the circuits was very easy to perform compared with other fault injection techniques such as clock glitching. However, fault injection targeting a specific flip-flop is much more difficult because it requires deep knowledge of the chip. Thus, the easy access of fault injection via scan chains seems not helpful enough for such an attack.

From the above analysis, scan-based attacks on cryptographic engines are the most practical security hazard induced by the existence of scan chains and the external access to scan chains via debug interfaces such as JTAG. Figure 3.4 illustrates a simplified view of how the scan-based attack can be performed against symmetric-key and public-key cryptographic process, more details of which can be found in [40].

Both symmetric-key and public-key algorithms usually perform the same operations for multiple iterations. Each iteration takes the key and intermediate results from last iteration as the input and outputs intermediate results that serve as input to the next iteration. For symmetric-key algorithms like Data Encryption Standard (DES) or Advanced Encryption Standards (AES), each bit of the intermediate result of one iteration depends on multiple key bits, while the intermediate result of each iteration of public-key algorithms like RSA or elliptic curve cryptography (ECC) depends on a single key bit.

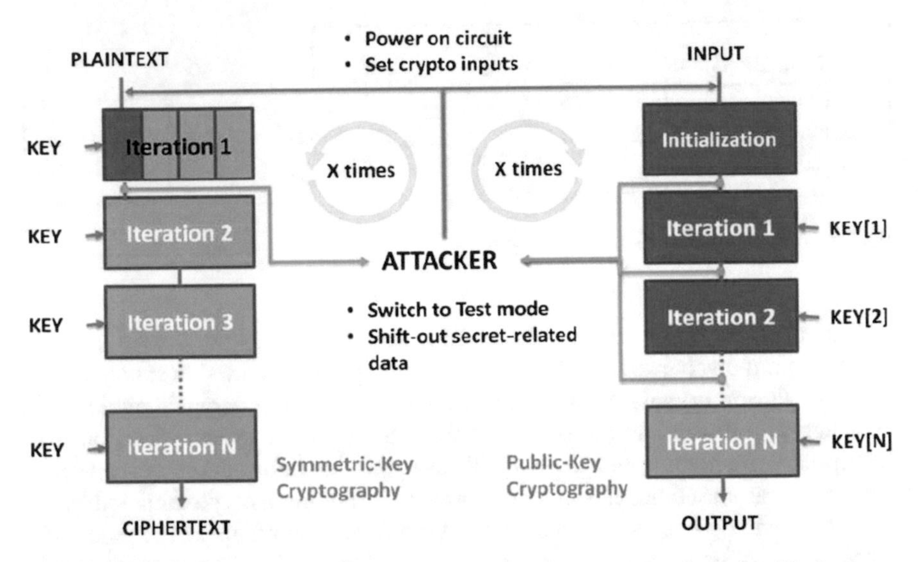

Fig. 3.4 Illustration of scan-based attacks on symmetric-key cryptography and public-key cryptography [14]

It is generally difficult to figure out the key by observing the plain-text and the cipher-text since there are many iterations in between. However, it is relatively easier to infer the key by observing the input data and the intermediate results of one iteration. Therefore, a scan-based attack on symmetric-key cryptographic engines focuses on retrieving the intermediate value resulted from the first iteration. The attacker chooses a plain-text and uses the flow shown in Fig. 3.3 to collect the intermediate value after the first iteration. The attacker repeats this process X times so that enough information can be collected to reveal the key. Attacking public-key cryptographic engines is slightly different because the attacker has to collect intermediate values in different iterations to figure out different bits of the key. X pairs of plain-text and intermediate result are collected for deriving the first bit of the key. Then the other bits of the key are revealed by shifting out intermediate states of further iterations.

With knowledge of the cryptographic algorithms, the attackers can simulate the intermediate results based on hypothetical keys. By comparing the retrieved intermediate results with the simulated ones, the attackers can confirm the correct hypothesis and thus figure out the key. For example, in public-key cryptography, the intermediate result of the first iteration depends on the first bit of the key, which can be 0 or 1. The attacker can simply simulate the intermediate results of a plain-text based on both hypotheses and see which one matches the actual intermediate result. If both match the intermediate result, then the attacker cannot tell the key bit based on this (plain-text, intermediate result) pair and need to use other pairs to determine the key bit. The attack on symmetric-key cryptography is similar though the attacker needs to simulate more hypotheses at once since the intermediate result depends on more key bits.

The first scan-based attack in the literature [43] was proposed to break a DES block cipher. By loading 64 pairs of known plain-texts with one-bit difference in the functional mode and then scanning out the internal states in the test mode, they first determine the positions of all scan elements in the scan chain. Then only three chosen plain-texts are applied to recover the first-round keys (48 bits). And similar attacks can be performed at the second and third rounds to recover the rest of key bits. Later, the same authors proposed a differential scan-based attack on the AES engine [44]. After that, scan-based attacks have been found effective on stream ciphers [31], RSA [41], and ECC [32]. Publications also show that the scan-based attacks can be performed in the presence of the advanced DfT structures such as partial scan, X-masking, and X-tolerant architecture thus making the attacks rather practical to perform [12, 13, 15, 16, 18, 28].

Most state-of-the-art scan-based attacks rely on the ability of switching from the functional mode to the test mode under the assumption that the data in the scan flip-flops can be preserved intact. Therefore, the designers can develop a countermeasure which injects random noise to the scan chain whenever there is a switch from functional mode to test mode, thwarting all these attacks. One simple form of this countermeasure is resetting the data in the scan elements whenever there is a switch from the functional mode to the test mode [22]. This solution can defend most of the scan-based attacks, but it also compromises the debugging capability since sometimes, the authorized users need this capability for debugging purposes. Moreover, a recent work proposed a new scan-based attack using only the test mode [1]. The initial attack analysis shows that only 375 test vectors are sufficient to reveal the 128-bit AES secret key with negligible time complexity. Whether this type of attack can succeed in the presence of advanced DfT structures is still under investigation.

3.4 Protection Against Hazards Induced by Debugging

3.4.1 Trade-Off Between Security and Debug

To prevent the security hazards induced by SoC debug components, one might simply attempt to disable the debug access after manufacturing tests and validation or customer configuration. This can be accomplished by blowing a fuse associated with the debug interface so as to disable access via the debug interface. The TI MSP430 Microcontroller is one example where the JTAG interface can be disabled [11]. The problem with this approach is that it compromises the capability to debug the SoC for purposes such as field return evaluation. To regain access to the debug interface after the fuse is blown, one must resort to complex and expensive techniques. Typically, using focused ion beam (FIB) modification to blow a counterpart fuse in the SoC can regain access to the debug interface. However, it has several problems. First, the equipment for performing FIB modification is expensive and complicated and the modification is often unsuccessful. Second, FIB modification often requires destructive de-encapsulation of the IC device and thus can prevent future evaluation. Third, FIB modification also results in an override of the customer configuration, therefore preventing subsequent access to such configuration information for further analysis. Finally, FIB modification is also relatively temporary due to metal migration, which at some point reconnects the blown counterpart fuse and returns the IC device to a state where the debug interface is inaccessible [9].

Even if the access to debug interfaces is disabled, much of the on-chip DfD instrumentation still remains after production. Though it is possible to disable the on-chip instrumentation, it will change the power/performance/energy profile of the production system from what has been used for validation. And again, the instrumentation is critical to field return evaluation or making in-field patches.

Though we should not permanently disable all debug access after production, restrictions should be in place in accordance with the life stage of the SoC. Design trade-off must be made between the debugging capabilities and security of an SoC along its product life cycle. After an SoC is designed, it has to be manufactured, tested, assembled, built into a product and shipped to customers. As it progresses toward a later stage of its product life cycle, the needs for protection gradually outweigh the needs for debug access, and therefore the SoC should be configured in a way that more restrictions are imposed on the debug access and fewer debugging features are allowed. Such configurations should be irreversible under most circumstances. In some situations such as field return evaluation, the protection of the device needs to be lowered temporarily. However, this feature should not redefine the default protection. Furthermore, this feature ought to be available to trusted entities only with restrictive authentication. Following these principles, a pragmatic designer would identify the security and debug requirements for each stage or mode and implement corresponding access control of the debug components (most likely the debug interface) based on authentication. We will give a historical review of the solutions proposed by both academia and industry along this line.

3.4.2 Authentication-Based Debug Access Control

One of the first approaches utilizes key-based locking/unlocking mechanism for controlling the access to JTAG functionality [33]. The user can unlock JTAG by shifting in the correct key that matches the secret key stored on the chip boundary. Otherwise, JTAG bypasses all the data from TDI to TDO. The process of shifting in the key is un-encrypted, which is vulnerable to eavesdropping on the JTAG communication. In addition, it is often too restrictive to have only two levels of access. In [10], authors proposed to reuse the flip-flops in the boundary cells as a linear shift feedback register to generate the key to reduce the area overhead for key storage.

The methods proposed in [23, 29] are similar in that they use key-based locking/unlocking mechanism to secure the JTAG scan chains. They allow the user to freely shift the contents of the scan chains without the correct key; however, the bits themselves are in a random order. The order is generated by a random number generator, and the correct order can be only restored using a secret key. Instead of restricting the access to the debug interface, this approach restricts the debug access by giving incorrect outputs. It still allows users to supply data to the scan chain, which could be exploited for fault injection. It is also a binary security mechanism since it either gives full access to the scan chains or scrambles the order.

To prevent eavesdropping attacks, a simple authentication scheme based on cryptograpic hash and shared key can suffice. The SoC device and the authorized user shall share a private key. For each test instruction (or debug command) to be issued, the user can use a cryptographic hash algorithm such as Secure Hash Algorithm (SHA) to generate a signature based on the test instruction and the shared key altogether. Then the test instruction and signature are sent to the SoC. The SoC can run the same hash on the received test instruction and the shared key to generate a signature and verify whether it matches the received signature. This approach eliminates the eavesdropping attack; however, it is still vulnerable to replay attack, in which an attacker monitors the message communication, duplicates, and replays the whole message (test instruction + signature) to spoof the SoC.

This replay hazard can be thwarted by the challenge/response authentication-based scheme proposed in [11], which utilizes SHA-256 cryptographic hash engine, shared secret key and random number generator to generate a challenge per communication session and verify the response. The SoC first generates a random number and sends it to the user as a challenge. The user uses the cryptographic hash to generate the signature based on the test instruction, shared key and challenge altogether, and then sends the instruction and signature as the response to the challenge. The SoC device can verify whether received signature is generated based on the same random number and the correct shared key, and thus verify the authenticity of the user. In addition, the authors proposed that a set of instructions should be public while others that can get access to sensitive information should be private. The public instructions do not require keys while each private instruction must have an independent secret key, thus making up different security levels of debug access.

The security of authentication based on shared secret keys largely relies on the confidentiality of the key. Reliable key management is a challenging problem. If each device of an SoC product family shares the same key, compromising the key of a single device would render the protection of all devices useless. If each device is assigned a unique key, the device provider needs to maintain a database recording the key corresponding to each device unique ID, which is costly and thus undesirable.

In [8], the authors proposed a three-entity authentication scheme, utilizing a separate secure server to authenticate the user for improved security. The device will generate a challenge every time upon the user's request to access the debug port. The user is connected to the secure server and relays the challenge to the server. The secure server's role is user authorization and verification upon reception of a challenge and generation of response to a given challenge upon successful verification. The challenge/response algorithm is based on ECC. The device owns a public-key and the secure server holds the private key. This approach offers a higher level of security by hiding the key from the user, and supports various security policies for user authentication and authorization on the server. However, the need for a server to authenticate the user on each debug interface access requires continuous communication with the server, disabling debug access and lowering overall availability when network is unavailable.

To improve the availability over [8], authors of [34] proposed a user authentication scheme in which the server issues to the authenticated user a credential with which to authenticate oneself to a particular device. The device verifies the user submitted credentials and opens the debug port to the users with valid credentials. This approach eliminates the need for networking with the server after a credential has been issued. The same authors later proposed an improved solution by incorporating the the maximum number of authentication allowed for one credential and mechanisms to deal with expired credentials [35].

The authors of [36, 37] formalized the concept of multilevel secure JTAG architecture and provided detailed hardware implementation specifications for enforcing the multilevel policy. Each debug instruction is assigned an access level and each user will be assigned a permission level after authentication. A user with a permission level P_i can execute any instruction with an access level A if $A \leq P_i$. The hardware implementation of such an architecture is composed of two primary components, the secure authentication module (SAM) and the access monitor (AM). SAM's functions are to provide an unlocking communication protocol, to set the user level, and to allow modification of access levels. The AM prevents potentially harmful data from being loaded into the scan chain. This work depicts a practical implementation scheme for multilevel secure debug access, where SAM serves as the security policy decision point and the AM acts as the security policy enforcement point.

Besides the academic proposals, industrial solutions for securing the debug access have also been offered. The ARM TrustZone architecture extension [2] is a system-wide approach to SoC security. JTAG port is a security-related block and thus is part of the architecture extension. Trustzone imposes hardware-based restrictions on JTAG operations, such as restricted or nonrestricted debug access, which can be configured by blowing fuses. Nonrestricted debug access is only used in the development phase of the product. The device delivered to customers is in the restricted mode, in which only the basic, noninvasive debugging functions of JTAG are available. However, the noninvasive debugging functions of JTAG might still be exploited as a backdoor to on-chip secrets. Moreover, Trustzone does not address the need of offering different levels of debug access to different users at different phases.

Freescale Semiconductor introduced the secure JTAG controller since i.MX31 and i.MX31L product families [5]. Freescale secure JTAG controller provides several configurations determined by a set of fuses. The secure JTAG controller in the latest i.MX6 processor allows four different JTAG security modes [20]:

- Mode 1: No JTAG—Maximum security. All JTAG features are permanently blocked.
- Mode 2: No Debug—High security. All security sensitive JTAG features are permanently disabled.

- Mode 3: Secure JTAG—Medium security. JTAG access is only permitted with challenge/response-based authentication mechanism.
- Mode 4: JTAG Enabled—Low security. All JTAG features enabled.

The fuse blowing for configuring security modes is an irreversible process. Once the fuse is blown, it is impossible to change the fuse back to its original state. To address the requirement of field return debug, another fuse is used to override the configuration by secure JTAG fuses. Once the fuse associated with field return is blown, maximum debugging access is obtained. The field return fuse can only be blown when a signed ROM image is loaded. In field return mode, the on-chip secrets will also be hidden from the debug access.

It is to be noted that these authentication-based protection mechanisms introduce high area overhead compared with the debug interface such as JTAG itself. For example, the key-based locking mechanism in [33] has about over 100 % area overhead. Challenge/response authentication based on shared keys between the user and the device such as [11] results in over 500 % area overhead. The three-entity authentication based on public-key cryptography such as [8] has over 2000 % area overhead. This is because the area of JTAG implementation is pretty small. Hence, in those publications, the area overhead is usually calculated considering large SoCs with cryptographic engines and microprocessors. For large SoC designs, the area overhead is usually small thus acceptable. However, for small SoCs and microcontrollers, how to design compact authentication hardware is an open question.

Also note that most of the proposed solutions focus on protecting the debug interfaces using authentication. This is based on the assumption that the traditional attacks would originate externally, without consideration of attack vectors from supply chain such as hardware Trojans and malicious third-party IPs. However, as there are emerging concerns about the supply chain security, the protection mechanisms should be applied not only to the debug interfaces but also to the on-chip DfD instrumentation.

3.4.3 Limitations and Challenges

The existing solutions are at best pragmatic workarounds to combat the security hazards induced by DfD components. A more fundamental approach is to incorporate the debug access as part of the security requirement and architecture definition. Besides confidentiality and integrity, there is another less frequently emphasized security requirement called *availability*. It requires that an asset must be accessible to an entity that requires such access per correct system functionality. The debug requirement can be viewed as an availability requirement [38]. Such a perspective to view security and debug requirements as an integral one would be more helpful to developing a comprehensive solution to address the trade-off between them.

The processes for defining the security and debug requirements and architectures are complex and involving multiple stake-holders. An effective solution for addressing security and debug trade-off should address a comprehensive set of aspects. An incomplete set of key aspects highlighted in [38] include:

- **Centralized Architecture**: The current security and debug architectures are largely decentralized, which makes it difficult to implement them correctly. The policy decision making should be centralized so that it can be effectively introspected for possible violations of the requirements.
- **Late Variability**: The debug requirements are often subject to late changes, which can happen during SoC integration or even after a silicon stepping. An effective solution should allow easy adaptation to the changing requirements and quick validation of the security impacts of the DfD changes.
- **Reusability**: The current solutions are ad-hoc and thus error-prone. A systematic design methodology with reusable components are essential for a viable solution.

The authors of [38] had a preliminary proposal of a centralized, firmware-controlled framework to fulfill the requirements of secure post-silicon debug. However, this still remains an open research area that requires extensive efforts.

3.5 Summary

The capabilities to debug an SoC at post-silicon stages are essential for SoC development. SoC debug circuitry, while offering increased observability and controllability of the internal states of the circuit, can be a backdoor for security attacks. Design decisions must be made regarding the trade-off of the debugging capabilities and the security protection at different stages of the SoC product life cycle. In this chapter, we review the common SoC debug architectures and give a comprehensive analysis of the known security hazards induced by SoC debug access. We review the published solutions for preventing debug access from untrusted entities while preserving debugging functionality, most of which implement access control mechanisms based on authentication of trusted entities.[1]

References

1. Ali, S., Sinanoglu, O., Saeed, S., Karri, R.: New scan-based attack using only the test mode. In: 2013 IFIP/IEEE 21st International Conference on Very Large Scale Integration (VLSI-SoC), pp. 234–239 (2013)
2. ARM: Designing with trustzone hardware requirements. ARM *whitepaper* (2005)
3. ARM: Coresight technical Introduction. ARM *whitepaper* (2013)

---

[1]Freescale and the Freescale logo are trademarks of Freescale Semiconductor, Inc., Reg. U.S. Pat. & Tm. Off. All other product or service names are the property of their respective owners. ARM and Cortex are trademark(s) or registered trademarks of ARM Ltd or its subsidiaries. 2014 Freescale Semiconductor, Inc.

4. Ashfield, E., Field, I., Harrod, P., Houlihane, S., Orme, W., Woodhouse, S.: Serial wire debug and the coresighttm debug and trace architecture (2006)
5. Ashkenazi, A.: Security features in the i.mx31 and i.mx31l multimedia applications processors. Freescale Semiconductor Inc. (2006)
6. Bennetts, B.: IEEE 1149.1 JTAG and boundary scan tutorial. http://www.asset-intertech.com/Products/Boundary-Scan-Test/e-Book-JTAG-Tutorial (2012)
7. Biham, E., Shamir, A.: Differential fault analysis of secret key cryptosystems. In: Proceedings of the 17th Annual International Cryptology Conference on Advances in Cryptology. CRYPTO '97, pp. 513–525. Springer, London (1997)
8. Buskey, R., Frosik, B.: Protected JTAG. In: 2006 International Conference on Parallel Processing Workshops. ICPP 2006 Workshops, pp. 8–414 (2006)
9. Case, L., Ashkenazi, A., Chhabra, R., Covey, C., Hartley, D., Mackie, T., Muir, A., Redman, M., Tkacik, T., Vaglica, J., et al.: Authenticated debug access for field returns. https://www.google.com.ar/patents/US20100199077 (2010). US Patent App. 12/363,259
10. Chiu, G.M., Li, J.M.: A secure test wrapper design against internal and boundary scan attacks for embedded cores. IEEE Trans. Very Large Scale Integr. (VLSI) Syst. **20**(1), 126–134 (2012)
11. Clark, C.: Anti-tamper JTAG TAP design enables DRM to JTAG registers and P1687 on-chip instruments. In: 2010 IEEE International Symposium on Hardware-Oriented Security and Trust (HOST), pp. 19–24 (2010)
12. Da Rolt, J., Das, A., Di Natale, G., Flottes, M., Rouzeyre, B., Verbauwhede, I.: A scan-based attack on elliptic curve cryptosystems in presence of industrial design-for-testability structures. In: 2012 IEEE International Symposium on Defect and Fault Tolerance in VLSI and Nanotechnology Systems (DFT), pp. 43–48 (2012)
13. Da Rolt, J., Das, A., Di Natale, G., Flottes, M.L., Rouzeyre, B., Verbauwhede, I.: A new scan attack on RSA in presence of industrial countermeasures. In: Proceedings of the Third International Conference on Constructive Side-Channel Analysis and Secure Design. COSADE'12, pp. 89–104. Springer, Berlin (2012)
14. Da Rolt, J., Das, A., Di Natale, G., Flottes, M.L., Rouzeyre, B., Verbauwhede, I.: Test versus security: past and present. IEEE Trans. Emerg. Top. Comput. **2**(1), 50–62 (2014). doi:10.1109/TETC.2014.2304492
15. Da Rolt, J., Di Natale, G., Flottes, M.L., Rouzeyre, B.: Are advanced DFT structures sufficient for preventing scan-attacks? In: VLSI Test Symposium (VTS), 2012 IEEE 30th, pp. 246–251 (2012)
16. DaRolt, J., Di Natale, G., Flottes, M.L., Rouzeyre, B.: Scan attacks and countermeasures in presence of scan response compactors. In: European Test Symposium (ETS), 2011 16th IEEE, pp. 19–24 (2011)
17. Dishnet: In house made with locking script. http://www.satcardsrus.com/dish_net%203m.htm (2012)
18. Ege, B., Das, A., Gosh, S., Verbauwhede, I.: Differential scan attack on AES with x-tolerant and x-masked test response compactor. In: 2012 15th Euromicro Conference on Digital System Design (DSD), pp. 545–552 (2012)
19. Freescale: Introduction to HCS08 background debug mode (2006)
20. Freescale: i.mx 6solox applications processor reference manual (2014)
21. Greenemeier, L.: iphone hacks annoy AT&T but are unlikely to bruise apple. Scientific American (2007)
22. Hely, D., Bancel, F., Flottes, M.L., Rouzeyre, B.: Test control for secure scan designs. In: Test Symposium, 2005. European, pp. 190–195 (2005)
23. Hely, D., Flottes, M.L., Bancel, F., Rouzeyre, B., Berard, N., Renovell, M.: Scan design and secure chip [secure IC testing]. In: On-Line Testing Symposium, 2004. IOLTS 2004. Proceedings. 10th IEEE International, pp. 219–224 (2004)
24. Homebrew development wiki JTAG-hack. http://dev360.wikia.com/wiki/JTAG-Hack (2012)
25. IEEE standard for in-system configuration of programmable devices: IEEE Std 1532–2001, pp. 1–130 (2001)

26. IEEE standard test access port and boundary scan architecture. IEEE Std 1149.1-2001, pp. 1–212 (2001)
27. Josephson, D., Poehhnan, S., Govan, V.: Debug methodology for the Mckinley processor. In: Test Conference, 2001. Proceedings. International, pp. 451–460 (2001)
28. Kapur, R.: Security vs. test quality: are they mutually exclusive? In: Test Conference, 2004. Proceedings. ITC 2004. International, pp. 1414– (2004)
29. Lee, J., Tehranipoor, M., Patel, C., Plusquellic, J.: Securing scan design using lock and key technique. In: 20th IEEE International Symposium on Defect and Fault Tolerance in VLSI Systems, 2005. DFT 2005, pp. 51–62 (2005)
30. Ley, A.: Doing more with less—an IEEE 1149.7 embedded tutorial: standard for reduced-pin and enhanced-functionality test access port and boundary-scan architecture. In: Test Conference, 2009. ITC 2009. International, pp. 1–10 (2009). doi:10.1109/TEST.2009.5355572
31. Liu, Y., Wu, K., Karri, R.: Scan-based attacks on linear feedback shift register based stream ciphers. ACM Trans. Des. Autom. Electron. Syst. **16**(2), 20:1–20:15 (2011)
32. Nara, R., Togawa, N., Yanagisawa, M., Ohtsuki, T.: Scan-based attack against elliptic curve cryptosystems. In: Design Automation Conference (ASP-DAC), 2010 15th Asia and South Pacific, pp. 407–412 (2010)
33. Novak, F., Biasizzo, A.: Security extension for IEEE Std 1149.1. J. Electron. Test. **22**(3), 301–303 (2006)
34. Park, K., Yoo, S.G., Kim, T., Kim, J.: JTAG security system based on credentials. J. Electron. Test. **26**(5), 549–557 (2010)
35. Park, K.Y., Yoo, S.G., Kim, J.: Debug port protection mechanism for secure embedded devices. J. Semicond. Technol. Sci. **12**(2), 241 (2012)
36. Pierce, L., Tragoudas, S.: Multi-level secure JTAG architecture. In: 2011 IEEE 17th International On-Line Testing Symposium (IOLTS), pp. 208–209 (2011)
37. Pierce, L., Tragoudas, S.: Enhanced secure architecture for joint action test group systems. IEEE Trans. Very Large Scale Integr. (VLSI) Syst. **21**(7), 1342–1345 (2013)
38. Ray, S., Yang, J., Basak, A., Bhunia, S.: Correctness and security at odds: post-silicon validation of modern SoC designs. In: Design Automation Conference (DAC), 2015 52nd ACM/EDAC/IEEE, pp. 1–6 (2015)
39. Rearick, J., Eklow, B., Posse, K., Crouch, A., Bennetts, B.: IJTAG (internal JTAG): a step toward a DFT standard. In: Test Conference, 2005. Proceedings. ITC 2005. IEEE International, pp. 8–815 (2005)
40. Rolt, J.D., Natale, G.D., Flottes, M.L., Rouzeyre, B.: A novel differential scan attack on advanced DFT structures. ACM Trans. Des. Autom. Electron. Syst. **18**(4), 58:1–58:22 (2013)
41. Ryuta, N., Satoh, K., Yanagisawa, M., Ohtsuki, T., Togawa, N.: Scan-based side-channel attack against RSA cryptosystems using scan signatures. IEICE Trans. Fundam. Electron. Commun. Comput. Sci. **93**(12), 2481–2489 (2010)
42. Tang, S., Xu, Q.: In-band cross-trigger event transmission for transaction-based debug. In: Design, Automation and Test in Europe, 2008. DATE '08, pp. 414–419 (2008)
43. Yang, B., Wu, K., Karri, R.: Scan based side channel attack on dedicated hardware implementations of data encryption standard. In: Test Conference, 2004. Proceedings. ITC 2004. International, pp. 339–344 (2004)
44. Yang, B., Wu, K., Karri, R.: Secure scan: A design-for-test architecture for crypto chips. IEEE Trans. Comput.-Aided Des. Integr. Circuits Syst. **25**(10), 2287–2293 (2006)

Chapter 4
IP Trust: The Problem and Design/Validation-Based Solution

Raj Gautam Dutta, Xiaolong Guo and Yier Jin

4.1 Introduction

A rapidly growing third-party intellectual property (IP) market provides IP consumers with more options for designing electronic systems. It also reduces the development time and expertise needed to compete in a market where profit windows are very narrow. However, one key issue that has been neglected is the security of electronic systems built upon third-party IP cores. Historically, IP consumers have focused on IP functionality and performance than security. The prejudice against the development of robust security policies is reflected in the IP design flow (see Fig. 4.1), where IP core specification usually includes functionality and performance measurements.

This lack of security assurance on third-party IPs is a major threat for the semiconductor industry. For example, a large number of side-channel-based attacks have been reported, which extract sensitive information from systems that were purportedly mathematically unbreakable [1–7]. The emergence of hardware Trojans (malicious logic) embedded in third-party IP cores has largely re-shaped the IP transaction market and there are currently no comprehensive detection schemes for identifying these Trojans. Some Trojan detection methods such as side-channel fingerprinting combined with statistical analysis have shown promising results, but most of the post-silicon stage Trojan detection methods rely on golden models which may not be available given the existence of untrusted IP cores.

R.G. Dutta · X. Guo · Y. Jin (✉)
University of Central Florida, Orlando, FL, USA
e-mail: rajgautamdutta@knights.ucf.edu

X. Guo
e-mail: guoxiaolong@knights.ucf.edu

Y. Jin
e-mail: yier.jin@eecs.ucf.edu

S. Bhunia et al. (eds.), *Fundamentals of IP and SoC Security*,
DOI 10.1007/978-3-319-50057-7_4

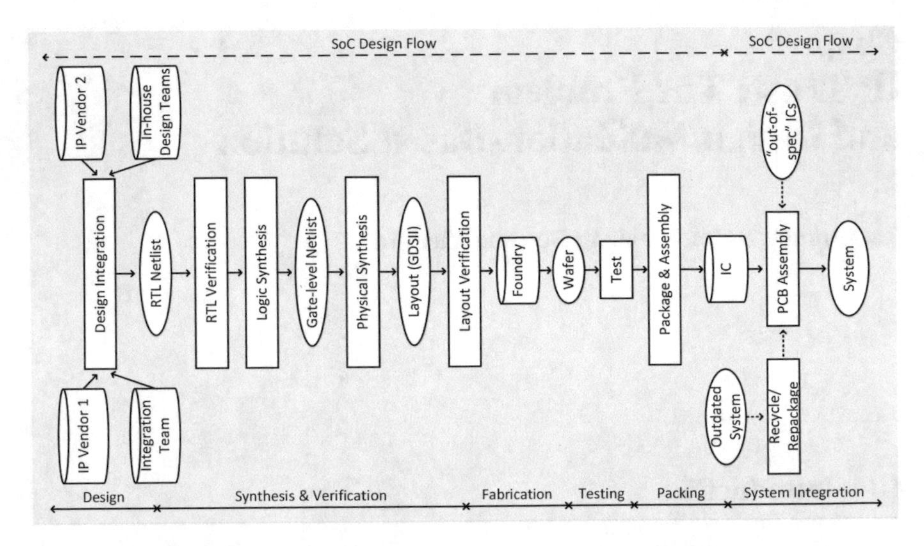

Fig. 4.1 Semiconductor supply chain: IC design flow

Upon the request for trusted IP cores, various IP protection and certification methods at the pre-silicon stage have been recently developed. In this chapter, most of these approaches will be introduced including hardware locking/encryption, FPGA bitstream protection, theorem proving, and equivalence checking.

The rest of the chapter is organized as follows: Sect. 4.2 discusses various hardware locking/encryption methods for preventing various threats to IP cores. For a better explanation, we divide these methods into three categories (i) combinational logic locking/encryption, (ii) finite state machine locking/encryption, and (iii) locking using reconfigurable components. In this section, we also discuss FPGA bitstream protection methods. Section 4.3 primarily discusses the existing equivalence checking and theorem proving methods for ensuring trustworthiness of soft IP cores. Finally, Sect. 4.4 concludes the chapter.

4.2 Design for IP Protection

Design for IP protection encompasses methods for authentication and prevention of IP from piracy, reverse engineering, overbuilding, cloning, and malicious tampering. Authentication approaches include IP watermarking [8–10] and IP fingerprinting [11–13], which can be used by IP owners for detecting and tracking both legal and illegal usages of their designs. However, these methods cannot prevent reverse engineering of designs and insertion of malicious logic. These limitations are overcome by the prevention methods. Most of the currently existing prevention approaches can be grouped under hardware locking/encryption (see Fig. 4.2). Hardware locking/encryption methods can be further divided into (i) combinational logic

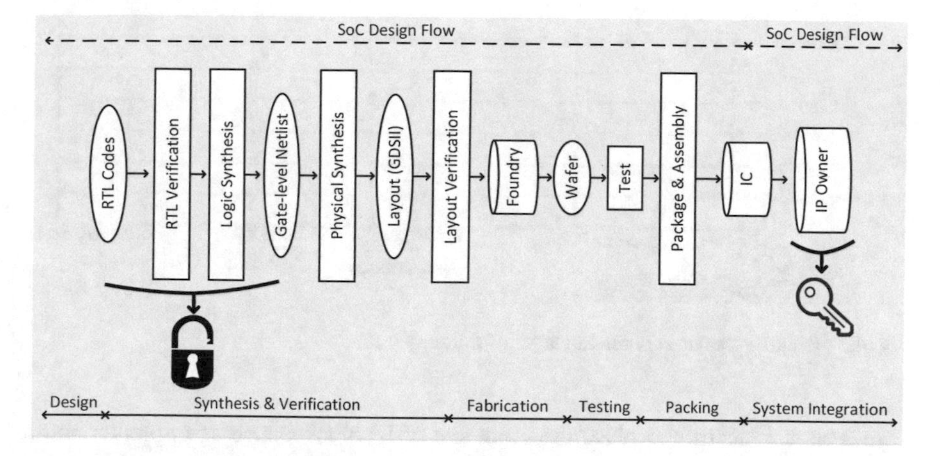

Fig. 4.2 Different stages at which hardware locking/encryption methods are applied

locking [14–18], (ii) finite state machine (FSM) locking [19–28], and (iii) locking using reconfigurable components [29–31]. The combinational logic locking method includes cryptographic algorithm for locking and logic encryption. The FSM locking methods include hardware obfuscation techniques and active hardware metering. The FPGA protection methods focus on securing the bitstream. In the rest of this section we will describe the threat model and each prevention method in details.

4.2.1 Threat Model

Security threats to an IP/IC vary depending on the location of the adversary in the supply chain (see Fig. 4.3). Below we briefly explain these threats to an IP/IC:

- *Cloning*: Adversary creates an exact copy or clone of the original product and sell it under a different label. To carry out cloning of ICs, an attacker should be either a manufacturer, system integrator, or a competing company equipped with necessary tools.
- *Counterfeiting*: When cloned products are sold under the label of the original vendor, without their authorization, it is called counterfeiting. This can be performed by an attacker in a manufacturing facility or by companies having capability to manufacture replicas of the original chip.
- *IC Overbuilding*: Another threat to an IC designer is overbuilding. In overbuilding, manufacturer or system integrator fabricates more IC than authorized.
- *IP Piracy*: In IP piracy, a system integrator steals the IP to claim its ownership or sell it illegally.
- *Reverse Engineering*: By analyzing an existing IC, manufacturers, system integrators, or companies having reverse engineering capabilities, can bypass security

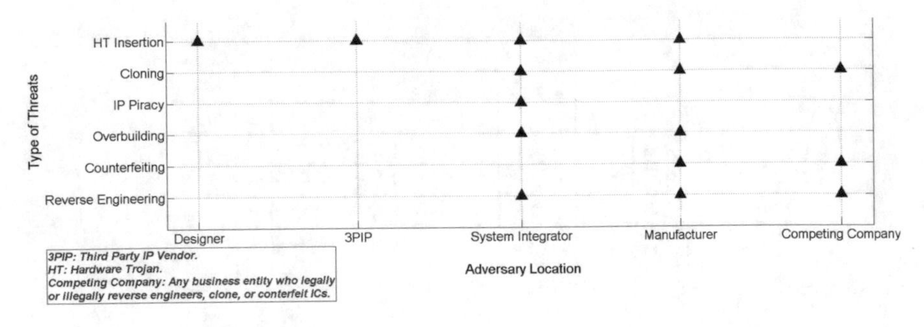

Fig. 4.3 Security threats at different stages of the supply chain

measures to learn and reuse the elements of IP blocks, such as implementation strategies, optimization algorithms, and design details. Consequently, adversaries can recreate the IP cores and sell them at much lower cost. Note that reverse engineering can also be used for IP piracy detection but it is out of the scope of this chapter.

- *Malicious Tampering*: Third-party IP providers can perform malicious tampering of IP by inserting hardware Trojans into the IP cores. Such attacks can also be performed by system integrators who have access to the IP core, or by manufacturer who can manipulate the lithographic masks by adding, deleting, or modifying gates/wires.

To protect an IP/IC from such threats, many prevention methods have been proposed, which are described in Sects. 4.2.2 and 4.2.3.

4.2.2 Combinational Logic Locking/Encryption

Combinational logic locking augments a combinational logic network by adding a group of lock inputs. Only if the correct key (generated at random) to the lock inputs is applied, the augmented network can produce the correct functionality [14, 15]. The locking mechanism is carried out by adding XOR gates on non-critical paths of the IC and by providing control to the key registers. Combinational locking and public-key cryptography was used in [14] for protecting the IC design. In [15], an attack was demonstrated to break the IP/IC protection method of [14] by "sensitizing" the key values to the output. The attacker on obtaining the modified netlist identified the unknown key by observing the output, provided the other key bits did not interfere with the "sensitized" path. In order to prevent key-leakage by "key-sensitization," key gates were inserted in such a way that propagation of a key value was possible only if certain conditions were forced on other key inputs. As these key inputs were not accessible to an attacker, they could not "sensitize" key values, thereby leaving the attacker the only option of brute force attacks. An algorithm was presented in [15], which used key gates interference graphs for insertion of key gates in such a way that attackers needed exponential number of brute force attempts to decipher the key. Compared to random insertion, this procedure incurred less area overhead as it required less number of XOR/XNOR as key gate. Another limitation of the combinational logic locking method of [14] was that inappropriate key input did not affect the output of the circuit.

Logic encryption was proposed in [16, 17], which used conventional fault simulation techniques and tools to guide the XOR/XNOR gate insertions and produced wrong output with 50 % Hamming distance between the correct and wrong outputs for an invalid key. This method masked the functionality and the implementation of a design by inserting key gates into the original design. To prevent collusion attack, physical unclonable functions (PUFs) were used to produce unique user keys for encrypting each IC. Instead of encrypting the design file by a cryptographic algorithm, the *logic encryption* method encrypted the hardware functionality. The performance overhead of this method was smaller than random key gate insertion method as it used smaller number of XOR/XNOR gates to achieve the 50 % Hamming distance.

Another combinational locking method was proposed in [18], which protected ICs from illegal overproduction and potential hardware Trojans insertion by minimizing rare signals of a circuit. The method made it harder for an attacker to exploit rare signals of a circuit to incorporate a hardware Trojan. An encryption algorithm modified the circuit, but preserved its functionality. The algorithm uses a probability-based method [32] to identify signals with low controllability. Among the identified signals, candidate signals for encryption were the ones with an unbalanced probability (signals with probability below 0.1 or above 0.9). For encryption, AND/OR gates were inserted in paths with large slack time and unbalanced probability. The type of gates to be inserted depended on the value of probability on the signal. When the probability of the signal was close to 0, an OR gate was included and the corresponding key value was 0 and when the probability was close to 1, an AND gate was included and the corresponding key value was 1. However, this method could not create multiple encryption key for the same design and hence, all the IP consumers of the design used the same key. Due to this limitation, it was not effective for preventing IP piracy.

4.2.3 Finite State Machine Locking/Encryption

Finite state machine (FSM) locking obfuscates a design by augmenting its state machine with a set of states. The modified FSM transit from the obfuscated states to the normal operating states after applying the specific input sequence (aka obfuscation key). The obfuscation method approximately transform the hardware design by preserving its functionality.

The FSM-based obfuscation method protects an IP/IC from reverse engineering, IP piracy, IC overproduction, and hardware Trojan insertion. Several variations of this method have been proposed in the literature [14–31]. In this chapter, we discuss three of its variations: (i) obfuscation by modifying gate-level netlist, (ii) obfuscation by modifying RTL code, and (iii) obfuscation using reconfigurable logic.

Obfuscation by Modifying Gate-Level Netlist

One of the methods for obfuscating a hardware design is to insert an FSM in the gate-level netlist [19–23]. The method in [19] obfuscated and authenticated an IP core by incorporating structural modification in the netlist. Along with the state transition function, large fan-in and fan-out nodes of the netlist were modified such that the design produces undesired output until a specific vector was given at the primary inputs. The circuit was re-synthesized after the modification to hide the structural changes. The FSM, which was inserted into the netlist to modify the state transition graph (STG), was connected to the primary inputs of the circuit. Depending on an initialization key sequence, the FSM operated the IP core in either the normal mode or the obfuscated mode. The maximum number of unique fan-out nodes (N_{max}) of the netlist were identified using an *iterative ranking algorithm*. The output of the FSM and the modified (N_{max}) nodes were given to an XOR gate. When the FSM output was 0, the design produced correct behavior. Although the method required less area and power overheads, it did not analyze security of the design. New methods were proposed to overcome this limitation [20–22]. A metric was developed to quantify the mismatch between the obfuscated design and the original design [20]. Also, the effect of obfuscation on security of the design was evaluated. Certain modifications were made to the methodology of [19] such as embedding "modification kernel function" for modifying the nodes selected by the *iterative ranking algorithm* and adding an authentication FSM, which acted as a digital watermark. These modifications prevented attacks from untrusted parties in the design flow with knowledge of the initialization sequence.

The obfuscation scheme of [20] was extended in [21, 22] to prevent insertion of trigger-activated hardware Trojans. The new method also ensured that such Trojans, when activated in the obfuscated mode, did not affect normal operation of the circuit. To incorporate these changes, the obfuscation state space was divided into (i) initialization state space and (ii) isolation state space [21]. On applying the correct key at power on, the circuit transitioned from the initialization state space to normal state space. However, an incorrect key transitioned the circuit to isolation state space from which the circuit could not return to the normal state space. Due to the extreme rareness of the transition condition for normal state space, it was assumed that the attacker was stuck in the isolation state space. Also, the insertion of a Trojan with wrong observability/controllability in the obfuscation mode would increase its detection probability at post-manufacturing testing. To further increase the probability of Trojan detection, the state space of the obfuscated mode was made larger than the normal mode. The proposed methodology was robust enough to prevent reverse engineering of modified netlist with large sequential circuits. Also, the area and power overheads of the method were relatively low.

However, the methodology of [20] could not protect an evaluation version of firm IP core. In [23], this problem was overcome by embedding a FSM in the IP netlist. The FSM disrupted normal functional behavior of the IP after its evaluation period.

The number of cycles required to activate the FSM depended on the number of bits in the state machine. Also, the activation probability decreased if the number of trigger nodes were increased. This method helped in putting an expiry date on the evaluation copy of the hardware IP. To distinguish between legally sold version of an IP and its evaluation version containing the FSM, IP vendors either (i) used disabling key to deactivate the FSM or (ii) provided a FSM-free version. The method structurally and functionally obfuscated the FSM to conceal it during reverse engineering. Area overhead of this method was directly proportional to the size of the FSM. However, the overhead decreased with an increase in size of the original IP.

Obfuscation by Modifying RTL Code of Design

Apart from netlist, obfuscation can also be carried out in RTL code of the design [24–27]. A key-based obfuscation approach for protecting synthesizable RTL cores was developed in [27]. The RTL design was first synthesized into a technology independent gate-level netlist and then obfuscated using the method of [19]. Then, the obfuscated netlist was decompiled into RTL in such a way that the modifications made on the netlist were hidden and the high-level HDL constructs preserved. A simple metric was presented to quantify the level of such structural and semantic obfuscation. This approach incurred minimal design overhead and did not affect adversely the automatic synthesis process of the resultant RTL code. However, decompilation removed some preferred RTL constructs and hence made the design undesirable for certain preferred design constraints. This limitation was overcome in [26], where the RTL code was first converted into a control and data flow graph (CDFG) and then a key-activated "Mode-Control FSM" was inserted to obfuscate the design. The CDFG was built by parsing and transforming concurrent blocks of RTL code. Small CDFGs were merged to build larger ones with more nodes. This helped in better obfuscation of the "Mode-Control FSM," which operated the design either in normal or obfuscated mode. This FSM was realized in the RTL by modifying a set of host registers. To further increase the level of obfuscation, state elements of the FSM were distributed in a non-contiguous manner inside one or more registers. After hosting the FSM in a set of selected host registers, several CDFG nodes were modified using the control signals generated from the FSM. The nodes with large fan-out cones were selected for modification, as they ensured maximum change in functional behavior at minimal design overhead. At the end of modifications on the CDFG, the obfuscated RTL was generated, which on powering up, initialized the design at the obfuscated mode. Only on the application of a correct input key, the "mode-control" FSM transited the system through a sequence of states to the normal mode. This method incurred low area and power overheads.

Another approach obfuscates the RTL core by dividing the overall functionality of the design into two modes, "Entry/Obfuscated mode" and "Functional mode," and encoding the path from the *entry/obfuscated mode* to the *functional mode* with a "code-word" [24]. The functionality of the circuit was divided by modifying the FSM representing the core logic. To modify the FSM, its states were extended and divided into *entry/obfuscated mode* and *functional mode* states. Only on the application of the right input sequence at the *entry mode*, the correct *Code-Word* was formed, which produced correct transitions in the *functional mode*. Unlike those methods where an invalid key disallowed entry to the *normal mode*, this method always allowed entry to the *functional mode*. However, the behavior in the functional mode depended on the value of the *Code-Word*. This *Code-Word* was not stored anywhere on chip, but it was formed dynamically during the *entry mode*. It was integrated into the transition logic to make it invisible to the attacker. In this method, the length of the *Code-Word* was directly related to the area overhead and security level required by the designer. A longer *Code-Word* meant higher level of security against brute force attacks, but at the cost of higher area overhead.

In [25], a key-based obfuscation method having two modes of operation, *normal mode* and *slow mode*, was developed to prevent IP piracy on sequential circuits. This method modified the state transition graph (STG) in such a way that the design operated in either mode depending on whether it was initialized with the correct key state. The key state was embedded in the power-up states of the IC and was known only to the IP owner. When the IP owner received the fabricated chip, power-up states were reset from the fixed initial state to the key state. As the number of power-up states was less in the design, chances of the IC being operational in the *normal mode* on random initialization were significantly reduced. Moreover, powering up the design with an incorrect initial state operated the IC in the *slow mode*, where it functioned slower than the *normal mode* without causing significant performance difference. This functionality prevented IP pirates from suspecting the performance degradation in the IC and the presence of *key state* in the design. To modify the STG, four structural operations were performed: (i) *retiming*, (ii) *resynthesis*, (iii) *sweep*, and (iv) *conditional stuttering*. In *retiming*, registers were moved in the sequential circuit using any one of the two operations. These operations included (i) adding a register to all outputs and deleting a register from each input of a combinational node and (ii) deleting a register from all outputs and adding a register to each input of a combinational node. *Resynthesis* restructured the netlist within the register boundaries whereas removal of redundant registers and logic that did not affect output was done using *sweep*. Both the *resynthesis* and the *retiming* operations preserved logical functionality of the design. *Conditional stuttering* involved addition of control logic to the circuit to stutter the registers under a given logic condition. On the other hand, inverse *conditional stuttering* removed certain control logic. *Stuttering* operations were done to obtain circuits which were cycle–accurate–equivalent. This method mainly focused on those real-time applications which were very sensitive to throughput. Unlike existing IC metering techniques, the secret key in this method was implicit, thus making it act as a hidden watermark. However, the area and power overheads of this method were higher than previous approaches.

An *active hardware metering* approach prevents overproduction of IC by equipping designers with the ability to lock each IC and unlock it remotely [28]. In this method, new states and transitions were added to the original finite state machine (FSM) of the design to create a *boosted finite state machine* (BFSM). This structural manipulation preserved the behavioral specification of the design. Upon activation, the BFSM was placed in the power-up state using an unique ID which was generated by the IC. To bring the BFSM into the functional initial state, the designer used an input sequence generated from the transition table. Black hole states were integrated with the BFSM to make the *active metering* method highly resilient against the brute force attacks. This method incurred low overhead and was applicable to industrial-size design.

Obfuscation Using Reconfigurable Logic

Reconfigurable logic was used in [29, 30] for obfuscation of ASIC design. In [29], reconfigurable logic modules were embedded in the design and their final implementation was determined by the end user after the design and the manufacturing process. This method assumed that the supply chain adversary has knowledge of the entire design except the exact implementation of the function and the internal structure of reconfigurable logic modules. The lack of knowledge prohibited an adversary from tampering the logic blocks. Combining this method with other security techniques provided data confidentiality, design obfuscation, and prevention from hardware Trojans. In the demonstration, a code injection Trojan was considered which, when triggered by a specific event, changed input–output behavior or leaked confidential information. The Trojan was assumed to be injected at the instruction decoded unit (IDU) of a processor during runtime and it was not detected by non-lock stepping concurrent checking methods, code integrity checker, and testing. To prevent such a Trojan attack, instruction set randomization (ISR) was done by obfuscating the IDU. For obfuscation, reconfigurable logic was used, which concealed opcode check logic, instruction bit permutation, or XOR logic. This method prevented the Trojan from monitoring an authentic computation. Moreover, Trojans which were designed to circumvent this method no longer remained stealthy and those trying to duplicate the IDU or modify it caused significant performance degradation. It was shown that the minimum code injection Trojan with a 1 KB ROM resulted in an area increase of 2.38 % for every 1 % increase in the area of the LEON2 processor.

To hide operations within the design and preserve its functionality, the original circuit was replaced with PUF-based logic and FPGA in [30]. PUF was also used to obfuscate signal paths of the circuit. The architecture for signal path obfuscation was placed in a location where most flip-flops were affected most number of times. To prevent this technique from affecting critical paths, wire swapping components (MUX'es with PUF as select input) were placed between gates with positive slack. The PUF-based logic and signal path obfuscation techniques were used simultaneously to minimize delay constraints of the circuit and maximize its security under user-specified area and power overheads. Two types of attacks were considered: (i) adversary can read all flip-flops, but can only write to primary inputs; (ii) adversary can read and write to all flip-flops of the circuit [30]. It was assumed that the adversary has complete knowledge of circuit netlist, but not of input–output mapping of any PUFs. For preventing the first type of attack, FPGA was used after the PUF and PUF were made large to accept a large challenge. To prevent the second attack, a PUF was placed in a location that was difficult to control directly using the primary inputs of the circuit. By preventing this attack, reverse engineering was made difficult for an adversary. These two methods were demonstrated on ISCAS 89 and ITC 99 benchmarks and functionality of circuits were obfuscated with area overhead upto 10 %.

In [31], obfuscation of DSP circuit was done using high-level transformation, key-based FSM, and a reconfigurator. High-level transformation does structural obfuscation of the DSP circuit at HDL or netlist level by preserving its functionality. This transformation was chosen based on the DSP application and performance requirements (e.g., area, speed, power, or energy). For performing high-level transformation on the circuit, reconfigurable secure switches were designed in [31]. These switches were implemented as multiplexers, whose control signals were obtained from a FSM designed using ring counters. Securities of these switches were directly related to the design of the ring counters. Another FSM, called the *obfuscated FSM*, was incorporated in the DSP circuit along with the reconfigurator. A configuration key was given to the *obfuscated FSM* for operating the circuit correctly. This key consisted of two parts: an *L-bit* initialization key and a *K-bit* configure data. The initialization key was used to activate reconfigurator via the *obfuscated FSM*, whereas *configure data* was used by the reconfigurator to control the operation of the switches. As configuration of the switches required correct initialization key and configure data, attacks targeting either of them could not affect the design. An adversary attempting to attack a DSP circuit, obfuscated with this method, had to consider the length of the configuration key and the number of input vectors required for learning the functionality of each variation mode. Structural obfuscation degree (SOD) and functional obfuscation degree (FOD) were used as metrics for measuring simulation-based attack and manual attack (visual inspection and structural analysis). SOD was estimated for manual attacks, whereas FOD estimated obfuscation degree of simulation-based attacks. A higher value of SOD and FOD indicated a more secure design. The area and the power overheads of this method were low.

4.2.4 Protection Methods for FPGA IP

Field-programmable gate arrays (FPGAs) have been widely used for many applications since 1980s. They provide considerable advantage in regards to design cost and flexibility. Due to accelerated time-to-market, designing a complete system on FPGA is becoming a daunting task. To meet the demands, designers have started using/reusing third-party intellectual property (IP) modules, rather than developing a system from scratch. However, this has raised the risk of FPGA IP piracy. Protection methods [33–36] have been proposed to mitigate this issue. In [35], a protection scheme was proposed which used both public-key and symmetric-key cryptography. To reduce area overhead, the public-key functionality was moved to a temporary configuration bitstream. Using five basic steps, the protection scheme enabled licensing of FPGA IP to multiple IP customers. However, this scheme restricted system integrators to the use of IP from a single vendor. The scope of the FPGA IP protection method of [35] was extended in [36], where system integrators could use cores from multiple sources. In [33], implementation of the protection methods of [35, 36] was carried out on commercially available devices. For securely transporting the key,

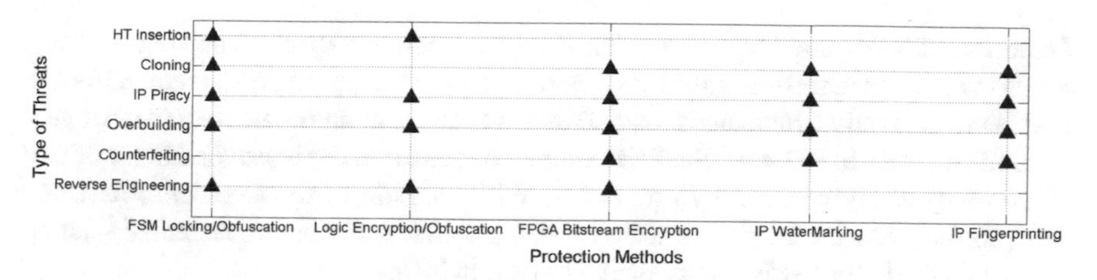

Fig. 4.4 Different protection methods

[33] used symmetric cryptography and trusted third-party provider. Use of symmetric cryptography also reduced the size of temporarily occupied reconfigurable logic for building the IP decryption key.

A practical and feasible protection method for SRAM-based FPGA was given in [34]. This approach allowed licensing of IP cores on a per-device basis and it did not require contractual agreement with trusted third-parties, large bandwidth, and complicated communication processes. The IP instance was encrypted for each system integrator and decryption key was generated using the license for the chips. This procedure ensured that the licensed IP core was used only on the contracted devices. Moreover, it helped to prevent IP core counterfeiting by tracking the unique fingerprint embedded in the licensed IP instance of the vendor. The proposed scheme did not require an external trusted third-party (TTP) and was applicable on IP cores designed for commercial purposes. It also helped in secure transaction of IP cores and prevented sophisticated attackers from cloning, reverse engineering, or tampering contents of the IP core.

A summary of all the above-described protection methods is shown in Fig. 4.4.

4.3 IP Certification

Recently, pre-silicon trust evaluation approaches have been developed to counter the threat of untrusted third-party resources [37–39]. Most of these methods try to trigger the malicious logic by enhancing functional testing with extra test vectors. Toward this end, authors in [37] proposed a method for generating "Trojan Vectors" to activate hardware Trojans during functional testing. To identify suspicious circuitry, unused circuit identification (UCI) [39] method analyzed the RTL code to find lines of code that were never used. However, these methods assume that the attacker uses rarely-occurring events as Trojan triggers. This assumption was voided in [40], where hardware Trojans were designed using "less-rare" trigger events.

Due to the limitations of enhanced functional testing methods in security evaluation, researchers started looking into formal solutions. Although at its early stage, formal methods have already shown their benefits over testing methods in exhaustive security verification [41–44]. A multi-stage approach was used in [41] for detection of hardware Trojans by identifying suspicious signals. The stages of this method included assertion-based verification, code coverage analysis, redundant circuit removal, equivalence analysis, and use of sequential automatic test pattern generation (ATPG). In [42–44], the PCH framework ensured the trustworthiness of soft IP cores by verifying security properties. With the help of the Coq proof assistant [45], formal security properties were proved in PCH. A review of currently existing formal methods for hardware security is given in [46].

Interactive Theorem Prover

Theorem provers are used to prove or disprove properties of systems expressed as logical statements. Since 1960s, several automated and interactive theorem provers have been developed and used for proving properties of hardware and software systems. However, verifying large and complex systems using theorem provers require excessive effort and time. Moreover, automated theorem provers require more developmental effort than proof assistants. Despite these limitations, theorem provers have currently drawn a lot of attentions in verification of security properties on hardware designs. Among all the formal methods, they have emerged as the most prominent solution for providing high-level protection of the underlying designs.

The proof-carrying hardware (PCH) framework uses interactive theorem prover and SAT solvers for verifying security properties on soft IP cores. This approach was used for ensuring the trustworthiness of RTL and firm cores [42, 47–49]. It is inspired from the proof-carrying code (PCC), which was proposed in [50]. Using the PCC mechanism, untrusted software developers/vendors certify their software code. During the certification process, software vendor develops *safety proof* for the safety policies provided by software customers. The vendor then provides the user with a PCC binary file, which includes the formal proof of the safety properties encoded with the executable code of the software. The customer becomes assured of the safety of the software code by quickly validating the PCC binary file in a proof checker. Efficiency of this approach in reducing validation time at the customer end led to its adoption in different applications.

Using the concept of PCC, authors in [48, 49, 51, 52] developed the PCH framework for dynamically reconfigurable hardware platforms. In this framework, authors used runtime combinational equivalence checking (CEC) for verifying equivalence between the design specification and the design implementation. A Boolean satisfiability (SAT) solver was used to generate resolution proof for unsatisfiability of the combinational miter circuit, represented in a conjunctive normal form (CNF). The proof traces were combined with the bitstream into a proof-carrying bitstream by the vendor and given to the customer for validation. However, the approach did not consider exchange of a set of security properties between the customer and the vendor. Rather it considers safety policy, which included agreements on a specific bitstream format, on a CNF to represent combinational functions, and the propositional calculus for proof construction and verification.

Another PCH framework was proposed in [42, 47], which overcame the limitations of the previous framework and expanded it for verification of security properties on soft IP cores. The new PCH framework was used for security property

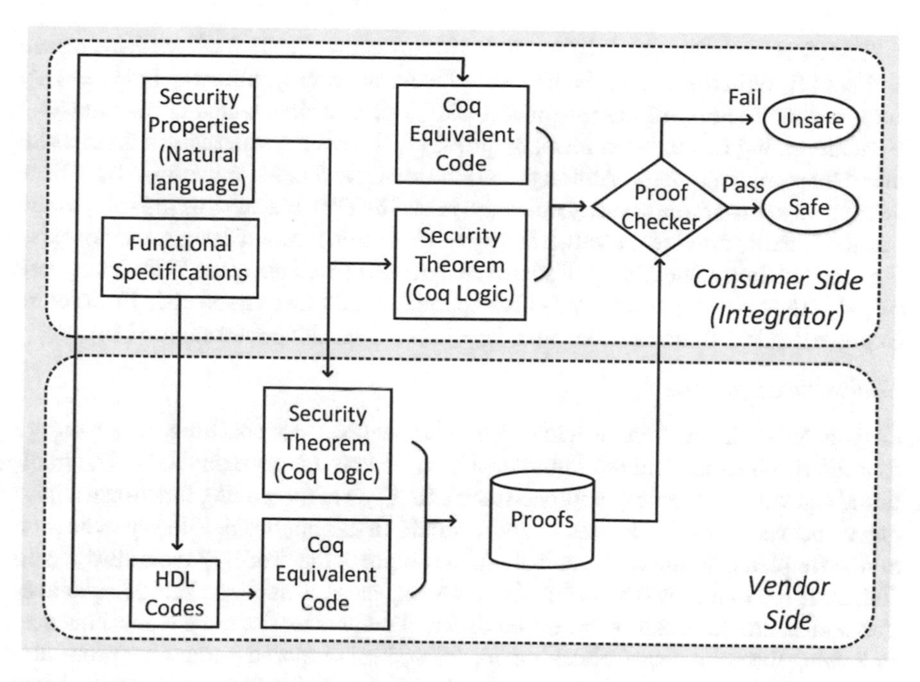

Fig. 4.5 Working procedure of the PCH framework [46]

verification on synthesizable IP cores. In the framework, Hoare-logic style reasoning was used to prove the correctness of the RTL code and implementation was carried out using the Coq proof assistant [45]. As Coq supported automatic proof checking, the security proof validation effort of IP customers was reduced. Moreover, usage of the Coq platform by both IP vendors and IP consumers ensures that same deductive rules could be used for validating the proof. However, Coq does not recognize commercial hardware description languages (HDLs) and security properties expressed in a natural language. To solve this problem, semantic translation of HDLs and informal security specifications to calculus of inductive construction (CIC) was done. Based on this PCH framework, a new trusted IP acquisition and delivery protocol was proposed (see Fig. 4.5), in which IP consumers provided both functional specifications and a set of security properties to IP vendors. IP vendors then developed the HDL code based on the functional specifications. The HDL code and security properties were then translated to CIC. Subsequently, proofs were constructed for security theorems and the transformed HDL code. The HDL code and proof for security properties were combined into a trusted bundle and delivered to the consumer. Upon receiving the trusted bundle, IP consumers first generate the formal representation of the design and security properties in CIC. The translated code, combined with formal theorems and proofs, was quickly validated using the proof checker in Coq platform.

The PCH framework was also extended to support verification of gate-level circuit netlist [44]. With the help of the new gate-level framework, authors in [44] formally analyzed the security of design-for-test (DFT) scan chains, which is the industrial standard testing method, and formally proved that a circuit with scan chain can violate data secrecy property. Although various attack and defense methods have been developed to thwart the security concerns raised by DFT scan chains [53–58], methods for formally proving the vulnerability of scan chain inserted designs did not exist. For the first-time vulnerability of such a design was proved using the PCH framework of [44]. The same framework was also applied in built-in-self-test (BIST) structure to prove that BIST structure can also leak internal sensitive information [44].

Equivalence Checking

Orthogonal to the theorem proving approach is equivalence checking, which ensure that the specification and the implementation of a circuit are equivalent. The traditional equivalence checking approach uses a SAT solver for proving functional equivalence between two representations of a circuit. In this approach, if the specification and the implementation were equivalent, the output of the "xor" gate was always zero (false). If the output was true for any input sequence, it implied that the specification and the implementation produced different outputs for the same input sequence. Following the equivalence checking approach, [38] proposed a four-step procedure to filter and locate suspicious logic in third-party IPs. In the first step, easy-to-detect signals were removed using functional vectors generated by a sequential ATPG. In the next step, hard-to-excite and/or propagate signals were identified using a full-scan N-detect ATPG. To narrow down the list of suspected signals and identify the actual gates associated with the Trojan, a SAT solver was used in the third step for equivalence checking of the suspicious netlist containing the rarely triggered signals against the netlist of the circuit exhibiting correct behavior. At the final step, clusters of untestable gates in the circuit were determined using the region isolation approach on the suspected signals list.

However, traditional equivalence checking techniques could result in state space explosion when large IP blocks were involved with significantly different specifications and implementations. They also could not be used on complex arithmetic circuits with larger bus widths. An alternative approach was to use computer symbolic algebra for equivalence checking of arithmetic circuit. These circuits constituted a significant portion of datapath in signal processing, cryptography, multimedia applications, error root causing codes, etc. Due to this, their chances of malfunctioning were very high. The new equivalence checking approach allowed verification of such large circuits and it did not cause state space explosion.

4.4 Conclusion

In this chapter, we analyzed existing prevention and certification methods for soft/firm hardware IP cores. The prevention methods largely consisted of various hardware locking/encryption schemes. These methods protected IP cores from piracy, overbuilding, reverse engineering, cloning, and malicious modifications. On the other hand, formal methods, such as theorem proving and equivalence checking, helped validate the trustworthiness of IP cores. These methods can help certify the trustworthiness of IP cores. Meanwhile, after a thorough analysis of all these proposed IP validation/protection methods, we realized that a single method is not sufficient to eliminate all the threats to IP cores. Combination of these methods becomes a necessity in order to ensure the security of IP cores and further secure the modern semiconductor supply chain.

Acknowledgements This work was supported in part by the National Science Foundation (CNS-1319105).

References

1. Kocher, P.: Advances in Cryptology (CRYPTO'96). Lecture Notes in Computer Science, vol. 1109, pp. 104–113 (1996)
2. Kocher, P., Jaffe, J., Jun, B.: Advances in Cryptology–CRYPTO'99, pp. 789–789 (1999)
3. Quisquater, J.J., Samyde, D.: Smart Card Programming and Security. Lecture Notes in Computer Science, vol. 2140, pp. 200–210 (2001)
4. Gandolfi, K., Mourtel, C., Olivier, F.: Cryptographic Hardwarc and Embedded Systems (CHES) 2001. Lecture Notes in Computer Science, vol. 2162, pp. 251–261 (2001)
5. Chari, S., Rao, J.R., Rohatgi, P.: Cryptographic Hardware and Embedded Systems—Ches 2002. Lecture Notes in Computer Science, vol. 2523, pp. 13–28. Springer, Berlin (2002)
6. Messerges, T.S., Dabbish, E.A., Sloan, R.H.: IEEE Trans. Comput. **51**(5), 541 (2002)
7. Tiri, K., Akmal, M., Verbauwhede, I.: Solid-State Circuits Conference, 2002. ESSCIRC 2002. Proceedings of the 28th European, pp. 403–406 (2002)
8. Fan, Y.C., Tsao, H.W.: Electr. Lett. **39**(18), 1316 (2003)
9. Torunoglu, I., Charbon, E.: IEEE J. Solid-State Circuits **35**(3), 434 (2000)
10. Kahng, A.B., Lach, J., Mangione-Smith, W.H., Mantik, S., Markov, I.L., Potkonjak, M., Tucker, P., Wang, H., Wolfe, G.: IEEE Trans. Comput.-Aided Des. Integr. Circuits Syst. **20**(10), 1236 (2001)
11. Lach, J., Mangione-Smith, W.H., Potkonjak, M.: IEEE Trans. Comput.-Aided Des. Integr. Circuits Syst. **20**(10), 1253 (2001)
12. Qu, G., Potkonjak, M.: Proceedings of the 37th Annual Design Automation Conference, pp. 587–592 (2000)
13. Chang, C.H., Zhang, L.: IEEE Trans. Comput.-Aided Des. Integr. Circuits Syst. **33**(1), 76 (2014)
14. Roy, F.K.J.A., Markov, I.L.: Design, Automation and Test in Europe (DATE), vol. 1 (2008)
15. Rajendran, J., Pino, Y., Sinanoglu, O., Karri, R.: Design Automation Conference (DAC), 2012 49th ACM/EDAC/IEEE, pp. 83–89 (2012)
16. Rajendran, J., Pino, Y., Sinanoglu, O., Karri, R.: Design, Automation Test in Europe Conference Exhibition (DATE), vol. 2012, pp. 953–958 (2012). doi:10.1109/DATE.2012.6176634

17. Rajendran, J., Zhang, H., Zhang, C., Rose, G., Pino, Y., Sinanoglu, O., Karri, R.: IEEE Trans. Comput. **99** (2013)
18. Dupuis, S., Ba, P.S., Natale, G.D., Flottes, M.L., Rouzeyre, B.: Conference on IEEE 20th International On-Line Testing Symposium (IOLTS), IOLTS '14, pp. 49–54 (2014)
19. Chakraborty, R., Bhunia, S.: IEEE/ACM International Conference on Computer-Aided Design 2008. ICCAD 2008, pp. 674–677 (2008). doi:10.1109/ICCAD.2008.4681649
20. Chakraborty, R., Bhunia, S.: IEEE Trans. Comput.-Aided Des. Integr. Circuits Syst. **28**(10), 1493 (2009). doi:10.1109/TCAD.2009.2028166
21. Chakraborty, R.S., Bhunia, S.: J. Electr. Test. **27**(6), 767 (2011). doi:10.1007/s10836-011-5255-2
22. Chakraborty, R., Bhunia, S.: IEEE/ACM International Conference on Computer-Aided Design—Digest of Technical Papers, 2009. ICCAD 2009, pp. 113–116 (2009)
23. Narasimhan, S., Chakraborty, R., Bhunia, S.: IEEE Des. Test Comput. **99**(PrePrints) (2011). http://doi.ieeecomputersociety.org/10.1109/MDT.2011.70
24. Desai, A.R., Hsiao, M.S., Wang, C., Nazhandali, L., Hall, S.: Proceedings of the Eighth Annual Cyber Security and Information Intelligence Research Workshop, CSIIRW '13, pp. 8:1–8:4. ACM, New York (2013). doi:10.1145/2459976.2459985
25. Li, L., Zhou, H.: 2013 IEEE International Symposium on Hardware-Oriented Security and Trust, HOST 2013, Austin, TX, USA, June 2–3, pp. 55–60 (2013). doi:10.1109/HST.2013.6581566
26. Chakraborty, R., Bhunia, S.: 23rd International Conference on VLSI Design, 2010. VLSID'10, pp. 405–410 (2010). doi:10.1109/VLSI.Design.2010.54
27. Chakraborty, R., Bhunia, S.: IEEE International Workshop on Hardware-Oriented Security and Trust, 2009. HOST '09, pp. 96–99 (2009). doi:10.1109/HST.2009.5224963
28. Alkabani, Y., Koushanfar, F.: USENIX Security, pp. 291–306 (2007)
29. Liu, B., Wang, B.: Design. Automation and Test in Europe Conference and Exhibition (DATE), pp. 1–6 (2014). doi:10.7873/DATE.2014.256
30. Wendt, J.B., Potkonjak, M.: Proceedings of the 2014 IEEE/ACM International Conference on Computer-Aided Design, ICCAD'14, pp. 270–277. IEEE Press, Piscataway (2014). http://dl.acm.org/citation.cfm?id=2691365.2691419
31. Lao, Y., Parhi, K.: IEEE Trans. Very Large Scale Integr. (VLSI) Syst. **99**, 1 (2014). doi:10.1109/TVLSI.2014.2323976
32. Natale, G.D., Dupuis, S., Flottes, M.L., Rouzeyre, B.: Workshop on Trustworthy Manufacturing and Utilization of Secure Devices (TRUDEVICE13) (2013)
33. Maes, R., Schellekens, D., Verbauwhede, I.: IEEE Trans. Inf. Forensics Secur. **7**(1), 98 (2012)
34. Zhang, L., Chang, C.H.: IEEE Trans. Inf. Forensics Secur. **9**(11), 1893 (2014)
35. Guneysu, T., Moller, B., Paar, C.: IEEE International Conference on Field-Programmable Technology, ICFPT, pp. 169–176 (2007)
36. Drimer, S., Güneysu, T., Kuhn, M.G., Paar, C.: (2008). http://www.cl.cam.ac.uk/sd410/
37. Wolff, F., Papachristou, C., Bhunia, S., Chakraborty, R.S.: IEEE Design Automation and Test in Europe, pp. 1362–1365 (2008)
38. Banga, M., Hsiao, M.: IEEE International Symposium on Hardware-Oriented Security and Trust (HOST), pp. 56–59 (2010)
39. Hicks, M., Finnicum, M., King, S.T., Martin, M.M.K., Smith, J.M.: Proceedings of IEEE Symposium on Security and Privacy, pp. 159–172 (2010)
40. Sturton, C., Hicks, M., Wagner, D., King, S.: 2011 IEEE Symposium on Security and Privacy (SP), pp. 64–77 (2011)
41. Zhang, X., Tehranipoor, M.: 2011 IEEE International Symposium on Hardware-Oriented Security and Trust (HOST), pp. 67–70 (2011)
42. Love, E., Jin, Y., Makris, Y.: IEEE Trans. Inf. Forensics Secur. **7**(1), 25 (2012)
43. Jin, Y., Yang, B., Makris, Y.: IEEE International Symposium on Hardware-Oriented Security and Trust (HOST), pp. 99–106 (2013)
44. Jin, Y.: IEEE Computer Society Annual Symposium on VLSI (ISVLSI) (2014)
45. INRIA: The coq proof assistant (2010). http://coq.inria.fr/

46. Guo, X., Dutta, R.G., Jin, Y., Farahmandi, F., Mishra, P.: Design Automation Conference (DAC), 2015 52nd ACM/EDAC/IEEE (2015) (To appear)
47. Love, E., Jin, Y., Makris, Y.: 2011 IEEE International Symposium on Hardware-Oriented Security and Trust (HOST), pp. 12–17 (2011)
48. Drzevitzky, S., Kastens, U., Platzner, M.: International Conference on Reconfigurable Computing and FPGAs, pp. 189–194 (2009)
49. Drzevitzky, S.: International Conference on Field Programmable Logic and Applications, pp. 255–258 (2010)
50. Necula, G.C.: POPL'97: Proceedings of the 24th ACM SIGPLAN-SIGACT Symposium on Principles of Programming Languages, pp. 106–119 (1997)
51. Drzevitzky, S., Platzner, M.: 6th International Workshop on Reconfigurable Communication-Centric Systems-on-Chip, pp. 1–8 (2011)
52. Drzevitzky, S., Kastens, U., Platzner, M.: Int. J. Reconfig. Comput. **2010** (2010)
53. Yang, B., Wu, K., Karri, R.: Test Conference, 2004. Proceedings. ITC 2004. International, pp. 339–344 (2004)
54. Nara, R., Togawa, N., Yanagisawa, M., Ohtsuki, T.: Proceedings of the 2010 Asia and South Pacific Design Automation Conference, pp. 407–412 (2010)
55. Yang, B., Wu, K., Karri, R.: IEEE Trans. Comput.-Aided Des. Integr. Circuits Syst. **25**(10), 2287 (2006)
56. Sengar, G., Mukhopadhyay, D., Chowdhury, D.: IEEE Trans. Comput.-Aided Des. Integr. Circuits Syst. **26**(11), 2080 (2007)
57. Da Rolt, J., Di Natale, G., Flottes, M.L., Rouzeyre, B.: 2012 IEEE 30th VLSI Test Symposium (VTS), pp. 246–251 (2012)
58. Rolt, J., Das, A., Natale, G., Flottes, M.L., Rouzeyre, B., Verbauwhede, I.: Constructive Side-Channel Analysis and Secure Design. In: Schindler, W., Huss, S. (eds.) Lecture Notes on Computer Science, vol. 7275, pp. 89–104. Springer, Berlin (2012)

Chapter 5
Security of Crypto IP Core: Issues and Countermeasures

Debapriya Basu Roy and Debdeep Mukhopadhyay

5.1 Introduction

The value of information in modern world has increased manifold in the last decade. Global spreading of Internet along with the recent advances in IoTs (Internet of Things) and PAN (Personalized Area Network) has forced the modern SoCs to handle large amount of sensitive information which are needed to be protected. Hence, cryptographic modules, protecting the sensitive information, are integral parts of modern SoCs. Most of the SoCs now contain HSM (Hardware Security Module) which is used for secure key generation and management for cryptographic operations along with secure crypto-processing. HSMs provide logical and physical protection for the digital keys which are used for encryption and authentication of secret information. Along with HSMs, a SoC may also contain dedicated hardware accelerator for some cryptographic operations to reduce the delay of the system. Generally, cryptographic operations involve computation intensive operations; hence usage of hardware accelerator is highly encouraged as it provides significant performance improvement of the SoC.

Both hardware accelerators and software routines generally implement standard cryptographic algorithms which are secure against theoretical attacks. The algorithms are standardized by *FIPS (Federal Information Processing Standards)* and *NIST (National Institute of Standards and Technology)* and we can consider them free of any theoretical or mathematical weakness which can be exploited by the adversaries. Example of such standard cryptographic algorithms are AES (Advanced Encryption Standard), RSA (Rivest Shamir Adleman Algorithm), ECC (Elliptic Curve Cryptography), etc. AES is the most popular symmetric key algo-

D.B. Roy (✉) · D. Mukhopadhyay
IIT, Kharagpur, India
e-mail: deb.basu.roy@cse.iitkgp.ernet.in

D. Mukhopadhyay
e-mail: debeep@cse.iitkgp.ernet.in

S. Bhunia et al. (eds.), *Fundamentals of IP and SoC Security*,
DOI 10.1007/978-3-319-50057-7_5

rithm whereas RSA and ECC are the popular asymmetric key algorithms. Implementation of these algorithms, both in hardware and software, can be considered as the crypto-IP core. They often form root of trust of the entire security module and hence need to be protected against malicious adversaries.

However, although these algorithms are proved to be secured in the classical sense where the attacker has the controllability of the input (plain-texts) and observability of the output (cipher-texts), in real life the scenarios are often different. This gap leads to the failure of the proofs when the adversary has access to much more information than what the classical cryptanalyst anticipated. Furthermore, the conventional designer engages several optimization techniques to improve the performance of a given algorithm. But in the field of cryptographic engineering it has been found repeatedly that naïve implementation optimizations had lead to catastrophic failures of the crypto-systems. Thus, performance of the crypto-systems is a delicate issue and the overhead of adding cryptographic layers should be as minimal as possible. Hence, a fresh approach is often required for developing cryptographic IP, where security is not an afterthought, but taken care from the beginning of the design cycle. In this chapter we will show that security cannot be guaranteed by just implementing a perfectly secure crypto-algorithm. A physical design of a crypto-algorithm can become vulnerable due to unintended leakages emitting from the design. This chapter provides a detailed analysis of such unintended leakages, popularly known as *side channel leakages*. We will also discuss what are the available countermeasures which can prevent these kinds of threats.

5.1.1 Implementation: An Issue for Security

In the recent past, it has been often found that security of a design is compromised due to the loopholes present in the design techniques. An example of this type of phenomenon is the recent *Heart Bleed Attack* on *OpenSSL* [1, 2] which affected lots of websites across the Internet. In OpenSSL, a *heartbeat* echoes back the data that the user has sent. The length of the data can be at most 64 KB. Thus a *heartbeat* comprises of two parts: data and length of the data. However, if an adversary sets the length as 64 KB and send just one byte of data, he will still get 64 KB of data as echo. This 64 KB of data comprised of one byte that the adversary has sent in heartbeat and rest of them are secret information which should not be disclosed to the adversary. Thus the reason of this *Heart Bleed Attack* is just a missing bound check in the code which does not affect the functionalities of OpenSSL, but compromises the security of the entire module. The web services affected by this *Heart Bleed Attack* includes *Gmail*, *Facebook* and *Yahoo* [3, 4]. Thus one can understand how much impact this kind of wrong engineering can have on security.

Optimization is another issue that need to be taken care of during design of a crypto-primitive. Modern CAD tools are powerful and are capable of analyzing a design and optimizing it if possible. However, optimizations may expose a design to several threats which may leak the key through some other stealthy channels. It is hence imperative for the designer of crypto-IPs to understand the possible threats and ensure that those leakages are properly mitigated. In particular in the following sequel, we shall consider the threats from side channel attacks, which leak information of intermediate states of a cipher which can be exploited to develop knowledge of the secret keys.

5.1.2 Side Channels

When a crypto-algorithm is cryptanalysed for any theoretical weakness, it is assumed that adversaries have access to the input (plain-text) and output (cipher-text) of the crypto-algorithm and the objective of the adversary is to find the secret key. In other words, the module executing crypto-algorithm is treated as black box to which the adversary can give plain-text as input and obtain cipher-text as output.

However, during the execution of any crypto-algorithm on a crypto-chip, an attacker has access to various other information other than the input (plain-text) and output (cipher-text). For example, the adversary can observe the power consumption, timing performance, electro-magnetic radiation or even acoustic behavior of the circuit. These information though seem harmless, they are highly alarming as they leak the internal state of the circuit, threatening compromise of sensitive information and defeating the very purpose of a crypto-system.

All these information that an adversary can observe from physical implementation of a crypto-algorithm are considered as side channel information, which essentially implies that we need to analyze the security of a crypto-system in a gray box model. Figure 5.1a shows the scenario when an adversary is carrying out theoretical cryptanalysis, whereas Fig. 5.1b shows the scenario when an adversary is equipped with the side channel information.

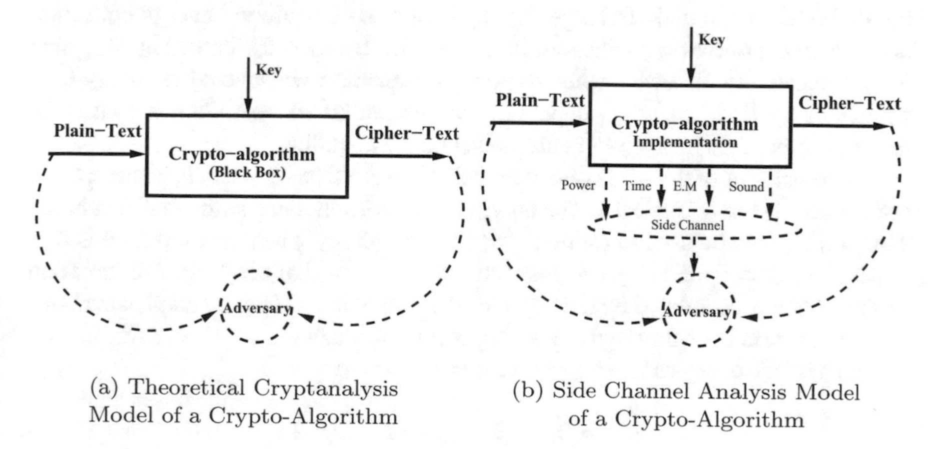

(a) Theoretical Cryptanalysis Model of a Crypto-Algorithm

(b) Side Channel Analysis Model of a Crypto-Algorithm

Fig. 5.1 Theoretical cryptanalysis and side channel analysis

Side channel analysis can be broadly classified into two different classes, namely passive and active side channel analysis, which are described as follows:

- Passive Side Channel Analysis: In passive side channel analysis, an adversary can only observe or record side channel information emitting from the system. For example, an adversary can observe power signatures or record electromagnetic radiations of the system to learn about the internal states of the circuits. These attacks, called as *power attack* and *electromagnetic attack*, are instances of passive side channel analysis.
- Active Side Channel Analysis: In active side channel analysis, an adversary cannot only observe the side channel information, but can also interfere with the circuit operation. An example of such active side channel analysis is *fault attack*. In this scenario an adversary intentionally injects faults during the circuit operation to obtain faulty outputs, which can be used to obtain the secret information.

In this chapter we are going to study in details about the different side channel analysis with more emphasis on power based side channel attack. We will start by describing various attack strategies and popular countermeasures. This chapter will also introduce the readers about the fault attack and side channel attack at the testing phase. In the next section we will introduce the rationals behind power based side channel analysis and corresponding attack strategies.

5.2 Power Analysis of Cryptographic Cores

Most modern VLSI circuits are made of CMOS (Complementary Metal Oxide Semiconductor) gates which have a power characteristic which depend on the transitions of data. This forms the basis for power analysis of cryptographic cores. Figure 5.2 shows a CMOS inverter which represents these class of underlying gates and shows the different charge and discharge paths which leads to different energy consumptions when the output capacitance gets charged or discharged, denoted by $E_{0\to1}$ and $E_{1\to0}$ respectively. Likewise when there is no transition we denote the energy consumptions by $E_{0\to0}$ and $E_{1\to1}$ depending on the output voltage. Consider an AND gate which is made by a similar complementary realization.

The transitions of the AND gate, denoted as $y = AND(a, b) = a \wedge b$, where a and b are bits are shown in Table 5.1. The energy levels are annotated in the fourth column. This column can be used to estimate the average energy when the output bit is 0 or 1, namely $E(q = 0)$ and $E(q = 1)$ respectively. We show that the power consumption of the device is correlated by the *value* of the output bit. It may be emphasized that this observation is central to the working of a DPA attack.

The average energies when $q = 0$ or $q = 1$ are:

$$E(q = 0) = (3E_{1\to0} + 9E_{0\to0})/12$$
$$E(q = 1) = (3E_{0\to1} + E_{1\to1})/4$$

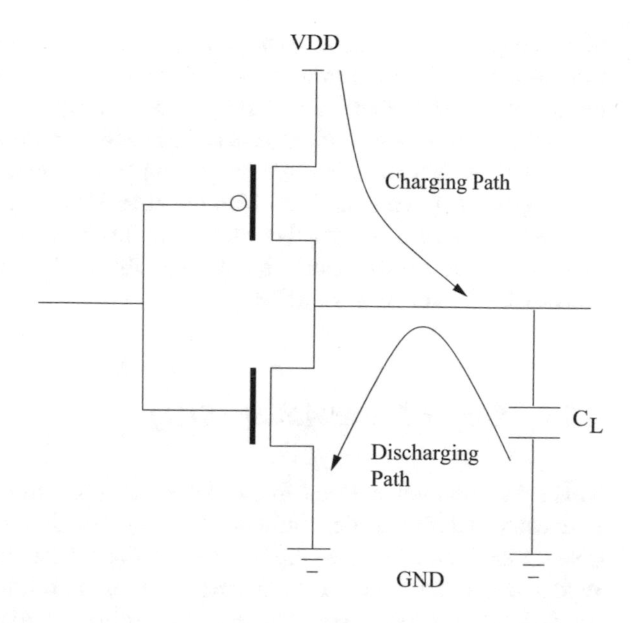

Fig. 5.2 Different power consumption for $0 \rightarrow 1$ and $1 \rightarrow 0$ transition for CMOS switch

Table 5.1 Transitions of an AND gate

a	b	q	Energy
$0 \rightarrow 0$	$0 \rightarrow 0$	$0 \rightarrow 0$	$E_{0 \rightarrow 0}$
$0 \rightarrow 0$	$0 \rightarrow 1$	$0 \rightarrow 0$	$E_{0 \rightarrow 0}$
$0 \rightarrow 0$	$1 \rightarrow 0$	$0 \rightarrow 0$	$E_{0 \rightarrow 0}$
$0 \rightarrow 0$	$1 \rightarrow 1$	$0 \rightarrow 0$	$E_{0 \rightarrow 0}$
$0 \rightarrow 0$	$0 \rightarrow 0$	$0 \rightarrow 0$	$E_{0 \rightarrow 0}$
$0 \rightarrow 0$	$0 \rightarrow 1$	$0 \rightarrow 1$	$E_{0 \rightarrow 1}$
$0 \rightarrow 0$	$1 \rightarrow 0$	$0 \rightarrow 0$	$E_{0 \rightarrow 0}$
$0 \rightarrow 0$	$1 \rightarrow 1$	$0 \rightarrow 1$	$E_{0 \rightarrow 1}$
$0 \rightarrow 0$	$0 \rightarrow 0$	$0 \rightarrow 0$	$E_{0 \rightarrow 0}$
$0 \rightarrow 0$	$0 \rightarrow 1$	$0 \rightarrow 0$	$E_{0 \rightarrow 0}$
$0 \rightarrow 0$	$1 \rightarrow 0$	$1 \rightarrow 0$	$E_{1 \rightarrow 0}$
$0 \rightarrow 0$	$1 \rightarrow 1$	$1 \rightarrow 0$	$E_{1 \rightarrow 0}$
$0 \rightarrow 0$	$0 \rightarrow 0$	$0 \rightarrow 1$	$E_{0 \rightarrow 0}$
$0 \rightarrow 0$	$0 \rightarrow 1$	$0 \rightarrow 0$	$E_{0 \rightarrow 1}$
$0 \rightarrow 0$	$1 \rightarrow 0$	$1 \rightarrow 0$	$E_{1 \rightarrow 0}$
$0 \rightarrow 0$	$1 \rightarrow 1$	$1 \rightarrow 1$	$E_{1 \rightarrow 1}$

Observe that if the four transition energy levels are different, then in general $|E(q = 0) - E(q = 1)| \neq 0$. This simple computation shows that if a large number of power traces are accumulated and divided into two bins: one for $q = 0$ and the other for $q = 1$ and when the means for the 0-bin and 1-bin are computed, the difference-of-mean (DOM) is expected to have a non-zero difference at some point. This forms the basis of Differential Power Analysis (DPA), where the variation of power of a circuit w.r.t. data is exploited. There are several types of power analysis. The most primitive form of power analysis is called as Simple Power Analysis or SPA. It is a technique that involves directly interpreting power consumption measurements with *cryptographic operations*. The objective of an SPA attack is to obtain the secret key in one or few traces. That makes an SPA quite challenging in practice. We first provide an overview on the same, before describing other forms of power analysis like Differential Power Attacks (DPA).

5.2.1 Simple Power Attack (SPA)

This subsection focuses on Simple Power Analysis (SPA), which is extremely easy to execute and if possible, could be extremely deadly for the security of the cryptocores. SPA is applicable to the implementations where value of key determines the type of operations to be executed. Attacker tries to exploit this operation dependency on the key bits through SPA. Generally, two different operations (for example multiplication and addition/squaring) have different power signature and hence it is easy to identify the type of operation from power traces. If the key value determines the operation type, an attacker can easily extract the key by classifying the operations from the power traces.

Apart from attacking, SPA can also be used to extract feature points from the power traces. Feature points are the points on the power traces which on being analyzed by DPA will give you the key. For example, while doing DPA on AES, we need to focus on the last round register update. By using SPA, we can pinpoint the last round register update and can record the power traces for only those points. This reduces both number of trace acquisition requirement and attack complexity.

We will provide example for both of the above instances. Using SPA, it is very easy to administer a successful attack on Elliptic Curve Cryptographic (ECC) operations, implemented with *Double-and-Add* algorithm. The next discussion focuses on this.

SPA on ECC

Elliptic curve cryptography (ECC) is a public key cryptography based on elliptic curves and finite field. The advantage of ECC over RSA lies in shorter key size and more security per key bit. Shorter key size of ECC also leads to a compact implementation on FPGA. Security of ECC depends upon the mathematical intractability of discrete logarithm of a point in elliptic curve with respect to a known base point. The most important operation on ECC is scalar multiplication operation which can be executed by Algorithm 1

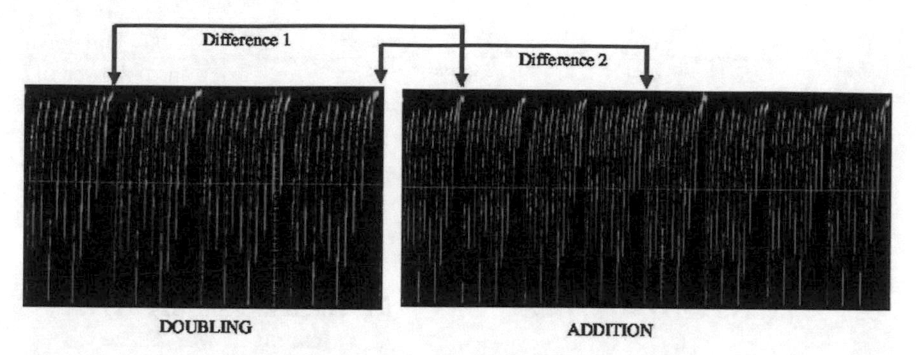

Fig. 5.3 Difference between doubling and addition (implemented on SASEBO-GII)

Data: Point P and scalar $k = k_{m-1}, k_{m-2}, k_{m-3} \dots k_2, k_1, k_0$,
where $k_{m-1} = 1$
Result: $Q = kP$
$Q = P$
for $i = m - 2$ *to* 0 **do**
  $Q = 2Q$ (Point Doubling)
  if k_i=*1* **then**
    $Q = Q + P$ (Point Addition)
  end
end

Algorithm 1: Double-and-Add Algorithm

As shown in Algorithm 1, ECC scalar multiplication operation involves two operations: point doubling and point addition. Point doubling happens for every key bit, whereas point addition happens only when the value of the key bit is 1. Thus we can see an operation dependency on the value of the key bits. So, if we can identify the point doubling and point addition operation from the power traces, we can easily extract the key and we can say that the implementation is vulnerable to SPA. Figure 5.3 shows the corresponding point doubling and point addition power traces for an elliptic curve implementation in binary curve. Point addition is more complex operation compared to point doubling and hence requires more time for completion. As we can in the Fig. 5.3, it is possible to distinguish between point doubling and point addition operation by just visual inspection. For more difficult situations, where visual inspection does not work, we can employ pattern matching methodologies to identify the differences between different operations.

SPA on AES: Extracting Feature Points

Apart from attacking, SPA can be used to identify operations like register update, multiplication, etc. If these operations are not key dependent, SPA cannot be used to extract the key and we need to shift our attack methodologies to more sophisticated

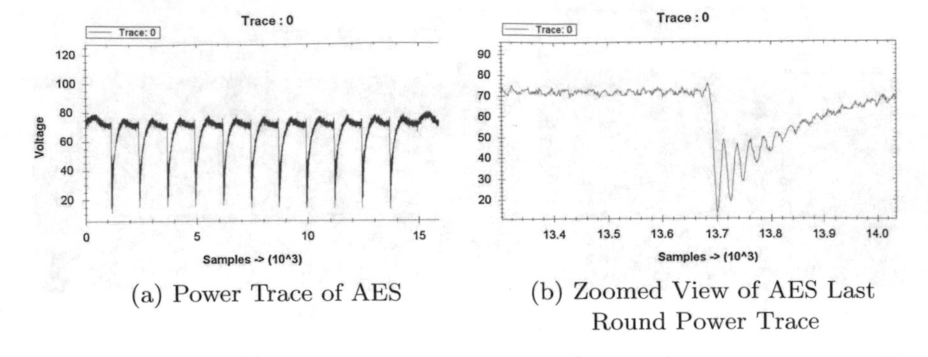

(a) Power Trace of AES

(b) Zoomed View of AES Last Round Power Trace

Fig. 5.4 Extracting feature points from AES

attack strategies like DPA. However, success of DPA depends upon selecting a particular region in a power trace where data dependency of the key can be exploited (basic principal of DPA). For example, AES can be attacked by DPA by targeting last round register update. SPA helps us to identify the last round register update and in turn enhances the success probability of DPA. Figure 5.4a shows the corresponding power trace of an AES implementation, obtained on SASEBO-GII board. From Fig. 5.4a, we can easily identify each round of AES algorithm. Moreover, we can now only focus on the last round register update operation, reducing the attack complexity. Figure 5.4b shows a zoomed view of last round register update.

Thus, we have seen how an adversary can exploit operation dependency on key bits through SPA. However, SPA is easy to prevent and can be easily avoided by very simple countermeasures. For example, SPA on ECC can be countered by employing Montgomery ladder [5], which removes operation dependency on key bits. But, this is not the case for Differential Power Attacks (DPA) where data dependency on the key bits are exploited by complex statistical analysis. In the next subsection we are going to focus on this.

5.2.2 *Differential Power Attacks*

Differential Power Attack (DPA) depends on the correlation of power consumption of the design with the underlying data. Differential power analysis depends largely on power simulation, which predicts the power variation of a device w.r.t. the change of internal values. The internal value is often computed from the ciphertext and an assumed portion of the key, and the *power model* maps the internal value to a power level. Actual values of the power are not important, but rather their relative ordering is crucial. Various forms of power models and variations have been suggested, but CMOS circuits which is the most common forms of technologies used in the present hardware design industry use the Hamming weight and Hamming distance power models to predict the variation of power. We present an overview on these models next. A more detailed description can be found in [5].

Hamming Weight Power Model

In this model, we assume that the power consumption of a CMOS circuit is proportional to the value at the tth time instance, say v_t. The model is oblivious of the state at the $(t-1)$th time instance, v_{t-1}. Thus, given the state v_t, the estimated power is denoted as $\mathrm{HW}(v_t)$, where HW is the Hamming Weight function. It may be argued that this is an inaccurate model, given the fact that the dynamic power of a CMOS circuit rather depends on the transition $v_{t-1} \rightarrow v_t$ than on the present state. However, the Hamming Weight model provides an idea of power consumption in several occasions. Consider situations where the circuit is pre-charged to a logic 1 or 0 (i.e., v_{t-1}=0 or 1), the power in that case is either directly or inversely proportional to the value v_t. In situations where the initial value v_t is also uniformly distributed, the dependence still holds directly or inversely owing to the fact that the transition power due to a $0 \rightarrow 1$ toggle or $1 \rightarrow 0$ switch are not same. This is because of different charge and discharge paths of a CMOS gate (Fig. 5.2). Thus assuming that the actual power is better captured by $P = HW(v_{t-1} \oplus v_t)$, there may exist a correlation between P and $HW(v_t)$ due to this asymmetric power consumption of CMOS gates.

Hamming Distance Power Model

This is a more accurate model of the power consumption of a CMOS gate. Here the power consumption of a CMOS circuit is assumed to be proportional to the Hamming Distance of the input and the output vector. In short, the power consumption is modeled as $P = HD(v_{t-1}, v_t) = HW(v_{t-1} \oplus v_t)$, where HD denotes the Hamming Distance between two vectors. This is a more accurate model as it captures the no of toggles in a net of the circuit. However, in order to use the model the attacker needs more knowledge than that for using the Hamming Weight model. The attacker here needs to know the state of the circuit in successive clock cycles. This model is useful for modeling the power consumption of registers and buses. On the contrary, they are incapable for estimating the power consumption due to combinatorial circuits as transitions of combinatorial circuits are unknown due to the presence of glitches.

DPA Using Difference of Means

We provide a discussion on the Difference-of-Mean (DoM) technique to illustrate a DPA on a block cipher like AES. The basic principle of the DoM method is based on the fact that the power consumption of a device is correlated with a target bit or a set of bits. In the simplest form, we obtain a large number of power consumption curves and the corresponding ciphertexts. The attacker then applies a divide and conquer strategy: he assumes a portion of the key which is required to perform the deciphering for a portion of the cipher for one round. Based on the guessed key he computes a target bit, typically the computation of which requires evaluation of an S-Box. Depending on whether the target bit is 0 or 1, the traces are divided into a 0-bin and 1-bin. Then the mean of all the traces in the 0-bin and 1-bin are computed and finally we compute Difference-of-Mean (DoM) of the mean curves. It is expected that for the correct key guess, there will be a time instance for which there is a non-negligible value, manifesting as a spike in the difference curve. The correlation of the power consumption of the device on the target bit is thus exploited to distinguish the correct key from the wrong ones.

Let us consider a sample run of the AES algorithm. We provide several runs of the AES algorithm with NSample randomly chosen plaintexts. Consider an iterated AES hardware where a register is updated by the output of the AES round every encryption. The power trace is stored in the array sample[NSample][NPoint], where NPoint is the length of the power trace corresponding to the power consumption after each round of the encryption. For each of the power traces we also store the corresponding ciphertexts in the array Ciphertext[NSample]. One can check that the power consumption is corresponding to the state of the register before the encryption, then updated by the initial key addition, followed by the state after each of the 9 rounds of AES, and finally the ciphertext after the 10th round.

The attack algorithm first *targets* one of the key bytes, key, for which one of the 16 S-Boxes of the last round is aimed at. We denote the corresponding ciphertext byte in the array Ciphertext[NSample] by the variable cipher. For each of the NSample plaintexts, the analysis partitions the traces, sample[NSample][NPoint] into a zero-bin and one-bin, depending on a target bit at the input of a target S-Box. For computing or estimating the target bit at the input of the target S-Box, the attack guesses the target key byte, and then computes the Inverse-SBox on the XOR of the byte cipher and the guessed key byte, denoted as key. One may observe that the ciphertexts for which the target byte in the ciphertext, cipher is same, always goes to the same bin. Thus the traces can be stored in a smaller array, sample[NCipher][NPoint], where NCipher is the number of cipher bytes (which is of course 256).

It essentially splits the traces into the two bins based on a target bit, say the LSB. The algorithm then computes the average of all the 0-bin and the 1-bin traces, and then computes the difference of the means, denoted as DoM. Let the number of traces present in the 0-bin be $count_0$ and the number of traces present in the 1-bin be $count_1$. Then the value of DoM is calculated according to the following equation:

$$DoM = \left| \frac{\sum_{i=0}^{count_0} sample[i][NPoint]_{cipher[0]=0}}{count_0} - \frac{\sum_{i=0}^{count_1} sample[i][NPoint]_{cipher[0]=1}}{count_1} \right| \quad (5.1)$$

It is expected that the correct key will have a significant Difference of Mean, compared to the wrong keys which have almost a negligible DoM. We then store the highest value of the DoM as the biasKey[NKey] for each of the key guesses. The key which has the highest bias value is returned as the correct key. An example of *DoM* on AES is shown in Fig. 5.5a.

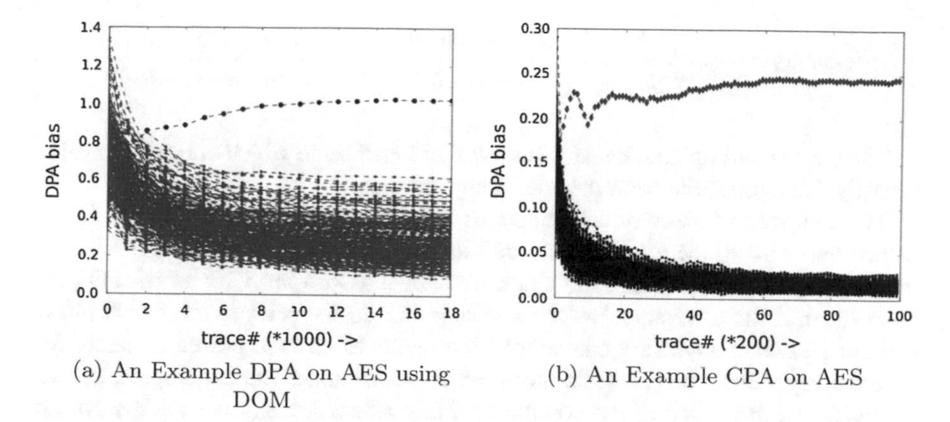

(a) An Example DPA on AES using DOM

(b) An Example CPA on AES

Fig. 5.5 Attack result of DPA on AES

DPA Using Correlation Analysis

Like in the DoM-based DPA attack, the Correlation Power Attack (CPA) also relies on targeting an intermediate computation, typically the input or output of an S-Box. These intermediate values are as seen previously computed from a known value, typically the ciphertext and a portion of the key, which is guessed. The power model is subsequently used to develop a *hypothetical* power trace of the device for a given input to the cipher. This hypothetical power values are then stored in a matrix for several inputs and can be indexed by the known value of the ciphertext or the guessed key byte. This matrix is denoted as **H**, the hypothetical power matrix. Along with this, the attacker also observes the actual power traces, and stores them in a matrix for several inputs. The actual power values can be indexed by the known value of the ciphertext and the time instance when the power value was observed. This matrix is denoted as **T**, the real power matrix. It may be observed that one of the columns of the matrix **H** corresponds to the actual key, denoted as k_c. In order to distinguish the key from the others, the attacker looks for *similarity* between the columns of the matrix **H** and those of the matrix **T**. The similarity is typically computed using the Pearson's Correlation coefficient.

The actual power value for all the NSample encryptions are observed and stored in the array trace[NSample][NPoint]. The attacker first scans each column of this array and computes the average of each of them, and stores in meanTrace[NPoint]. Likewise, the hypothetical power is stored in an array hPower[NSample][NKey] and the attacker computes the mean of each column and stores in meanH[NKey] by scanning each column of the hypothetical matrix. The attacker then computes the correlation value to find the similarity of the ith column of the matrix hPower and the jth column of trace. The correlation is computed as follows and stored in the array result[NKey][NPoint]:

$$result[i][j] = \frac{\sum_{k=0}^{NSample}(hPower[i][k] - meanH[i])(trace[j][k] - meanTrace[j])}{\sum_{k=0}^{NSample}(hPower[i][k] - meanH[i])^2 \sum_{k=0}^{NSample}(hPower[i][k] - meanH[i])^2}$$

The corresponding attack result is shown in Fig. 5.5b in which correct key value is easily distinguishable from the wrong key guesses.

Till now we discussed attack strategies in which attacker has the access to the device only during the attack phase, and does not have any control on the device prior the attack. These attack strategies are known as *non-profiling attacks*. On the other hand, if the adversary has the access to the device prior to the attack phase and during which adversary can control the inputs to observe power signature for different key values, he can build more sophisticated attack methodologies, known as *profiling attack*. One of the popular profiling attack strategy is *template attack*, which is described in the next subsection.

5.2.3 Template Attack

Template attack was introduced by Chari et al. in [6] as "the strongest form of side channel attack possible in an information theoretic sense." In the template attack, it is assumed that the attacker has an access to the device or clone of the device which he can use for profiling of the device. During profiling the attacker collect power traces for different key values and classes. In the attack phase, the adversary recovers the correct key from the design under attack using the estimated leakage distribution. Profiling attacks are more generic than their non-profiling counterparts in the sense that they require weaker assumption on the leakage model. But their applications are limited by the requirement of the access to a cloned device. The first template attack proposed in [6] consists of following steps:

- **Template Building**:

1. Collect T number of power traces for each class of key value. For example, an attacker can collect power traces for each hamming weight class of the key. Let us assume that there are L different classes of key.
2. Compute mean of the power traces for each L classes. Let us denote this mean power traces as $M = (M_1, M_2, M_3, \ldots, M_L)$.
3. Now, we need to identify high SNR region in the power traces. This is extremely important for the success of the template attack as it is crucial to reduce the trace length to reduce the execution time of template attack. There are several ways to reduce the trace length to a small (n) high SNR sample. In [6], the author proposed a method in which pairwise difference between mean power traces were taken and only the points with high differences were considered. Let us assume these points as $(P = P_1, P_2, \ldots, P_n)$. There are other possible methods also. For example, *NICV* [7] or analysis in principal subspaces [8] can also be used to find out the high SNR points.

4. In the next step, noise matrix N is calculated for each of the power traces. The noise matrix for ith trace is calculated as follows: $N_i = (T_i[0] - M_l[0], T_i[1] - M_l[1], \dots, T_i[n] - M_l[n])$, where l is the class of the corresponding trace. Then we compute covariance matrix between the N_is of a particular class, denoted as $\sum_{N_i}[u, v] = cov(N_i(P_u), N_i(P_v))$.
5. Thus, we have now built the template for each of the classes $(M_i, \sum_{N_i})$.

- **Attack Phase**:

1. Let us denote the trace under attack as t. We compute $n_{test_i} = M_i - t$ for each classes.
2. The probability that the trace under attack t belongs to particular class i can be computed by maximum likelihood test in multivariate gaussian distribution using following formula

$$p(n_{test_i}) = \frac{1}{(2\pi)^n |\sum_{N_i}|} e^{-\frac{1}{2} n_{test_i}{}^T (\sum_{N_i})^{-1} n_{test_i}} \tag{5.2}$$

Template attack, based on stronger adversary attack model, requires very few number of power traces compared to standard DPA. To protect crypto-systems against this kind of sophisticated profiling attack or standard non-profiling DPA attack, we need to develop efficient countermeasures which is the focus of the next section.

5.3 Countermeasures Against Power Analysis

As discussed in the previous section, countermeasure design to mitigate power attack is of extreme importance. A countermeasure design generally adds more overhead to the design, making the crypto-system more resource hungry. Hence, the objective is to design countermeasures which are lightweight but provide adequate security to thwart power-based side channel attacks.

5.3.1 *t-Private Circuit*

Private circuit [9] is a countermeasure which provides a theoretical framework to construct a crypto-circuit to prevent probing attack. Probing attack is the most powerful side channel attack where an adversary can observe exact value of fixed number of circuit's internal nets. Though t-private circuit is designed to prevent probing attack, it is equally applicable to other side channel attacks like power attack or electromagnetic radiation attack. Moreover, this countermeasure is based on sound theoretical proof, providing a formal method to develop a countermeasure. In this subsection we are going to give a brief description of *private circuit* with an analysis of its advantage and disadvantages. For theoretical proofs and analysis of circuit, we encourage the readers to go through [9].

Construction of t-Private Circuit

In t-private circuit approach, a circuit is transformed in such a way that any adversary, having capability of observing t nets, cannot get access to a single bit of sensitive information. The minimum number of probes required by an adversary to extract one bit of information is $t+1$. This subsection provides a brief description of such transformation.

- *Input Encoding*:- Any input bit a is transformed into a vector of $2t+1$ bits. The first $2t$ bits are random values $(a_1, a_2, \ldots, a_{2t})$ and the last bit (a_{2t+1}) is computed by the following way:

$$a_{2t+1} = a \oplus \bigoplus_{i=1}^{2t} a_i \ . \tag{5.3}$$

- *AND gate*: Inputs a and b of *AND* gate are transformed into vectors $\grave{a} = (a_1, a_2, \ldots, a_{2t+1})$ and $\grave{b} = (b_1, b_2, \ldots, b_{2t+1})$. Output of the *AND* gate is also a vector $\grave{c} = (c_1, c_2, \ldots, c_{2t+1})$, which is calculated by following steps:

  1. Generate random bits $r_{i,j}$, where $i \neq j$ and $1 \leq i \leq j \leq 2t+1$.
  2. Compute $r_{j,i} = (r_{i,j} \oplus a_i b_j) \oplus a_j b_i$, where $i \neq j$ and $1 \leq i \leq j \leq 2t+1$.
  3. Compute $c_i = a_i b_i \oplus \bigoplus_{j \neq i} r_{i,j}$, where $1 \leq i \leq 2t$ and $1 \leq j \leq 2t$.

- *NOT gate*: Input a is transformed into a vector $\grave{a} = (a_1, a_2, \ldots, a_{2t+1})$. Output $\bar{\grave{a}}$ is computed by inverting any bit of $\grave{a}$. E.g., $\bar{\grave{a}} = (a_1, a_2, \ldots, \overline{a_{2t+1}})$.
- *XOR gate*: Like *AND* gate; inputs a and b of *XOR* gate are transformed into vectors $\grave{a} = (a_1, a_2, \ldots, a_{2t+1})$ and $\grave{b} = (b_1, b_2, \ldots, b_{2t+1})$. Output $\grave{c} = (c_1, c_2, \ldots, c_{2t+1})$ is calculated in the following way. Perform:

$$c_i = a_i \oplus b_i, 1 \leq i \leq 2t \ . \tag{5.4}$$

Using these transformations, one can easily transform any digital circuit to a t-private circuit, because this set of gates is universal.

Let us consider an *AND* gate in t-private circuit for $t = 1$. Inputs of the *AND* gate are two vectors $\grave{a} = (a_1, a_2, a_3)$ and $\grave{b} = (b_1, b_2, b_3)$, encoded according to (5.3). Output $\grave{c} = (c_1, c_2, c_3)$ is calculated as follows:

$$c_1 = a_1 b_1 \oplus r_{1,2} \oplus r_{1,3} \tag{5.5}$$

$$c_2 = a_2 b_2 \oplus (r_{1,2} \oplus a_1 b_2) \oplus a_2 b_1 \oplus r_{2,3} \tag{5.6}$$

$$c_3 = a_3 b_3 \oplus (r_{1,3} \oplus a_1 b_3) \oplus a_3 b_1 \oplus (r_{2,3} \oplus a_2 b_3) \oplus a_3 b_2 \tag{5.7}$$

The t-private circuits are secure against an adversary capable of observing any t nets of the circuit at any given time instant. Construction of t-private circuits involves $2t$ number of random bits for each input bit of the circuit. Moreover, it requires $2t + 1$ random bits for each 2-input *AND* gate present in the circuit. The overall complexity of the design is $\mathcal{O}(nt^2)$ where n is the number of gates in the circuit.

The complexity of $\mathcal{O}(nt^2)$ is often considered impractical for several practical implementations. After the publication of [9], there have been many works which try to improve its result, in particular the area overhead. In [10], the authors have improved the complexity of private circuit from $\mathcal{O}(nt^2)$ to $\mathcal{O}(nt)$. Moreover, in their recent work [11], they have further improved it to $\lceil t/2 \rceil$ for private circuits. They have also provided theoretical analysis and improvement of private circuit in context of power-based side channel attack and glitch [12, 13].

Due to its high area requirement, it is not practical to use private circuit as a side channel countermeasure despite being theoretically secure. However, private circuit provides the basis of one of the most popular countermeasure Masking, which is described in the next subsection.

5.3.2 *Masking*

Masking is probably the most popular countermeasure against side channel analysis and has been studied in great details. The objective of the masking scheme is to randomize the intermediate results and is to make power consumption of the device independent of the sensitive data processed. The countermeasure is based on the fact that the power consumption of the devices are uncorrelated with the actual data as they are *masked* with a random value.

In masking every intermediate value which is related to the key is concealed by a random value m which is called as the *mask*. Thus, we transform the intermediate value v as $v_m = v * m$, where m is randomly chosen and varies from encryption to encryption. The attacker does not know the value m. The operation $*$ could be either exclusive or, modulo addition, or modulo multiplication. *Boolean masking* is a special term given to the phenomenon of applying $\oplus$ as the above operation to conceal the intermediate value. If the operation is addition or multiplication, the masking is often referred to as *arithmetic masking*.

Masking at the Algorithmic Level

In this class of masking, the intermediate computations of an algorithm are masked. Depending on the nature of the masking scheme, namely Boolean or arithmetic, the nonlinear functions of the given cipher are redesigned. In the subsequent discussion, we are going to focus on implementing masking scheme on nonlinear S-Box module. We will show how to employ masking scheme on an efficient design of S-Box and will highlight the corresponding design overhead.

Masking the AES S-Box

AES S-Box is the most critical element in the AES design. It is actually a substitution table which takes an eight bit input and produce an eight bit output. The output of the S-Box are generated by computing inverse of the input in $GF(2^8)$ and then applying an affine transformation. It can be implemented on hardware using a look-up tables or embedded memories which will store the 256 entries of the S-Box. However, S-Box output can be also be produced by doing an isomorphic transformation from $GF(2^8)$ to composite field $GF((((2)^2)^2)^2)$ and doing field operation in the composite field. This approach is highly efficient as it can be implemented with very few gates compared to table implementation of S-Box [14]. The isomorphic transformation function is linear and can be masked using *XORs* quite conveniently.

Now to compute the S-Box using composite field, we need to compute inverse in $GF(2^4)^2$, which can be computed as follows: Let the irreducible polynomial of an element in $GF(2^4)^2$ be $r(Y) = Y^2 + Y + \mu$ ($\mu \in GF(2^4)$). An element in the composite field can be represented as $\gamma = (\gamma_1 Y + \gamma_0)$ and let the inverse be $\delta = (\gamma_1 Y + \gamma_0)^{-1} = (\delta_1 Y + \delta_0) \bmod (Y^2 + Y + \mu)$.

Thus, the inverse of the element is expressed by the following equations:

$$\delta_1 = \gamma_1(\gamma_1^2\mu + \gamma_1\gamma_0 + \gamma_0^2)^{-1} \tag{5.8}$$

$$\delta_0 = (\gamma_0 + \gamma_1)(\gamma_1^2\mu + \gamma_1\gamma_0 + \gamma_0^2)^{-1} \tag{5.9}$$

Equivalently, we can write:

$$\delta_1 = \gamma_1 d' \tag{5.10}$$

$$\delta_0 = (\gamma_1 + \gamma_0)d' \tag{5.11}$$

$$d = \gamma_1^2\mu + \gamma_1\gamma_0 + \gamma_0^2 \tag{5.12}$$

$$d' = d^{-1} \tag{5.13}$$

Next we consider the masking of these operations. The masked values corresponding to the input is thus, $(\gamma_1 + m_h)Y + (\gamma_0 + m_l)$, of which the inverse is to be computed such that the output of the equations of Eq. 5.10 are also masked by random values, respectively m'_h, m'_l, m_d, m'_d.

Let us consider the masking of Eq. 5.10. Thus we have:

$$\delta_1 + m'_h = \gamma_1 d' + m'_h \tag{5.14}$$

Hence we need to compute $\gamma_1 d' + m'_h$. However, because of the additive masking, we have both γ_1 and d' masked. Thus, we can compute the *masked value* $(\gamma_1 + m_h)(d' + m'_d)$, and then add some *correction terms* to obtain $\gamma_1 d' + m'_h$.

To elaborate, the correction terms will be:

$$(\gamma_1 d' + m'_h) + (\gamma_1 + m_h)(d' + m'_d) = (\gamma_1 + m_h)m'_d + m_h d' + m'_h \tag{5.15}$$

One has to take care when adding correction terms that no intermediate values are correlated with values, which an attacker can predict. We thus mask d' in the correction term as follows: $(\gamma_1 + m_h)m'_d + m_h(d' + m'_d) + m_h m'_d + m'_h$.

Thus, the entire computation can be written as:

$$\begin{aligned}\delta_1 + m'_h &= (\gamma_1 + m_h)(d' + m'_d) + \\ &\quad (\gamma_1 + m_h)m'_d + m_h(d' + m'_d) + m_h m'_d + m'_h \\ &= f_{\gamma_1}((\gamma_1 + m_h), (d' + m'_d), m_h, m'_h, m'_d)\end{aligned} \tag{5.16}$$

We thus have:

$$f_{\gamma_1} = (\gamma_1 + m_h)m'_d + m_h(d' + m'_d) + m_h m'_d + m'_h \tag{5.17}$$

This equation can be significantly reduced if we consider reuse of mask value to reduce the corresponding circuit complexity. Using $m'_d = m_l$ and $m'_h = m_h$, we have the following:

$$\begin{aligned}f_{\gamma_1} &= (\gamma_1 + m_h)(d' + m_l) + \\ &\quad (\gamma_1 + m_h)m_l + m_h(d' + m_l) + m_h m_l + m_h\end{aligned} \tag{5.18}$$

Likewise, one can derive the remaining 2 equations (Eqs. 5.11 and 5.12) in the masked form. For Eq. 5.11, we have:

$$\begin{aligned}\delta_0 + m'_l &= (\gamma_1 + \gamma_0)d' + m'_l \\ &= (\gamma_1 d' + m'_h) + (\gamma_0 + m_l)(d' + m'_d) + (d' + m'_d)m_l \\ &\quad +(\gamma_0 + m_l)m'_d + m'_h + m'_l + m_l m'_d \\ &= (\delta_1 + m'_h) + (\gamma_0 + m_l)(d' + m'_d) + (d' + m'_d)m_l \\ &\quad +(\gamma_0 + m_l)m'_d + m'_h + m'_l + m_l m'_d \\ &= f_{\gamma_0}((\delta_1 + m'_h), (\gamma_0 + m_l), (d' + m'_d), m_l, m'_h, m'_l, m'_d)\end{aligned} \tag{5.19}$$

Again the resulting circuit complexity can be reduced by reusing the same mask. we can choose $m'_l = m_l$ and $m'_d = m_h = m'_h$. Thus we have:

$$f_{\gamma_0} = (\delta_1 + m_h) + (\gamma_0 + m_l)(d' + m_h) + (d' + m_h)m_l$$
$$+(\gamma_0 + m_l)m_h + m_h + m_l + m_l m_h \quad (5.20)$$

Thus continuing for Eq. 5.12 we have:

$$\begin{aligned} d + m_d &= \gamma_1^2 p_0 + \gamma_1\gamma_0 + \gamma_0^2 + m_d \\ &= (\gamma_1 + m_h)^2 p_0 + (\gamma_1 + m_h)(\gamma_0 + m_l) + (\gamma_0 + m_l)^2 + (\gamma_1 + m_h)m_l + (\gamma_0 + m_l)m_h \\ &\quad + m_h^2 p_0 + m_l^2 + m_h m_l + m_d \\ &= f_d((\gamma_1 + m_h), (\gamma_0 + m_l), p_0, m_h, m_l, m_d) \end{aligned} \quad (5.21)$$

We choose $m_h = m_d$ to reduce the gate-count. Thus,

$$\begin{aligned} f_d &= (\gamma_1 + m_h)^2 p_0 + (\gamma_1 + m_h)(\gamma_0 + m_l) + (\gamma_0 + m_l)^2 + (\gamma_1 + m_h)m_l + (\gamma_0 + m_l)m_h \\ &\quad + m_h^2 p_0 + m_l^2 + m_h m_l + m_h \end{aligned} \quad (5.22)$$

The masking of Eq. 5.13 involves performing the following operations on $d + m_d$ and obtained the masked inverse, $d' + m'_d$. Thus one needs to develop a circuit for $f_{d'}$ for which:

$$\begin{aligned} d' + m'_d &= f_{d'}(d + m_d, m_d, m'_d) \\ &= d^{-1} + m'_d \end{aligned} \quad (5.23)$$

Masking Eq. 5.13 involves masking an inverse operation in $GF(2^4)$. Hence the same masking operations as above can be applied while reducing the inverse to that in $GF(2^2)$. Thus, we can express an element in $GF(2^4)$ $\delta = \Gamma_1 Z + \Gamma_0$, where Γ_1 and $\Gamma_0 \in GF(2^2)$. Interestingly, in $GF(2^2)$ the inverse is a linear operation making masking easy! Thus we have, $(\Gamma + m)^{-1} = \Gamma^{-1} + m^{-1}$. This reduces the gate count considerably.

Masking at Gate Level

One of the most popular countermeasures to prevent DPA attacks at gate level is masking. Although there are various techniques to perform masking, the method of masking has finally evolved to the technique proposed in [15]. The principle of this masking technique is explained briefly, in reference to a 2 input and gate. The same explanation may be extended to other gates, like or, xor etc. The gate has two inputs a and b and the output is $q = a$ and b. The corresponding mask values for a, b and q are respectively m_a, m_b and m_q. Thus the masked values are: $a_m = a \oplus m_a, b_m = b \oplus m_b, q_m = q \oplus m_q$. Hence the masked and gate may be expressed as: $q_m = f(a_m, b_m, m_a, m_b, m_q)$. The design proposed in [15] proposes a hardware implementation for the function f for the masked and gate, which may be easily generalized to a masked multiplier. This is because the 2-input and gate is a special of case of an n-bit multiplier, as for the and gate we have $n = 1$. The masked multiplier (or masked and gate by assuming $n = 1$) is depicted in Fig. 5.6. The correctness of the circuit may be established by the following argument:

$$\begin{aligned} q_m &= q \oplus m_q \\ &= (ab) \oplus m_q \\ &= (a_m \oplus m_a)(b_m \oplus m_b) \oplus m_q \\ &= (a_m b_m \oplus b_m m_a \oplus a_m m_b \oplus m_a m_b \oplus m_q) \end{aligned}$$

However, it should be noted that the order of the computations performed is of extreme importance. The correct order of performing the computations are as follows:

$$q_m = a_m b_m + (m_a b_m + (a_m m_b + (m_a m_b + m_q)))$$

The ordering follows to ensure that the unmasked values are not exposed during the computations. Further it should be emphasized that one cannot reuse the mask values. For example one may attempt to make one of the input masks, m_a same as the output mask, m_q. While this may seem to be harmless, it can defeat the purpose of the masking. In the subsequent discussions, we will give examples of side channel attacks which take advantage of improper implementation of masking scheme.

Attack on Masking

Masking, being the most widely used countermeasure, has been constantly evaluated against sophisticated side channel analysis. There are several instances where a masked hardware has failed to prevent side channel attacks. Masking prevents first order (1o) differential power attacks and can be broken by second order differential attacks. However, in literature it has been shown that masking can be broken by first order differential attack if it is not implemented in a correct way. We will mainly focus on two different attack methodologies: *Attack due to glitches* and *collision-correlation attack.*

The Masked and Gate and Vulnerabilities Due to Glitches

The circuit for computing the masked AND gate is shown in Fig. 5.6. The same circuit can be applied for masking a $GF(2^n)$ multiplier as well. We observe that the masked multiplier (or, AND gate) requires four normal multipliers (or, AND gates) and four normal n-bit (or 1-bit) XOR gates. Also it may be observed that the multipliers (or, AND gates) operate pairwise on (a_m, b_m), (b_m, m_a), $(a_m m_b)$ and (m_a, m_b).

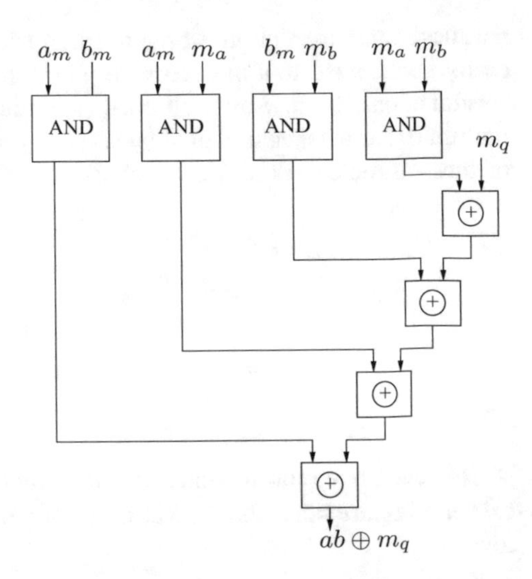

Fig. 5.6 Architecture of a masked multiplier

Each of the element of the pairs has no correlation to each other (if the mask value is properly generated) and are independent of the unmasked values a, b and q. One can obtain a transition table and obtain the expected energy for generating $q = 0$ and $q = 1$. The gate now has 5 inputs and thus there can be $4^5 = 1024$ transitions (like Table 5.1). If we perform a similar calculation as before for unmasked gates, we find that the energy required to process $q = 0$ and $q = 1$ are identical. Thus if we compute the mean difference of the power consumptions for all the possible 1024 transitions for the two cases: $q = 0$ and $q = 1$, we should obtain theoretically zero. Likewise the energy levels are also not dependent on the inputs a and b and thus supports the theory of masking and show that the masked gate should not leak against a first order DPA. However in this analysis we assume that the CMOS gates switch once per clock cycle, which is true in the absence of glitches.

But glitches are a very common phenomenon in digital circuits, as a result of which the CMOS gates switch more than once in a clock signal before stabilizing to their steady states. One of the prime reasons of glitches in digital circuits is different arrival times of the input signals, which may occur in practice due to skewed circuits, routing delays, etc. As can be seen in the circuit shown in Fig. 5.6, the circuit is unbalanced which leads to glitches in the circuit.

The work proposed in [16] investigates various such scenarios which causes glitches and multiple toggles of the masked AND gate. The assumption is that each of the 5 input signals toggle once per clock cycle and that one of the inputs arrive at different time instance than the others. Moreover we assume that the delay between the arrival time of two distant signals is more than the propagation time of the gate. As a special case, consider situations when only one of the five inputs arrives at a different moment of time than the remaining four inputs.

There exist ten such scenarios as each one of the 5 input signals can arrive either before or after the four other ones. In every scenario there exist $4^5 = 1024$ possible combinations of transitions that can occur at the inputs. However, in each of the ten scenarios where the inputs arrive at two different moments of time, the output of the masked and gate performs two transitions instead of one. One transition is performed when the single input performs a transition and another one is performed when the other four input signals perform a transition. Thus the Transition Table for such a gate in this scenario would consist of 2048 rows and we observe that the expected mean for the cases when $q = q_m \oplus m_q = 0$ is different from that when $q = 1$. Similar results were found in other scenarios as well. This bias of leakage in masked gates in presence of glitches can be exploited to apply successful attacks on masking countermeasure.

Collision -Correlation Attack:

The main objective of a collision-correlation attack is to find out the region in the power traces which handles same data, and use this knowledge to get access to secret information. To illustrate this attack methodology, let us consider an example. Assume that we are going to implement the attack on an masked AES implementation. Internal states of the AES design is denoted as $(x_0, x_1, \ldots, x_{15})$, where x_i= a data byte. Similarly, key is also denoted as $(k_0, k_1, \ldots, k_{15})$. S-Boxes of the AES implementation is masked and input masks for all the bytes is same. So each masked S-box has a input $x_i' = x_i + u$, u is the input mask. Output of the masked S-Box(S') is $S'(x_i') = y_i' = y_i + v$, v is the output mask and $y_i = S(x_i)$. Now we obtain N power traces for a same message M and identify the power traces for each masked S-Box access in the first round. Using any statistical distinguisher (for example Pearson's correlation coefficient), we can find out whether any of the S-Boxes are having *collision*, i.e., whether they are handling the same data or not.

Once we get a collision between the ith S-Box access and jth S-Box access, we can obtain following formulation.

$$x_i' = x_j'$$
$$x_i \oplus u = x_j \oplus u$$
$$m_i \oplus k_i = m_j \oplus k_j$$
$$m_i \oplus m_j = k_i \oplus k_j$$

This experiment is repeated for multiple times with different messages (m_i = ith byte of plain-text) to obtain sufficient number of collisions such that the guessing entropy of the key reduces. The attack methodology described here was first introduced in [17]. Consequently more advanced attack methodology with different statistical tools have been presented in [18, 19].

Generally when a crypto-algorithm is developed, side channel security is not considered as a required parameter. The crypto-algorithm is initially formulated to thwart theoretical cryptanalysis techniques. After the development of the algorithm, side channel protection is provided during the implementation by adding an external side channel countermeasure like masking. This approach leads to resource hungry designs and has a huge overhead. For example, gates required by masked AES is three times of original AES implementation.

Another alternative approach is to design crypto-algorithms which will be secure against side channel by its construction. An example of such technique is given in the next subsection.

5.3.3 DRECON: DPA Resistant Encryption by Construction

The scheme we present is called DRECON (DPA Resistant Encryption by CONstruction) [20], which attempts to design a complete block cipher with DPA prevention as a prerequisite. The construction is inspired from tweakable block ciphers [21], where in addition to the plaintext and key, the cipher takes a *tweak*. However, unlike the tweakable block ciphers in [21], the construction requires the tweak to be kept secret. The tweak is used to choose an sbox from a given pool of cryptographically strong sboxes, thus modifying the mapping between the plaintext and the ciphertext. Protection against DPA is obtained based on the assumption that the tweak is exclusively shared between the sender and the receiver and modified in every encryption. The construction is supported by information theoretic proofs of security to show that is secured against first order power attacks.

DRECON

The secret in DRECON comprises of the tuple (t, k), where t is called the tweak and k the key used in the block cipher. The key k is held constant for all encryptions, while the tweak t changes for each encryption, using a tweak generation algorithm. The tweak is used to select a function from the set $\mathcal{F}\{\mathsf{F}_1, \mathsf{F}_2, \cdots, \mathsf{F}_r\}$, where $\mathsf{F}_j : \mathbb{F}_2^n \mapsto \mathbb{F}_2^n$ and $(1 \leq j \leq r)$, are cryptographically strong sbox functions. For every application of the sbox on X, a function from $\mathcal{F}$ is selected depending on the value of the tweak (t) and applied to X. This sbox, known as the *tweaked-sbox*, is represented by $\vec{S}(\cdot, \cdot)$ and defined as follows:

$$\vec{S}(t, X) \leftarrow \mathsf{F}_t(X) \qquad \text{where } t \overset{R}{\leftarrow} \{1, 2, \cdots, r\}.$$

In a typical iterative block cipher, the first round key is added to plaintext before the sbox operation and the sbox operation has the form $S(x \oplus k)$. However, in DRECON, we choose to omit the key-whitening at the beginning and end of the encryption. Thus, each round except the last round consists of a substitution layer, diffusion layer and key addition layer. The last round consists of only substitution layer. Each of the sboxes of the substitution layers is replaced by the tweaked-sbox. For all round, the same tweaks are used while the sboxes in a round have differ-

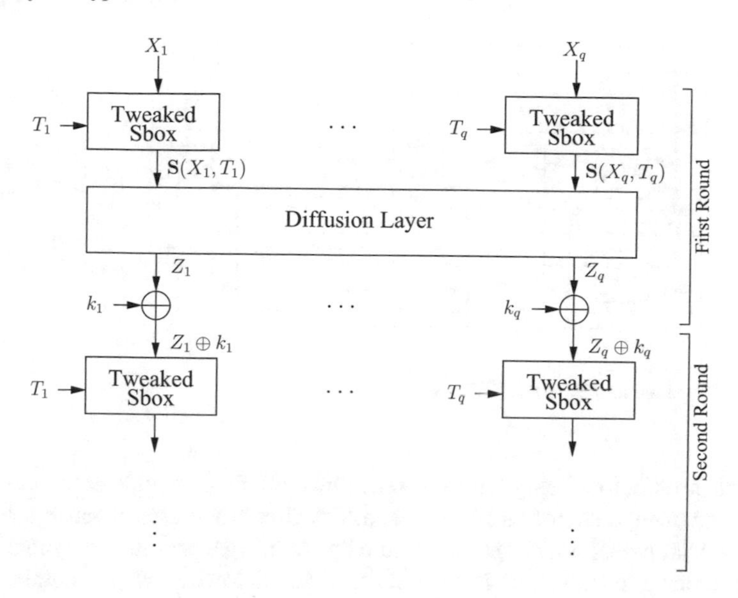

Fig. 5.7 First round of DRECON. The same structure is repeated for all rounds except the last round which consists of only substitution layer

ent tweaks. The first round of DRECON is shown in Fig. 5.7. It may be noted that DRECON requires no key-whitening at the beginning and end of the block cipher since the tweaked-sboxes provide the required randomization of the input and output respectively.

Tweak Generation Algorithm

From the master tweak agreed upon by the sender and receiver, tweaks need to be generated for each encryption. The tweak generation needs to produce uniformly random tweaks in the range of 1 to r in order to select one of the r sboxes (for DRECON-AES $r = 16$ or 256). Further, the algorithm needs to be secure against power attacks as is discussed in detail in [22].

Any mask generation function (MGF) or stream cipher implemented in a secure manner can be used as a tweak generator. However, given the fact that the adversary has no control or knowledge of the input and output of the tweak generator, lightweight solutions can be developed by balancing registers and minimizing the combinatorial logic, which can otherwise leak [23]. A possible construction for a tweak generation algorithm makes use of an LFSR as shown in Fig. 5.8. The design uses two pairs of shift registers (S and $\overline{S}$), each comprising of 128 flip-flops. The flip-flops in $\overline{S}$ are a complement of the flip-flops in S. To obtain such a state, the master tweak is used to seed S and the complement of the master tweak is used to seed $\overline{S}$. Further, the feedback obtained from a 128 degree primitive polynomial is

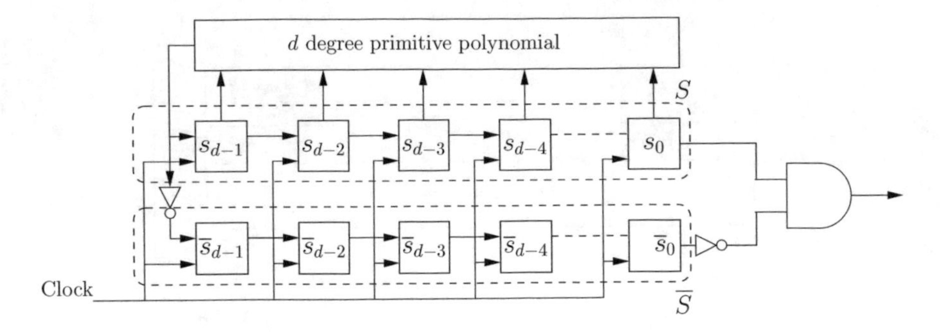

Fig. 5.8 Tweak generation for DRECON

complemented before being fed back to $\overline{S}$. Since all clocks toggle at the same time, the leakage from the registers is minimized. The alternate source of leakage, from the combinatorial paths, is also kept minimum by choosing a primitive polynomial with small number of coefficients. For DRECON-AES, the primitive polynomial chosen is $\alpha^{128} \oplus \alpha^{95} \oplus \alpha^{57} \oplus \alpha^{45} \oplus \alpha^{38} \oplus \alpha^{36} \oplus 1$.

Hardware Implementation of DRECON on FPGAs

The adapted $n \times n$ AES algorithm with DRECON is called $n \times n$ DRECON-AES where possible values of n are 4 and 8. The DRECON-AES has the following properties. Each round of DRECON-AES has the same structure as that of $n \times n$ AES except the *AddRoundKeys* of the first and the last round are omitted. The *ShiftRows* operation of the last round is also omitted, and thus last round is left with only the *SubBytes* operation (Fig. 5.9). Further, each $n \times n$ bit sbox is replaced by a $n \times n$ bit tweaked-sbox. Each tweaked-sbox is a set of 2^n ($r = 2^n, n = 4$ or 8) non-linear functions having the equal cryptographic strength.

The implementation of DRECON on an FPGA device is shown in Table 5.2. DRECON 4×4 and 8×8 has been implemented and compared with a standard

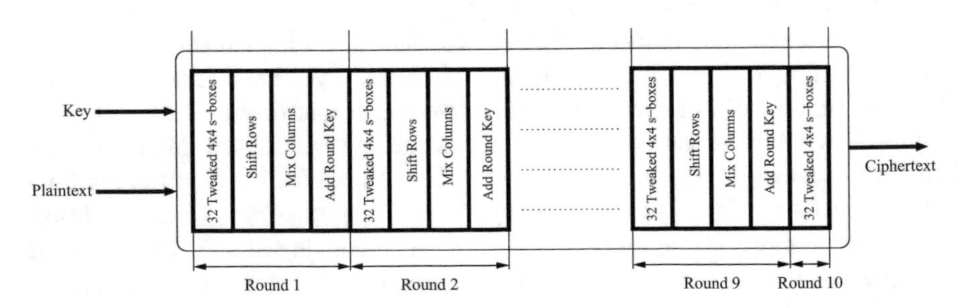

Fig. 5.9 4×4 AES adapted for DRECON

Table 5.2 Comparing resource requirements for 4×4 DRECON-AES with masking on an FPGA (XC5VLX50-2FF324)

Implementation	Slices	LUTs	Registers	Clock cycles	Clock period (ns)
4×4 AES	1120	3472	1270	11	11.14
Masked 4×4 AES	3427	10589	1765	11	23.83
4×4 DRECON-AES	1379	3868	1583	11	10.3

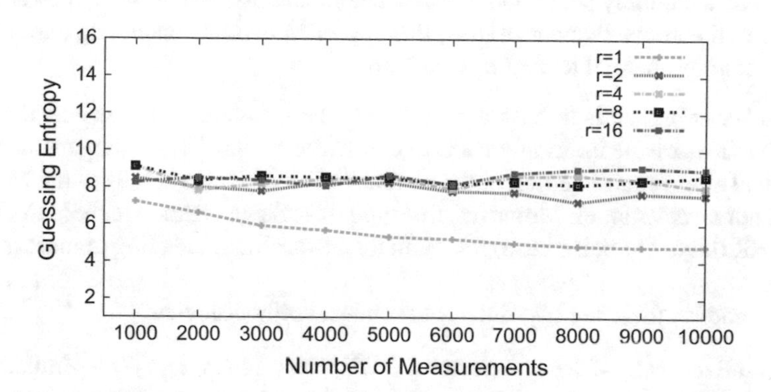

Fig. 5.10 Guessing Entropy versus number of measurements for different size of tweaks

masking scheme. However, it may be stressed DRECON is not a countermeasure, but a design principle in the block cipher to resist against DPA.

Practical Evaluation: Practical evaluation of 4×4 DRECON-AES has been performed on the above design targetting an implementation on SASEBO-GII side-channel evaluation board [24]. Since block RAMs are difficult to attack [25], we implemented with distributed RAMs in order to get a faster comparison. We used CPA as the attack algorithm where the target is computed by randomly guessing the value of the tweak. The attack was repeated with different tweak sizes and the efficiency was measured using the guessing entropy metric [26]. When the tweak is not present (the size of the pool of sboxes $\mathcal{F}$ is one, i.e., $h_t = 0$ and $r = 1$), as shown in Fig. 5.10, the guessing entropy decreases as the number of measurements increases, and with 10000 traces the guessing entropy becomes almost 5. This is significantly below the guessing entropy of random guess. When $r = 16$ (i.e., $h_t = 4$, the size of the input to the sboxes), the guessing entropy is around 8.5 all the time, implying that the scheme is as secure as a random guess of K. Figure 5.10 presents the results.

5.3.4 Evaluation of Side Channel Leakage for a Crypto Core

When a crypto-chip is designed, it is imperative to analyze the chip for both functionality and side channel security. Now there are two ways to test a design for side channel security. The first one is to carry out actual attacks on the chip to see whether it is possible for an adversary to get the secret information. However, this approach poses several problems like:

1. The evaluator needs to have a deep understanding of the hardware implementation of the crypto-algorithm.
2. He needs to carry out several different attacks with different attack models to become absolutely sure about the side channel resistance of the crypto-chip.
3. Due to the above two constraints, this approach is time consuming and not suitable for commercial testing mechanisms.

Another approach is to have a simple test where instead of doing actual attacks, we try to measure the information leakage from the design. This testing methodology is fast and does require neither understanding of the intrinsic details of the hardware design nor attack model. Moreover, this type of leakage testing can help to identify high SNR (signal to noise ratio) region in the power traces, leading to more efficient attacks.

In literature, there are two different such test methodologies:

- **Normalized Inter-Class Variance (NICV) Test**: NICV test [7] is similar to statistical *F-test*, where we try to find high SNR value in given power traces. The steps involved in computing NICV are as follows:

  1. Let us assume that plain-text input to the design can be separated into n number of different classes.(For example, a plaintext byte can be divided into 9 different classes depending upon its hamming weight)
  2. Collect power traces and separate them into the n classes depending upon the input plain-text value
  3. Compute the mean of each class and then compute the variance of these mean curves. Let us denote this variance as *mean class variance (MCV)*
  4. Compute the variance of all the power traces. Let us denote this variance as *all trace variance (ATV)*
  5. $NICV = \frac{MCV}{ATV}$

  Generally NICV is used to find out favorable points in the power traces to execute power attack. It can also be useful for profiling and training phase of template attack. An NICV plot, obtained during power analysis on block cipher *SIMON* is shown in Fig. 5.11a.

- **Test Vector Leakage Assessment (TVLA)**: TVLA test is similar to the statistical *T-test*. Like NICV, this test can be used to find out high SNR region in the power traces. Moreover, this test provides yes-no answer to the question whether the device is side channel secure or not. This testing method is fast and can be used in high speed commercial testing of the crypto-chips.

  The steps in TVLA test are as follows:

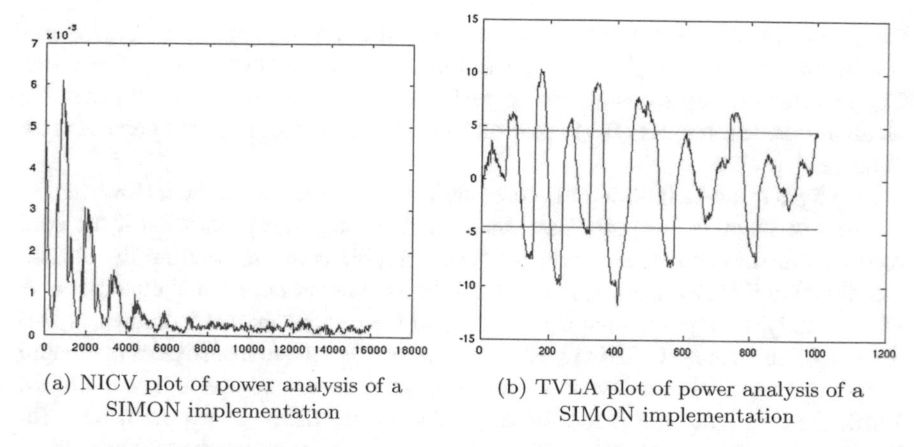

(a) NICV plot of power analysis of a SIMON implementation

(b) TVLA plot of power analysis of a SIMON implementation

Fig. 5.11 Different leakage evaluation methodology

1. Create two different Dataset (Q_1 and Q_2), each with n instances of plain-text and key. One comprising of same plain-text and the other one contains random plain-texts. Both the data-set has same key.
2. Now obtain n power traces for each dataset and compute the TVLA metric according to the following formula

$$TVLA = \frac{mean(Q_1) - mean(Q_2)}{\sqrt{\frac{Var(Q_1)}{n} + \frac{Var(Q_2)}{n}}} \tag{5.24}$$

3. For any sample point in the power trace, if $|TVLA| \geq 4.5$, we consider the device not secure against side channel attack.

This methodology can be applied on each bit of the plain-text also, i.e., it is possible to generate TVLA plot for each bit of the plain-text. A TVLA plot of a SIMON implementation is shown in Fig. 5.11b.

In this chapter we have given a brief introduction of power based side channel attack and the countermeasures to mitigate the resulting threats. Next we are going to give readers brief idea about fault analysis which can also threaten the security of crypto-systems.

5.4 Fault Attack Resistant Crypto Cores

With the growing importance of secured communication the importance of cryptographic cores have increased. Hardware solutions of these cores are developed using ASIC (Application Specific Integrated Circuit) libraries or on FPGA (Field

Programmable Gate Array) platforms. Recently the enhancement of the FPGAs has lead to the use of these platforms for in-house development of cryptographic IPs. The fact that the entire design can be performed in the laboratory, without relying on an untrusted third party fab makes such design flows ideal from the point of view of security [27].

In a System-on-Chip (SoC) the cores are pretested and pre-verified. However the test and verification is mostly functional and the integrator is satisfied if the core meets its functionalities. However for cryptographic cores, apart from the normal functionality it is also important to model the core under not normal conditions. A related study of cryptographic algorithms and the designs thereof is known as Differential Fault Analysis (DFA) [28]. This analysis technique investigates the nature of induced faults when a device is stressed beyond its normal operating conditions. While the core integrator is mostly satisfied when the faults are not permanent, for the attacker a single transient fault is enough to get the complete key for even standard ciphers like AES-128 [29].

5.4.1 General Principle of DFA of Block Ciphers

In this subsection, we study the basic principle of DFA, which shall be subsequently applied for the AES algorithm. As apparent from the name, DFA combines the concepts of differential cryptanalysis with that of fault attacks. DFA is applicable to almost any secret key crypto-system proposed so far in the open literature. DFA has been used to attack many secret key crypto-systems, including DES, IDEA, and RC5 [30].

There has been considerable amount of work about DFA of AES. Some of the DFA proposals are based on theoretical model [31–38], while others launched successful attacks on ASIC and FPGA devices using previously proposed theoretical models [37, 39–42]. The key idea of DFA is composed of three steps as shown in Fig. 5.12. (1) Run the cryptographic algorithm and obtain non-faulty ciphertexts. (2) Inject faults, i.e., rerun the algorithm with the same input, but in unexpected environmental conditions, and obtain faulty ciphertexts (3) Analyze relationship between the non-faulty and faulty ciphertexts to significantly reduce the key space.

5.4.2 Fault Analysis of AES

Probably AES is the block cipher which has been studied most against DFA. There has been several works on DFA of AES, using various types of fault models, like (1) bit faults, (2) random byte faults, (3) multiple byte faults. In the following sequel, we provide an understanding of DFA on AES.

We consider the attack based on byte level faults. We assume that certain bits of a byte is corrupted by the induced fault and the induced difference is confined within

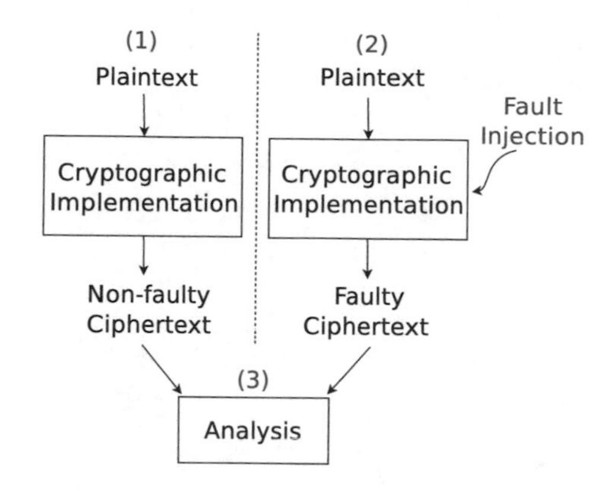

Fig. 5.12 Three steps of DFA

a byte. The fact that the fault is induced in the penultimate round implies that apart from using the differential properties of S-box (as used in the bit-level DFA on last round of AES), the attacker also uses the differential properties due to the diffusion properties of the Mix-Columns operation of AES. AES diffusion is provided using a 4×4 MDS matrix in the Mix-Columns. Due to this matrix multiplication, if one byte difference is induced at the input of a round function, the difference is spread to 4 bytes at the round output. Figure 5.13 shows the flow of fault.

The induced fault has generated a single byte difference at the input of the 9th round Mix-Columns. Let f be the byte value of the difference and the corresponding 4-byte output difference is $(2f, f, f, 3f)$, where 2, 1, and 3 are the elements of the first row of the Mix-Columns matrix. The 4-byte difference is again converted to (f_0, f_1, f_2, f_3) by the non-linear S-box operation in tenth round. The Shift-Rows operation will shift the differences to 4 different locations. The attacker has access to the fault-free ciphertext C and faulty ciphertext C^*, which differs only in 4 bytes. Now, we can represent the 4-byte difference $(2f, f, f, 3f)$ in terms of the tenth round key K^{10} and the fault-free and faulty ciphertexts by the following equations:

$$\begin{aligned} 2f &= S^{-1}(C_{0,0} \oplus K^{10}_{0,0}) \oplus S^{-1}(C^*_{0,0} \oplus K^{10}_{0,0}) \\ f &= S^{-1}(C_{1,3} \oplus K^{10}_{1,3}) \oplus S^{-1}(C^*_{1,3} \oplus K^{10}_{1,3}) \\ f &= S^{-1}(C_{2,2} \oplus K^{10}_{2,2}) \oplus S^{-1}(C^*_{2,2} \oplus K^{10}_{2,2}) \\ 3f &= S^{-1}(C_{3,1} \oplus K^{10}_{3,1}) \oplus S^{-1}(C^*_{3,1} \oplus K^{10}_{3,1}) \end{aligned} \tag{5.25}$$

These four equations can be represented in the form $A = B \oplus C$ where A, B, and C are bytes in F_{2^8}, having 2^8 possible values each. Now a uniformly random choice of (A, B, C) is expected to satisfy the equation with probability $\frac{1}{2^8}$. Therefore, in this case 2^{16} out of 2^{24} random choices of (A, B, C) will satisfy the equation.

This fact can be generalized. Consider we have M such related equations. These M equations consist of N uniformly random byte variables. The probability that a random choice of N variables satisfy all the M equations simultaneously is $(\frac{1}{2^8})^M$.

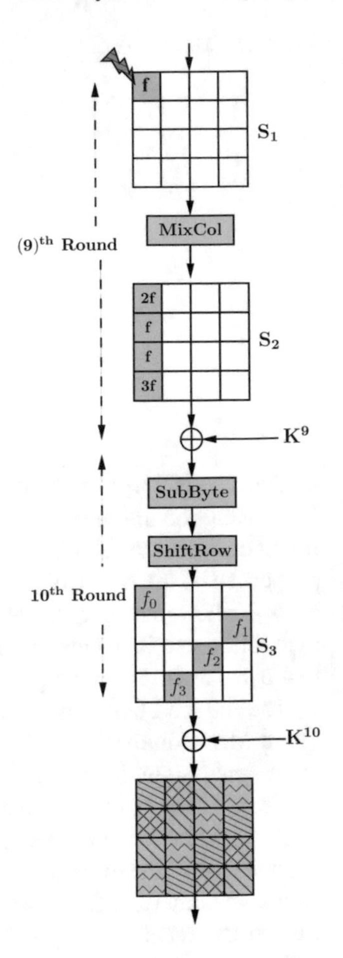

Fig. 5.13 Differences across the last 2 rounds

Therefore, the reduced search space is given by $(\frac{1}{2^8})^M \cdot (2^8)^N = (2^8)^{N-M}$. For our case, we have four equations which consist of five unknown variables: f, $K^{10}_{0,0}$, $K^{10}_{1,3}$, $K^{10}_{2,2}$, and $K^{10}_{3,2}$. Therefore, the four equations will reduce the search space of the variables to $(2^8)^{5-4} = 2^8$. That means out of 2^{32} hypotheses of the 4 key bytes, only 2^8 hypotheses will satisfy the above 4 equations. Therefore, using one fault the attacker can reduce the search space of the 4 key byte to 2^8. Using two such faulty ciphertexts one can uniquely determine the key quartet. For one key quartet one has to induce 2 faults in the required location. For all the 4 key quartets, i.e., for the entire AES key an attacker thus needs to induce 8 faults. Therefore using 8 faulty ciphertexts and a fault-free ciphertext, it is expected to uniquely determine the 128-bit key of AES.

The attack can further be improved. It was shown in [36] that instead of inducing fault in 9th round, if we induce fault in between 7th and 8th round Mix-Columns, we can determine the 128-bit key using only 2 faulty ciphertexts. Figure 5.14 shows the spreading of faults when it is induced in such a fashion. The single byte difference at

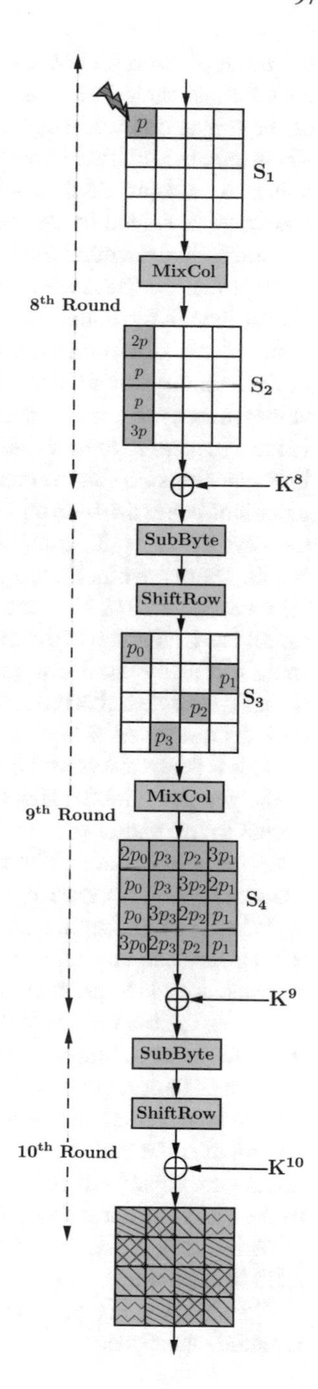

Fig. 5.14 Differences across the last three rounds

the input of 8th round Mix-Columns is spread to 4 bytes. The Shift-rows operation ensures that there is one disturbed byte in each column of the state matrix. Each of the 4-byte difference again spreads to 4 bytes at 9th round Mix-Columns output. Therefore, the relation between the fault values in the 4 columns of difference-state matrix S_4 is equivalent to 4 faults at 4 different columns of 9th-round input-state matrix as explained in the previous attack. This implies that using 2 such faults we can uniquely determine the entire AES key.

Note that the exact working of the DFA proposed in [36] is slightly different from above, though the underlying principle is the same. The attack maintains a list $\mathcal{D}$ for each column of the difference matrix S_4 assuming a one-byte fault in the input of the penultimate round Mix-Columns. The size of the table $\mathcal{D}$ is thus 4×255 4-byte values, as the input fault can occur in any byte of a column and can take 255 non-zero values. Assuming that the fault occurs in the difference matrix S_3 in the first column, then equations similar to equation (5.25) can be written, with the left hand side of the equations being a 4-byte tuple $(\Delta_0, \Delta_1, \Delta_2, \Delta_3)$. It is expected that the correct guess of the keys $K^{10}_{0,0}, K^{10}_{1,3}, K^{10}_{2,2}$, and $K^{10}_{3,2}$ should provide a 4-byte tuple which belongs to the list $\mathcal{D}$. There are other wrong keys which also pass this test, and analysis shows that on an average 1036 elements pass this test with a single fault. Repeating the same for all the 4-columns of the difference matrix S_4 reduces the AES key to $1036^4 \approx 2^{40}$ (note that as the fault is assumed to be between 7th and 8th round each column of S_3 has a byte disturbed). However, if 2 faults are induced then with a probability of 0.98 the unique AES key is returned.

This is the best-known DFA of AES to date when the attacker does not have access to the plaintext and the attacker needs to determine the key uniquely. However with access to the plaintexts, the attacker can still improve the attack by performing the DFA using only fault and a further reduced brute force guess. Also it is possible to reduce the time complexity of the attack further from 2^{32} to 2^{30}.

When the attacker has access to the plaintexts in addition to the ciphertexts [43], the attacker can do brute-force on the possible keys. The objective of this attack or its extensions is to perform the attack using only one fault. While a unique key may not be obtainable with a single fault, the AES key size can reduce to such a small size that a brute force search can be easily performed. It may be noted that reducing the number of fault requirements from 2 to 1 should not be seen in terms of its absolute values. In an actual fault attack, it is very unlikely that the attacker can have absolute control over the fault injection method and hence may need more trials. Rather these attacks are capable of reducing the number of fault requirements by half compared to the attacks proposed in [36].

This attack is comprised of two phases: the first phase reducing the key space of AES to around 2^{36} values, while the second phase reducing it to around 2^8 values.

Consider the Fig. 5.14, where from the first column of S_4 we get following 4 differential equations:

$$\begin{aligned}2p_0 &= S^{-1}(C_{0,0} \oplus K^{10}_{0,0}) \oplus S^{-1}(C^*_{0,0} \oplus K^{10}_{0,0})\\ p_0 &= S^{-1}(C_{1,3} \oplus K^{10}_{1,3}) \oplus S^{-1}(C^*_{1,3} \oplus K^{10}_{1,3})\\ p_0 &= S^{-1}(C_{2,2} \oplus K^{10}_{2,2}) \oplus S^{-1}(C^*_{2,2} \oplus K^{10}_{2,2})\\ 3p_0 &= S^{-1}(C_{3,1} \oplus K^{10}_{3,1}) \oplus S^{-1}(C^*_{3,1} \oplus K^{10}_{3,1})\end{aligned} \tag{5.26}$$

In the above 4 differential equation we only guess the 2^8 values of p_0 and get the corresponding possible 2^8 hypotheses of the key quartet by applying the S-box difference distribution table. Therefore, one column of S_4 will reduce the search space of one quartet of key to 2^8 choices. Similarly, solving the differential equations from all the 4 columns we can reduce the search space of all the 4 key quartets to 2^8 values each. Hence, if we combine all the 4 quartets we get $(2^8)^4 = 2^{32}$ possible hypotheses of the final round key K^{10}. We have assumed here that the initial fault value was in the $(0, 0)$th byte of S_1. If we allow the fault to be in any of the 16 locations, the key space of AES is around 2^{36} values. This space can be brute-force-searched within practical time and hence shows that effectively one fault is sufficient to reduce the key space to practical limits.

The search space of the final round key can be further reduced if we consider the relation between the fault values at the state matrix S_2, which was not utilized in the previous attacks. This step serves as a second phase, which is coupled with the first stage on all the 2^{32} keys (for an assumed location of the faulty byte). Hence using only one faulty ciphertext one can reduce the search space of AES-128 key to 256 choices. However, the time complexity of the attack is 2^{32} as we have to test all the hypothesis of K^{10}. The time complexity of the attack can however be improved to 2^{30} by exploiting a property of the key-schedule.

A Practical DFA on an AES-Key Schedule Core

In FPGA implementations, the key schedule operation is often done prior to the AES encryption and the round keys are stored in RAM which is being read out for each of the 10 rounds of AES-128 implemented in iterative fashion for the first nine rounds. An AES round includes all the four transformation operations in a combinatorial logic. The S-box in `SubBytes` module is implemented in a look-up table fashion. Figure 5.15a shows the block diagram of 32-bit AES-128 key schedule. The four 32-bit registers *R0, R1, R2, R3* hold the four words of the round keys (word represents a data of 32 bit width). As the design is 32-bit, therefore only one word of the AES round key is generated in each clock cycle. In first four clock cycles, the *select1* and *select0* lines will load the initial AES key into the four registers. In the subsequent clock cycles the *select0* line will load the value of output register W_i to *R0, R1, R2, R3*. The value of W_i will be stored in one of the four registers depending on the write enable signals WR0–WR3 of these registers. *SW* and *RW* in figure represent the `SubWord` and `RotWord` operations of AES key schedule [44]. In each cycle one

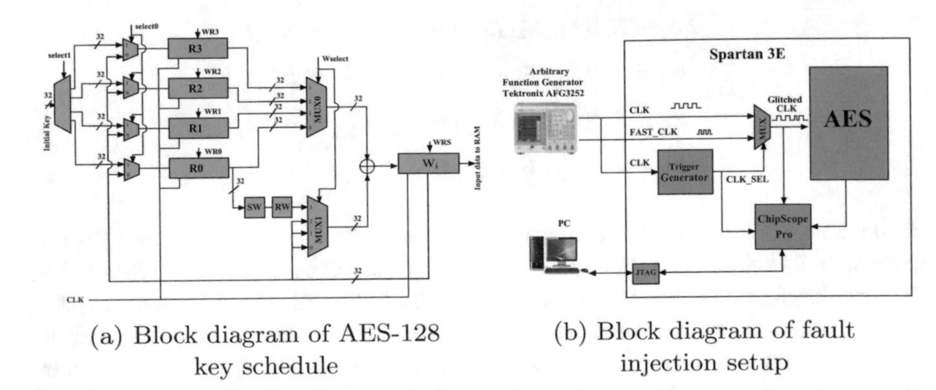

(a) Block diagram of AES-128 key schedule

(b) Block diagram of fault injection setup

Fig. 5.15 Fault injection in AES hardware implementation

word of round key is generated and stored in register W_i and ultimately stored in the RAM. Therefore, in the first 44 clock cycles, all the ten round keys are generated.

Figure 5.15b shows the fault injection setup. We use two clocks *CLK* and *FAST_CLK*, generated from Tektronix AFG3252 arbitrary function generator. One is the normal clock (*CLK*) and the other one is a fast clock. The trigger generator generates the *CLK_SEL* signal which initially being low selects *CLK*. At the beginning of eighth round, the trigger generator makes the *CLK_SEL* signal high for one clock pulse which selects *FAST_CLK* and thus generates glitch in the clock line. This creates setup time violation in the path $LP1 : R0 \rightarrow SW \rightarrow RW \rightarrow MUX1 \rightarrow XOR \rightarrow W_i$, which results in faulty data in register W_i depending on time period (glitch width) of the fast clock. This is the critical (longest) path in the key schedule module of AES-128 architecture. Let the other long paths in decreasing order of lengths and affected by timing violations be $LP2$, $LP3$ and so on which we will use later on. As the fault is generated during key schedule operation, this fault gets propagated to the subsequent round keys. The architecture of the AES-128 cipher is implemented using Verilog HDL in Xilinx Spartan-3E FPGA XC3S500E device with input clock *CLK* at an operating frequency of 36 MHz. We used ChipScope Pro 7.1 analyzer to observe the faulty output.

5.4.3 Experimental Results on Fault Models

In the experimental setup, the frequency of *CLK* is held constant at 36 MHz as the operating frequency of the AES while the *FAST_CLK* is increased in steps of 1 MHz from 36 MHz onwards to generate faults in the AES key schedule [45]. In each step we run 512 encryptions and collect the samples through ChipScope. Until 85 MHz no fault occurs in register W_i. The experimental observations to follow are specific to our hardware implementation of AES-128. The specific observations of

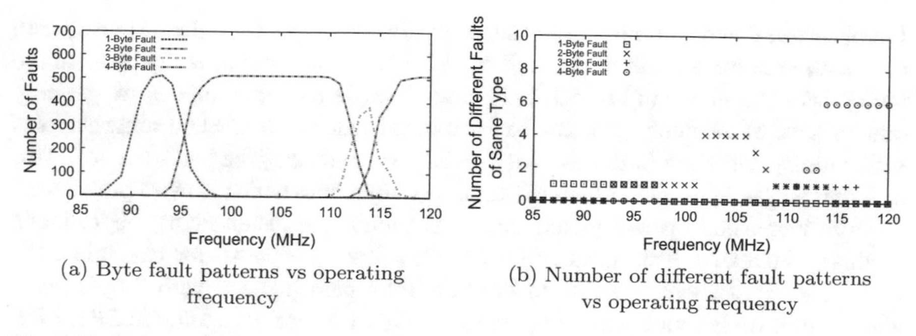

(a) Byte fault patterns vs operating frequency

(b) Number of different fault patterns vs operating frequency

Fig. 5.16 Fault patterns versus operating frequency

fault occurrences in other implementations may vary but the trend and comparisons across different types of faults are similar.

Fault Propagation with Increasing Frequency

From 87 MHz onwards faults start appearing whose distribution can be seen in Fig. 5.16a. Initially only single-byte faults occur. The number of samples with single-byte faults increases with the increase in *FAST_CLK* frequency. At 93 MHz, all the 512 samples in the trace are infected in one byte. The first fault appears in one of the bits (corresponding to $LP1$) of the 4-byte W_i register. When the *FAST_CLK* frequency is further increased, the next fault occurs in another bit (corresponding to $LP2$). If $LP1$ and $LP2$ are in the same byte then multiple 1-byte faults occur else it is a 2-byte fault. But the probability of fault occurrence in the same byte ($\frac{8-1}{31}$, i.e., $\frac{7}{31}$) is lesser than the probability of fault occurrence in other bytes ($\frac{32-8}{31}$, i.e., $\frac{24}{31}$). In the overlapping region of 1-byte and 2-byte fault in Fig. 5.16a, the next faulty bit occurs in a different byte which causes a 2-byte fault in some of the samples. The same argument applies to other overlapping regions between 2-byte fault and 3-byte fault and between 3-byte and 4-byte fault.

From this distribution of different faults it is obvious that initially only 1-byte faults occur and beyond a frequency range (in our case it is 119 MHz) all the samples have 4-byte faults. From the observed results in Fig 5.16a, it is seen that beyond some upper limit frequency all bytes can be corrupted by the glitch. Therefore, generating an all-byte fault is much easier than generating a 1-byte.

Number of Different Types of Faults with Increasing Frequency

In the previous discussion, we observed the distribution of different types of faults from 1-byte to 4-byte, with change in the operating frequency. However, the number of different instances of the same type of fault does not follow these distributions.

We observed that the number of instances of different types of faults increases with the increase in fault-width of the fault model, i.e., the number of faulty bytes in the fault model. As shown in Fig. 5.16b, in case of single byte fault model, we get only one instance of the faults. This can be attributed to the fact that when some samples suffer timing violation in the two long paths $LP1$ and $LP2$, the paths occur in two different output bytes rather than in the same output byte which is more probable.

With increasing operating frequency, the number of instances of 2-byte faults gradually increases. This means that the next few long paths, for example $LP3$, $LP4, \ldots, LPk_2$, (where k_2 refers to the final long path infected with 2-byte fault) which suffer timing violation lie in the same two bytes corresponding to $LP1$, $LP2$. Also from Fig. 5.16b, the number of different instances of 2-byte faults increases as the operating frequency increases beyond 101-MHz. As the *FAST_CLK* frequency is increased further, we see that the number of instances of 2-byte faults reduces to one and 3-byte faults increases. This happens because all the instances of timing violations $LP1$, $LP2, \ldots, LPk_2$, occur simultaneously and each of the separate individual occurrence of previous fault instances disappear from the samples. Here k_i is a integer number, where $2 \leq i \leq 4$. At the same time some new long paths $LP(k_2 + 1), \ldots,$ LPk_3, suffered from timing violation in some of the samples. Each of these long paths leads to a faulty byte which is different from the faulty bytes contained in the 2-byte fault. If frequency is increased in such a way, we observe that after certain frequency range only 4-byte fault instances exist. For even higher levels of frequencies, we see increasingly more instances of 4-byte faults. This is observed since more and more paths $LP(k_3 + 1)\ \ldots,$ $LP(k_4)$, suffer timing violations leading to more affected faulty bytes and this observation is seen in almost all samples as we keep on increasing the *FAST_CLK* frequency. Also from Fig. 5.16b, the number of different types of 4-byte faults increases beyond 113 MHz and the number of such different instances is the highest amongst all the different fault patterns. To sum up, we see that 1-byte faults are seen to exist within a limited operating frequency window (only single instance been seen). Beyond this frequency window only multi-byte faults occur and after a certain maximum operating frequency, only 4-byte faults and that too with different instances exist in abundance. The existence of such numerous different instances of 4-byte faults in the experiments makes this class of multi byte faults the most effective to perform DFA of AES Key schedule as we can have numerous faulty ciphertexts at a particular frequency and hence multiple equations for many unknown variables in terms of faulty ciphertexts as revealed in the next subsection.

5.4.4 Proposed DFA on AES-192 Key Schedule

From the experimental results in the previous subsection it is observed that 4-byte fault in the AES key schedule can be easily injected using methods like clock glitching. In this subsection, we present DFA on AES-192 and AES-256 key schedule [45]. The challenge in 4-byte fault model compared to single byte fault model is that it induces more number of unknown variables in the differential equations that need to be solved in order reduce the search space of the key. Especially when the fault is induced in the key schedule, the challenge increases manifold due to the diffusion in the key schedule of AES. When it comes to AES-192 and AES-256, we need to find two round keys in-order to get the master key which makes the job of the attacker more difficult as he cannot directly apply the technique of AES-128 which only retrieves the final round key. In this work we propose new technique which shows that 4-byte fault model can also be used against AES-192 and AES-256 with relatively lesser number of fault inductions compared to DFA using single byte fault model in [46].

The proposed attack on AES-192 requires only two faulty ciphertexts to reduce the search space to 2^{32}. The flow of fault in AES-192 key schedule is shown in Fig. 5.17a. Figure 5.17b shows the corresponding flow of fault in AES-192 states. Here, a, b, c, d, e, f, g, and h are the fault values. We have two different faulty ciphertexts C_1^* and C_2^*, based on above figures, corresponding to which we have two sets of values (a_1, b_1, c_1, d_1) and (a_2, b_2, c_2, d_2) of (a, b, c, d).

Consider the first row of S_0, we can represent these values of a in terms of fault-free and faulty ciphertexts and the final round key K^{12}. Therefore, we get a set of four equations corresponding to four key bytes of K^{12}. Using two faulty ciphertexts we get two sets of equations. This is also true for the other three rows of S_0. It may be observed that each set of equations consists of two variables of (a, b, c, d, e, f, g, h) and four key bytes of K^{12}. For example, equations from the first row will have (a, e) and first row bytes of K^{12}. We solve these equations in row-wise fashion.

For the first row equations, we guess a_1, e_1, e_2, and get the values of four key bytes from the first set of equations (of C_1). We test each of these values by the second set of equations (of C_2). Only, the right candidates of the four key bytes will satisfy both

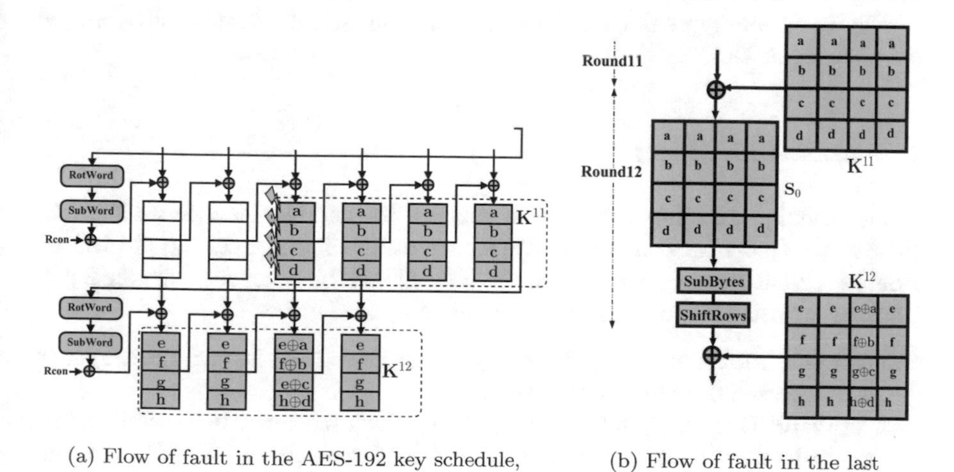

(a) Flow of fault in the AES-192 key schedule, from the first column of K^{11}

(b) Flow of fault in the last two rounds of AES-192

Fig. 5.17 Flow of faults in AES-192

the sets of equations [29, Sect. 3.1]. We apply, the same technique for the rest of the three rows of S_0 and uniquely determine the values of corresponding key quartets. Thus, combining all the four quartets we get K^{12}.

It may be observed in Fig. 5.17a, that in order to get the master key we need K^{12} and the last two columns of K^{11}. For the last column of K^{11}, we consider the relations between the fault values in Fig. 5.17b. The two sets of fault values (a, b, c, d) and (e, f, g, h) in Fig. 5.17a, are related by the following equations:

$$e = SB(K^{11}_{1,3}) \oplus SB(K^{11}_{1,3} \oplus b) \quad f = SB(K^{11}_{2,3}) \oplus SB(K^{11}_{2,3} \oplus c)$$
$$g = SB(K^{11}_{3,3}) \oplus SB(K^{11}_{3,3} \oplus d) \quad h = SB(K^{11}_{0,3}) \oplus SB(K^{11}_{0,3} \oplus a),$$

where $K^r_{i,j}$ is the (i, j) byte of the r-th round key K^r. We have two faulty ciphertexts corresponding to which we get two sets of equations. In these two sets of equations the two sets of values of a, b, c, d, e, f, g, h are already determined while retrieving K^{12}. Therefore, using the two sets of values of the variables we can uniquely determine the four key bytes $(K^{11}_{1,3}, K^{11}_{2,3}, K^{11}_{3,3}, K^{11}_{0,3})$. Hence, using two faulty ciphertexts the attack reduces the search space of the AES-192 key to 32 bits which is in practical search limits.

5.4.5 Countermeasures Against DFA

In this subsection we are going to focus on different countermeasures to protect the crypto-system against fault attack. We will mainly discuss one such countermeasure known as **Concurrent Error Detection (CED) for DFA** [47].

CEDs uses four types of redundancy: information, time, hardware, and hybrid, as shown in Fig. 5.18.

Information Redundancy

In this technique, the input message is encoded to generate a few check bits, and these bits are propagated along with the input message. These check bits are derived from the output message and compared with the check bits encoded from the input. Three information redundancy techniques are discussed below:

1. Parity-1: One can use single bit parity for the entire 128-bit state, and the parity bit is checked once for the entire round [48].
2. Parity-16: One parity bit can be generated for each input byte. While some parity-16 techniques depend on the S-box implementations [49, 50], a general parity formation is proposed in [51].
3. Robust Code: The parity code suffers from nonuniform fault coverage [52], e.g., parity-16 cannot detect an even number of faulty bits in a byte. Robust

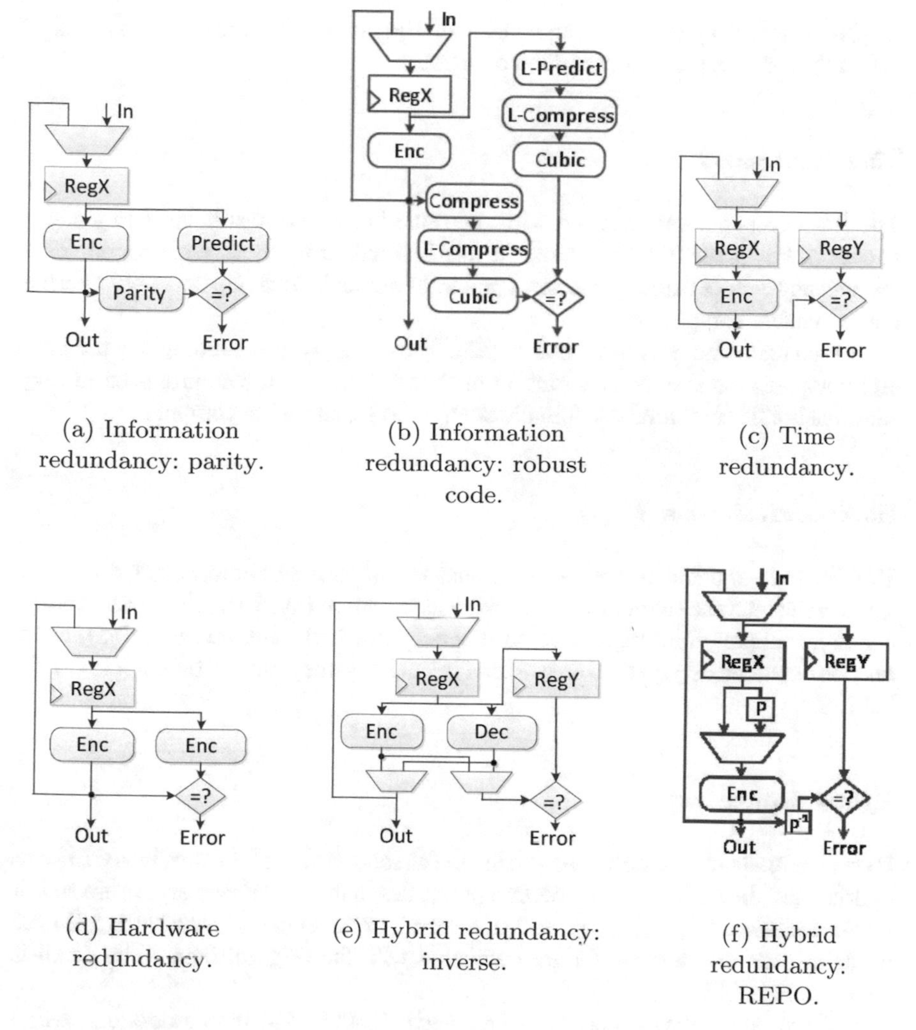

(a) Information redundancy: parity. (b) Information redundancy: robust code. (c) Time redundancy.

(d) Hardware redundancy. (e) Hybrid redundancy: inverse. (f) Hybrid redundancy: REPO.

Fig. 5.18 Four types of CEDs

code provides uniform fault coverage for all types of faults [52]. It uses a prediction circuit at round input to predict a nonlinear property of the round output as shown in Fig. 5.18b. The prediction circuit has a linear predictor (L-Predict), linear compressor (L-Compress), and a cubic function (Cubic). The L-Predict will take the round key and the round input and generate a 32-bit output. The L-Compress and cubic function will reduce the 32-bit data into 28 bits. There are three components at the round output to extract the nonlinear property of the output: the compressor (Compress), the linear compressor, and the cubic function. Each byte of the compressor output $L(j)$ is equivalent to the component-wise

XOR of four bytes of the same column. The output of the linear predictor $L_l(j)$ is the same as the output of the compressor.

Time Redundancy

The function is computed twice with the same input, and results are compared as shown in Fig. 5.18c. One redundant cycle is required to check each round. Time redundancy cannot detect permanent and transient faults that appear in both normal and redundant computations.

A time redundancy is proposed in [53]. The design simply recomputes the input and compares the results. A variation of the time redundancy is proposed in [54]. The function is computed on both clock edges to speed up the computation.

Hardware Redundancy

The circuit is duplicated, and both original and duplicated circuits are fed with the same inputs and the outputs are compared as shown in Fig. 5.18d. Hardware redundancy technique offers high fault coverage against both random faults [53], but it may be bypassed by an attacker who can inject the same faults in both copies of the hardware.

Hybrid Redundancy

Hybrid redundancy techniques combine the characteristics of the previous CED categories, and they often explore certain properties in the underlying algorithm and/or implementation. In [55], an operation, a round, or the entire encryption is followed by its inverse, and the results are compared with the original input. The detail is shown in Fig. 5.18e.

In Recomputing with Permuted Operands (REPO) [56] the authors discover a special invariance of AES and use it to detect faults (Fig. 5.18f). First, the data is computed usually. Then, the same data is permuted and computed. After the results are inverse permuted, the result should be the same as without any permutation. Redundant rounds are inserted in the encryption. In each redundant round, the input data is permuted and AES computes the permuted data. Then, the round output is inverse permuted and compared with the original output. Any mismatch shows that faults are detected. REPO provides close to 100 % fault coverage to both permanent and transient faults.

Thus in this section we have provided a brief discussion on fault analysis along with the countermeasures to protect the crypto-system against it. In the next section we are going to discuss about the vulnerabilities that may arise in the crypto-system due to conventional chip testing mechanisms.

5.5 Testing and Validation of Crypto Cores

Reliability of devices has become a serious concern, with the increase of complexity of ICs and the advent of deep sub-micron process technology. The growth of applications of cryptographic algorithms and their requirement for real time-processing has necessitated the design of crypto-hardware. But along with the design of such devices, testability is a key issue. What makes testability of these circuits more challenging compared to other digital designs, is the fact that popular design for testability (DFT) methodologies, such as scan chain insertion, can be used as a double-edged sword. Scan chain insertion, which is a desirable test technique owing to its high fault coverage and small hardware overhead, open "side channels" for cryptanalysis [57, 58]. Scan chains are used to access intermediate values stored in the flip-flops, thereby ascertaining the secret information, often the key. Conventional scan chains fail to solve the conflicting requirements of effective testing and security [57]. So, one of the solutions that have been suggested is to blow off the scan chains from the crypto ICs, before they are released into the market. But such an approach is unsatisfactory and directly conflicts the paradigm of DFT. In order to solve this problem of efficiently testing cryptographic ICs several research works have been proposed.

5.5.1 Working Principle of Scan-Chain-Attacks on Block Ciphers

In this subsection, we outline the working principle of scan chain-based-attacks on block ciphers. Similar attacks can be performed on other category of ciphers as well.

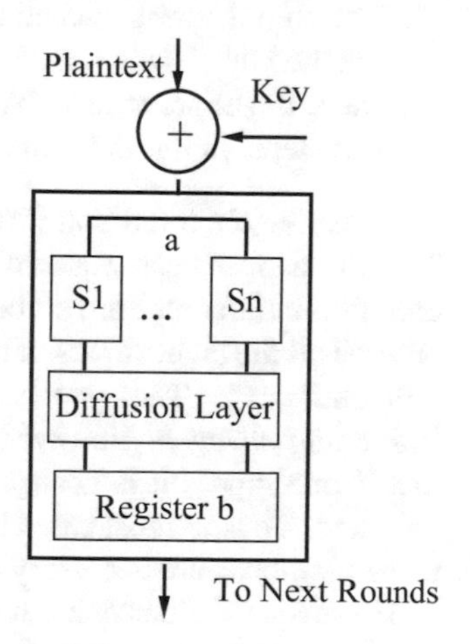

Fig. 5.19 Generic Structure of a block cipher

The security of the block cipher is obtained due to the properties of the round function as shown in Fig. 5.19, and the number of rounds in the cipher. However, when the design is prototyped on a hardware platform, and a scan chain is provided to test the design, the attacker uses the scan chain to control the input patterns, and observe the intermediate values in the output patterns. The security of the block cipher is thus threatened, as the output after few rounds is revealed to the adversary. The attacker then analyzes the data and applies conventional cryptanalytic methods on a much lessened cipher [59].

We next summarize the scan based attack, w.r.t. Fig. 5.19. Without loss of generality, let us assume that the S-boxes are byte-wise mappings, though the discussion can be easily adapted for other dimensions. The attack observes the propagation of a disturbance in a byte through a round of the cipher. If one byte of the plaintext is affected, say p_0, then one byte of a, a_0 gets changed (see figure). The byte passes through an S-box and produces an output, which is diffused in the output register, b_0.

The diffusion layer of AES-like ciphers are characterized by a property called as, *branch number*, which is the minimum total number of disturbed bytes at the input and output of the layer. For example, the MixColumns step of AES has a branch number of 5, indicating that if b_1 input bytes are disturbed at the input of MixColumns, resulting in b_2 bytes at the output which get affected, then $b_1 + b_2 \geq 5$.

Depending upon the branch number of the diffusion layer, the input disturbance spreads to say t-number of output bits in the register b. The attacker tries to exploit this property to first ascertain the correspondence of the flip-flops of register b, with the output bits in the scan-out pattern. Next the attacker applies a one-round differential attack to determine the secret key.

1. The attacker first resets the chip and loads the plaintext p and the key k, and applies one normal clock cycle. The XOR of p and k is thus transformed by the S-boxes and the diffusion layers, and is loaded into the register b.
2. The chip is now switched to the test mode and the contents of the flip-flops are scanned out. The scanned out pattern is denoted by TP_1.
3. Next, the attacker disturbs one byte of the input pattern and repeats the above two steps. In this case, the output pattern is TP_2.

It may be observed that if the attacker observes the difference between TP_1 and TP_2, the attacker can observe the positions of the contents of the register b. The ones in the difference are all because of the contents of register b. In order to better observe all the bit positions of register b, the attacker repeats the process with further differential pairs. There can be a maximum 256 possible differences in the plaintext byte being changed. However, the ciphers satisfy avalanche criteria, which states that if one input bit is changed, on an average at least half of the output bits get modified. Thus in most cases because of this avalanche criteria of the round, much fewer plaintexts are necessary to obtain the locations of all the registers b.

However, the attacker has only ascertained the location of the values of the register b in the scanned-out patterns. But, it is surely an unintended leakage of information.

For example, the difference of the scanned out patterns giving away the Hamming distance after one round of the cipher.

The attacker now studies the properties of the round structures. The S-box is a non-linear layer, with the property that all possible input and output pairs are not possible. As an example, for the present day standard cipher, the Advanced Encryption Standard (AES), given a possible input and output pair, on an average one value of the input to the S-box is possible. That is, if the input to the S-box, denoted by S is x, and the input and output differentials are α, β, then there is one solution on an average to the equation:

$$\beta = S(x) \oplus S(x \oplus \alpha)$$

The adversary can introduce a differential α through the plaintext. However, he uses the scanned out data to infer the value of β. In order to do so, he observes the diffusion of the unknown β through the diffusion layer. In most of the ciphers, like the AES, the diffusion layers are realized through linear matrices.

To be specific in case of AES, a differential $0 < \alpha \leq 255$ in one of the input bytes, is transformed by the S-box to β and then passes to the output after being transformed by the diffusion layer as follows:

$$\begin{pmatrix} \alpha\,0\,0\,0 \\ 0\,0\,0\,0 \\ 0\,0\,0\,0 \\ 0\,0\,0\,0 \end{pmatrix} \Rightarrow \begin{pmatrix} \beta\,0\,0\,0 \\ 0\,0\,0\,0 \\ 0\,0\,0\,0 \\ 0\,0\,0\,0 \end{pmatrix} \Rightarrow \begin{pmatrix} 2\beta\,0\,0\,0 \\ 3\beta\,0\,0\,0 \\ \beta\;0\,0\,0 \\ \beta\;0\,0\,0 \end{pmatrix}$$

Thus the attacker knows that the differential in the scanned-out patterns, TP_1 and TP_2, has the above property. That is there are 4 bytes in the register b, denoted by d_0, d_1, d_2, d_3, such that:

$$d_0 = 2d_2; d_1 = 3d_2; d_2 = d_3 \tag{5.27}$$

The attacker in the previous attack has ascertained the positions of the 32 bits of the register b in the scanned out pattern. But he does not know the correct pattern. The above property says that the correct pattern will satisfy the above property. If w is the number of ones in the XOR of TP_1 and TP_2, then there are 32_{C_w} possible patterns. Out of that, the correct one will satisfy the above property. The probability of a random string satisfying the above property is 2^{-24}. Thus, if $w = 24$ as an example, then the number of satisfying permutations is $32_{C_4} \times 2^{-24} \approx 1$. Thus, there is a single value which satisfies the above equations. This helps the attacker to get the value of β. The attacker already knows the value of α from the plaintext differential. Thus, the property of the S-box ensures that there is on an average one single value of the input byte of the S-box. Thus, the attacker gets the corresponding byte for a (see figure). The attacker then computes one byte of the key by XORing the plaintext, p_0 with the value of the byte a_0, that is, $k_0 = p_0 \oplus a_0$.

The remaining key bytes may be similarly obtained. In the literature, there are several reported attacks on the standard block ciphers, namely DES and AES [60], but all them follows the above general ideas of attacking through controllability and observability through scan chains.

Countermeasures

An interesting alternative and one of the best method was proposed in [61, 62] where a secure scan chain architecture with mirror key register was used to provide both testability and security. Figure 5.20 shows the diagram of the secure scan architecture. The design uses the idea of a special register called as the mirror key register (MKR), which is loaded with the key, stored in a separate register, during encryption. However during encryption, the design is in a secure mode and the scan chains are disabled. When the design is in the test mode, the design is in the insecure mode and the scan chains are enabled. During this time the MKR is detached from the key register. The transition from the insecure mode to the secure mode happens, by setting the Load_key signal high and the Enable_scan_in and Enable_scan_out signals low. However the transition from the secure mode to the insecure mode happens only through a power_off state and reversing the above control signals. It is expected that the power off removes the content of the MKR, and thus does not reveal the key to a scan-chain based attacker.

But this method has the following shortcomings:

- Security is derived from fact that switching off power destroys the data in registers. So, if the secret is permanently stored on-chip (example credit cards, cell-phone simcards, access cards) even after turning the power off the information exists inside the chip. This can be extracted from a device having such a scan chain in the insecure mode.
- At-speed testing or on-line testing is not possible with this scheme.
- The cryptographic device can be a part of a critical system that remains ON continuously (like satellite monitoring system). In such devices power off is not possible. Hence testing in such a scenario requires alternative solutions.

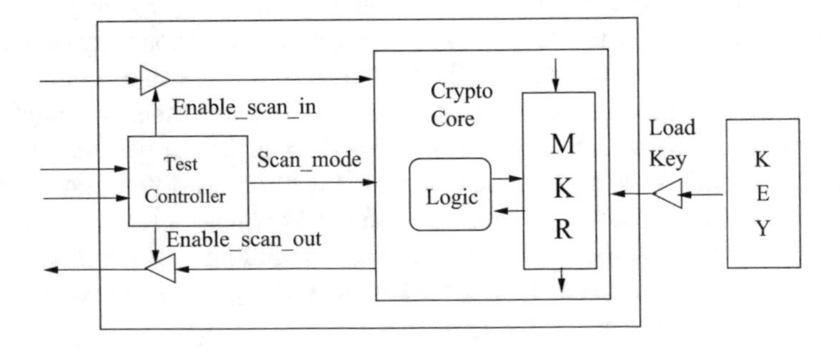

Fig. 5.20 Secure scan architecture with mirror key register [61]

- One of the most secured mode of operation of a block cipher like AES is Cipher Block Chaining (CBC) where the ciphertext at any instant of time depends on the previous block of ciphertext [60]. If testing is required at an intermediate stage then the device needs to be switched off. Thus for resuming data encryption all the previous blocks have to be encrypted again. This entire process has also to be synchronized with the receiver which is decrypting the data. Therefore such modes of block ciphers cannot be tested efficiently using this scheme.

5.6 Conclusions

The chapter shows the security issues involved in designing IP cores for cryptographic algorithms. It shows the threats that loom over the designs of even standard cryptographic algorithms when conventional design approaches are adopted through the availability of several side channel sources. It starts with the underlying principles of power analysis and discusses in details on several forms of power attacks. Mitigation schemes and also evaluation strategies for such attacks are also discussed alongwith. The chapter also discusses on fault attacks of crypto-cores, taking examples of the AES core. Laboratory results have been furnished throughout to demonstrate the practicality of such attack vectors. Various redundancy architectures to prevent such fault attacks are presented. Finally, a discussion on threats from adoption of conventional testing schemes have been provided along with a popular method to improve resistance against such menacing threats on the crypto-IC.

References

1. The Heartbleed Bug (2014)
2. Green, M.: Attack of the week: OpenSSL Heartbleed (2014)
3. Subramanian, N.: Websites affected by Heartbleed: change your Gmail, Facebook and Yahoo passwords right now (2014)
4. Team, M.: The Heartbleed Hit List: The Passwords You Need to Change Right Now (2014)
5. Mukhopadhyay, D., Chakraborty, R.S.: Hardware Security: Design, Threats, and Safeguards. CRC Press (2014)
6. Chari, S., Rao, J., Rohatgi, P.: Template attacks. In: Kaliski, B., Koç, K., Paar, C. (eds.) Cryptographic Hardware and Embedded Systems—CHES 2002. Lecture Notes in Computer Science, vol. 2523, pp. 13–28. Springer, Berlin (2003)
7. Bhasin, S., Danger, J., Guilley, S., Najm, Z.: NICV: normalized inter-class variance for detection of side-channel leakage. IACR Cryptol. ePrint Arch. **2013**, 717 (2013)
8. Archambeau, C., Peeters, E., Standaert, F.-X., Quisquater, J.-J.: Template attacks in principal subspaces. In: Goubin, L., Matsui, M. (eds.) Cryptographic Hardware and Embedded Systems—CHES 2006. Lecture Notes in Computer Science, vol. 4249, pp. 1–14. Springer, Berlin (2006)
9. Ishai, Y., Sahai, A., Wagner, D.: Private circuits: securing hardware against probing attacks. In: Proceedings of CRYPTO 2003, pp. 463–481. Springer (2003)
10. Park, J., Tyagi, A.: *t*-Private logic synthesis on FPGAs. In: 2012 IEEE International Symposium on Hardware-Oriented Security and Trust (HOST), pp. 63–68 (2012)

11. Park, J., Tyagi, A.: *t*-private systems: unified private memories and computation. In: Chakraborty, R., Matyas, V., Schaumont, P. (eds.) Security, Privacy, and Applied Cryptography Engineering. Lecture Notes in Computer Science, vol. 8804, pp. 285–302. Springer International Publishing (2014)
12. Gomathisankaran, M., Tyagi, A.: Glitch resistant private circuits design using HORNS. In: IEEE Computer Society Annual Symposium on VLSI, ISVLSI 2014, Tampa, FL, USA, July 9–11, 2014, pp. 522–527 (2014)
13. Park, J., Tyagi, A.: Towards making private circuits practical: DPA resistant private circuits. In: IEEE Computer Society Annual Symposium on VLSI, ISVLSI 2014, Tampa, FL, USA, July 9–11, 2014, pp. 528–533 (2014)
14. Wong, M., Wong, M., Hijazin, I., Nandi, A.: Composite field GF(((22)2)2) AES s-box with direct computation in gf(24) inversion. In: 2011 7th International Conference on Information Technology in Asia (CITA 11), pp. 1–6 (2011)
15. Trichina, E.: Combinational logic design for AES subbyte transformation on masked data. IACR Cryptol. ePrint Arch. **2003**, 236 (2003)
16. Mangard, S., Popp, T., Gammel, B.: Side-channel leakage of masked cmos gates. In: Menezes, A. (ed.) Topics in Cryptology CT-RSA 2005. Lecture Notes in Computer Science, vol. 3376, pp. 351–365. Springer, Berlin (2005)
17. Clavier, C., Feix, B., Gagnerot, G., Roussellet, M., Verneuil, V.: Improved collision-correlation power analysis on first order protected aes. In: Preneel, B., Takagi, T. (eds.) Cryptographic Hardware and Embedded Systems CHES 2011. Lecture Notes in Computer Science, vol. 6917, pp. 49–62. Springer, Berlin Heidelberg (2011)
18. Moradi, A.: Statistical tools flavor side-channel collision attacks. In: Pointcheval, D., Johansson, T. (eds.) Advances in Cryptology EUROCRYPT 2012. Lecture Notes in Computer Science, vol. 7237, pp. 428–445. Springer, Berlin (2012)
19. Moradi, A., Mischke, O.: How far should theory be from practice? In: Prouff, E., Schaumont, P. (eds.) Cryptographic Hardware and Embedded Systems CHES 2012. Lecture Notes in Computer Science, vol. 7428, pp. 92–106. Springer, Berlin (2012)
20. Hajra, S., Rebeiro, C., Bhasin, S., Bajaj, G., Sharma, S., Guilley, S., Mukhopadhyay, D.: DRECON: DPA resistant encryption by construction. In: Pointcheval, D., Vergnaud, D. (eds.) Progress in Cryptology - AFRICACRYPT 2014 - 7th International Conference on Cryptology in Africa, Marrakesh, Morocco, May 28–30, 2014. Proceedings. Lecture Notes in Computer Science, vol. 8469, pp. 420–439. Springer (2014)
21. Liskov, M., Rivest, R.L., Wagner, D.: Tweakable block ciphers. In: Yung, M. (ed.) CRYPTO. Lecture Notes in Computer Science, vol. 2442, pp. 31–46. Springer (2002)
22. Medwed, M., Standaert, F.-X., Großschädl, J., Regazzoni, F.: Fresh re-keying: security against side-channel and fault attacks for low-cost devices. In: Bernstein, D.J., Lange, T. (ed.) AFRICACRYPT. Lecture Notes in Computer Science, vol. 6055, pp. 279–296. Springer (2010)
23. Guilley, S., Sauvage, L., Flament, F., Vong, V.-N., Hoogvorst, P., Pacalet, R.: Evaluation of power constant dual-rail logics countermeasures against DPA with design time security metrics. IEEE Trans. Comput. **59**(9), 1250–1263 (2010)
24. Research Center for Information Security National Institute of Advanced Industrial Science and Technology. Side-channel Attack Standard Evaluation Board SASEBO-GII Specification (Version 1.01) (2009)
25. Shah, S., Velegalati, R., Kaps, J.-P., Hwang, D.: Investigation of DPA resistance of block RAMs in cryptographic implementations on fpgas. In: Prasanna, V.K., Becker, J., Cumplido, R. (eds.) ReConFig, pp. 274–279. IEEE Computer Society (2010)
26. Standaert, F.-X., Malkin, T., Yung, M.: A unified framework for the analysis of side-channel key recovery attacks. In: Joux, A. (ed.) EUROCRYPT. Lecture Notes in Computer Science, vol. 5479, pp. 443–461. Springer (2009)
27. Ali, S., Chakraborty, R.S., Mukhopadhyay, D., Bhunia, S.: Multi-level attacks: an emerging security concern for cryptographic hardware. In: DATE, pp. 1176–1179. IEEE (2011)
28. Biham, E., Shamir, A.: Differential fault analysis of secret key cryptosystems. In: Kaliski Jr., B.S. (eds.) CRYPTO. Lecture Notes in Computer Science, vol. 1294, pp. 513–525. Springer (1997)

29. Tunstall, M., Mukhopadhyay, D., Ali, S.S.: Differential fault analysis of the advanced encryption standard using a single fault. In: Ardagna, C.A., Zhou, J. (eds.) WISTP. Lecture Notes in Computer Science, vol. 6633, pp. 224–233. Springer (2011)
30. Biham, E., Shamir, A.: Differential fault analysis of secret key cryptosystems. In: Proceedings of Eurocrypt. Lecture Notes in Computer Science, vol. 1233, pp. 37–51 (1997)
31. Blömer, J., Seifert, J.-P.: Fault based cryptanalysis of the advanced encryption standard (AES). In: Financial Cryptography, pp. 162–181 (2003)
32. Giraud, C.: DFA on AES. In: IACR e-print archive 2003/008, p. 008. http://eprint.iacr.org/2003/008 (2003)
33. Moradi, A., Shalmani, M.T.M., Salmasizadeh, M.: A generalized method of differential fault attack against AES cryptosystem. In: CHES, pp. 91–100 (2006)
34. Mukhopadhyay, D.: An improved fault based attack of the advanced encryption standard. In: AFRICACRYPT, pp. 421–434 (2009)
35. Dusart, G.L.P., Vivolo, O.: Differential fault analysis on AES. In: Cryptology ePrint Archive, pp. 293–306 (2003)
36. Piret, G., Quisquater, J.: A differential fault attack technique against SPN structures, with application to the AES and Khazad. In: CHES, pp. 77–88 (2003)
37. Saha, D., Mukhopadhyay, D., Chowdhury, D.R.: A diagonal fault attack on the advanced encryption standard. IACR Cryptol. ePrint Arch. **581** (2009)
38. Tunstall, M., Mukhopadhyay, D., Ali, S.: Differential fault analysis of the advanced encryption standard using a single fault. In: WISTP, pp. 224–233 (2011)
39. Agoyan, M., Dutertre, J.-M., Naccache, D., Robisson, B., Tria, A.: When clocks fail: on critical paths and clock faults. In: CARDIS, pp 182–193 (2010)
40. Barenghi, A., Hocquet, C., Bol, D., Standaert, F.-X., Regazzoni, F., Koren, I.: Exploring the feasibility of low cost fault injection attacks on sub-threshold devices through an example of a 65 nm AES implementation. In: Proceedings of Workshop RFID Security Privacy, pp. 48–60 (2011)
41. Khelil, F., Hamdi, M., Guilley, S., Danger, J.L., Selmane, N.: Fault analysis attack on an AES FPGA implementation. In: ESRGroups, pp. 1–5 (2008)
42. Selmane, N., Guilley, S., Danger, J.-L.: Practical setup time violation attacks on AES. In: European Dependable Computing Conference, pp. 91–96 (2008)
43. Mukhopadhyay, D.: An improved fault based attack of the advanced encryption standard. In: Preneel, B. (ed.) AFRICACRYPT. Lecture Notes in Computer Science, vol. 5580, pp. 421–434. Springer (2009)
44. National Institute of Standards and Technology: Advanced Encryption Standard. NIST FIPS PUB **197** (2001)
45. Ali, S., Mazumdar, B., Mukhopadhyay, D.: A fault analysis perspective for testing of secured soc cores. IEEE Des. Test **30**(5), 63–73 (2013)
46. Kim, C.H.: Improved differential fault analysis on AES key schedule. IEEE Trans. Inf. Forensics Secur. **7**(1), 41–50 (2012)
47. Guo, X.: Fault Attacks and Countermeasures on Symmetric/Key Cryptographic Algorithms. Ph.D. thesis
48. Wu, K., Karri, R., Kuznetsov, G., Goessel, M.: Low cost concurrent error detection for the advanced encryption standard. In: ITC, pp. 1242–1248 (2004)
49. Bertoni, G., Breveglieri, L., Koren, I., Maistri, P., Piuri, V.: Error analysis and detection procedures for a hardware implementation of the advanced encryption standard. IEEE Trans. Comput. **52**(4), 492–505 (2003)
50. Mozaffari-Kermani, M., Reyhani-Masoleh, A.: A lightweight high-performance fault detection scheme for the advanced encryption standard using composite field. IEEE Trans. VLSI Syst. **19**(1), 85–91 (2011)
51. Mozaffari-Kermani, M., Reyhani-Masoleh, A.: Concurrent structure-independent fault detection schemes for the advanced encryption standard. IEEE Trans. Comput. **59**(5), 608–622 (2010)

52. Karpovsky, M., Kulikowski, K.J., Taubin, E., Member, S.: Robust protection against fault-injection attacks of smart cards implementing the advanced encryption standard. In: DNS, pp. 93–101 (2004)
53. Malkin, T., Standaert, F.-X., Yung, M.: A comparative cost/security analysis of fault attack countermeasures. In: FDTC, pp. 109–123 (2005)
54. Maistri, P., Leveugle, R.: Double-data-rate computation as a countermeasure against fault analysis. IEEE Trans. Comput. **57**(11), 1528–1539 (2008)
55. Karri, R., Wu, K., Mishra, P., Kim, Y.: Concurrent error detection schemes of fault based side-channel cryptanalysis of symmetric block ciphers. IEEE Trans. Comput.-Aid. Des. **21**(12), 1509–1517 (2002)
56. Guo, X., Karri, R.: Recomputing with permuted operands: a concurrent error detection approach. IEEE Trans. Comput.-Aid. Des. Integr. Circ. Syst. **32**(10), 1595–1608 (2013)
57. Kapoor, R.: Security vs. test quality: Are they mutually exclusive? In: ITC'04: Proceedings of the International Test Conference, Washington, DC, USA, 2004, p. 1413. IEEE Computer Society (2004)
58. Yang, B., Wu, K., Karri, R.: Scan based side channel attack on dedicated hardware implementations of data encryption standard. In: ITC'04: Proceedings of the International Test Conference, pp. 339–344, Washington, DC, USA, 2004. IEEE Computer Society (2004)
59. Mukhopadhyay, D., Chakraborty, R.: Testability of cryptographic hardware and detection of hardware trojans. In: 2011 20th Asian Test Symposium (ATS), pp. 517–524 (2011)
60. Stallings, W.: Cryptography and Network Security: Principles and Practice. Pearson Education (2002)
61. Wu, K., Yang, B., Karri, R.: Secure scan: a design-for-test architecture for crypto-chips. In: DAC'05: Proceedings of 42nd Design Automation Conference, pp. 135–140 (2005)
62. Yang, B., Wu, K., Karri, R.: Secure scan: a design-for-test architecture for crypto chips. IEEE Trans. Comput.-Aid. Des. Integr. Circ. Syst. **25**(10), 2287–2293 (2006)

Chapter 6
PUF-Based Authentication

Jim Plusquellic

6.1 Introduction

The internet-of-everything has created vast opportunities for the integration of microelectronic systems into nearly every aspect of our lives, but it has also expanded the attack surface of such systems, providing an ever-widening opportunity for malicious adversaries to steal private information, destroy property or worse, and subvert systems in a manner that results in the loss of human life [1–15]. These problems are becoming particularly acute with the proliferation of mobile computing and the debut of new information-sharing and control systems such as the health information exchange, embedded medical devices, smart grid, home automation, smart cars, smart cards, RFID, and sensor networks. Stronger, physical-layer security, and trust primitives are needed for modern electronic systems to counter the advantage made available to adversaries by the increasing proliferation, diversity, and complexity of software and hardware.

Physical-layer refers to components that are rooted in the hardware, and that provide support for secure execution of algorithms, and for secure generation and storage of secrets (keys). A *physical unclonable function* (PUF) is a physical-layer primitive that is designed to derive entropy (randomness) from variations in the structural and electrical characteristics of integrated circuits (ICs) [16]. Similar to DNA profiles among humans, no two ICs are (or can intentionally be manufactured to be) identical. PUFs measure and digitize small "analog" differences among identically designed ICs to generate unique and unclonable bitstrings. The random and persistent nature of the entropy source within ICs addresses important physical security requirements that relate to the generation and storage of keys. Most PUF designs use standard IC manufacturing processes, which benefits low-cost appli-

J. Plusquellic (✉)
University of New Mexico, Albuquerque, NM, USA
e-mail: jimp@ece.unm.edu

S. Bhunia et al. (eds.), *Fundamentals of IP and SoC Security*,
DOI 10.1007/978-3-319-50057-7_6

cations by eliminating the need for costly non-volatile memory (NVM). PUFs can be integrated into any type of system, including system-on-a-chip (SoC), an application specific integrated circuit (ASIC) or field programmable gate array (FPGA).

This chapter focuses on the design of authentication protocols which utilize physical-layer cryptographic primitives such as the PUF, and describes the benefits (and drawbacks) they offer over traditional software-based authentication protocols. PUF-based authentication protocols are less than 15 years old and many have not yet been fully vetted. Therefore, the development of low cost, secure protocols, and proofs of their attack resilience is still very much a moving target. We provide a high-level description of algorithmic security primitives and authentication protocols, and then present a snapshot of the current state of the art, fully acknowledging that the latter is rapidly evolving and still considered an open research problem by the hardware security and trust community.

6.2 Information Security and Cryptography

The term **information security** refers to vast array of mechanisms, protocols, and algorithms, which are designed to protect information from unauthorized access, modification, and destruction [17]. Information security has four primary objectives including confidentiality, data integrity, authentication, and non-repudiation [18]. *Confidentiality* refers to maintaining privacy or secrecy of information and is traditionally ensured using encryption techniques. *Data integrity* relates to a property of the data, that it has not been altered by an unauthorized party, and is typically implemented using secure hashing schemes. *Authentication* is a process that confirms the identity of an entity or the original source of data using corroborative evidence, and can be carried out using modification detection codes (MDCs), message authentication codes (MACs), and digital signatures. *Non-repudiation* refers to a process that associates an entity with a commitment or action, thereby preventing the entity from claiming otherwise, and is traditionally ensured using digital signature schemes.

The primary goal of **cryptography** is to provide a theoretical basis and practical specifications for techniques that meet these information security goals. A wide variety of cryptographic primitives have been developed to provide information security. Menezes et al. [18] propose a taxonomy which partitions cryptographic primitives into three basic categories, namely *unkeyed primitives*, *symmetric-key primitives,* and *public-key primitives*. *Unkeyed* primitives include cryptographic hash functions, one-way permutations, and random sequences. The keyed primitives include a wide variety of symmetric and public-key ciphers, MACs (which are keyed hash functions), signatures, and pseudo-random number generators (those relevant to authentication are described in the next section). Each primitive can be evaluated according to a set of criteria such as the level of security they provide as well as the performance and overhead associated with a particular implementation of the primitive.

Authentication protocols are implemented as an exchange of messages between two or more parties, usually over an unsecured network. Authentication utilizes cryptographic primitives as countermeasures to adversarial manipulation of the transmitted messages and as mechanisms to protect the interfaces of the communicating entities from information leakage and tracking. PUFs provide novel ways of designing protocols but cannot be used by themselves to implement all of the security requirements of the protocol. Sections 6.3 and 6.4 provide an overview of traditional security-related primitives commonly used in authentication protocols, as well as algorithms and evaluation metrics that are required when using PUFs for authentication. Once the groundwork of authentication has been established, we then describe several PUF implementations and PUF-based authentication protocols in Sects. 6.5 and 6.6.

6.3 Cryptographic Primitives for Authentication Protocols

A cryptographic protocol is a distributed algorithm defined by a sequence of steps precisely specifying the actions required of two or more entities to achieve a specific security objective [18]. All protocols make use of cryptographic primitives that provide specific security properties. In this section, we briefly describe the primitives most commonly used in authentication protocols.

6.3.1 Random Number Generation

Random numbers are important in many cryptographic protocols, e.g., session keys, nonce for authentication, randomized procedures, etc. Random numbers must be selected uniformly from a distribution, thereby ensuring that all possible values are equally likely, as a means of maximizing the difficulty of algorithmic and brute force attacks carried out by adversaries against the protocol. Requests that are common in cryptographic protocols include "select an element at random from the sequence $\{1, 2, \ldots, n\}$" or "generate a random string of symbols of length m over the alphabet G of n symbols." *Uniformly* refers to the probability that a given symbol is selected and by definition is equal to $1/n$ for an alphabet of n symbols, and $1/n^m$ for a string of symbols of length m.

Traditionally, deriving random numbers from physical sources was difficult and costly, spurring the development of software-based alternatives such as techniques based on *pseudorandom sequences* and *seed* parameters (PRNGs) [19]. NIST recommends several such cryptographically secure PRNGs, each based on different types of cryptographic primitives such as hash functions, MACs and block ciphers [20]. Although most are considered cryptographically secure, they each depend on a random seed with high entropy. An *entropy accumulator* can be used to derive the seed from a "non-ideal" physical source of randomness, whereby the input bit-stream produced by the non-ideal source is processed by the entropy accumulator into an *m-bit pool* of high entropy. The entropy accumulator can be a cryptographic hash function [19]. Alternatively, the physical layer nature of PUFs make them cost-effective and well suited as the physical source of randomness. Recent work shows that appropriate post-processing of PUF responses allow them to be used directly as TRNGs, i.e., without the need of PRNGs [21].

6.3.2 Cryptographic Hash Functions

As mentioned above, secure hash functions are used to realize a fundamental information security property, namely that related to the *integrity of data*. Compression is a defining characteristic of many-to-one hash functions, whereby binary strings of arbitrary length are mapped to strings of fixed length n. The *n-bit* hash output is a compact representation of the input string. The many-to-one property implies that *collisions* are possible, a condition in which two distinct input strings map to the same hash. *Cryptographic* hash functions (referred to as hash functions subsequently) add important security-related properties to traditional hash functions and have the following characteristics [22]:

- It is easy to compute the hash for any input string.
- It is computationally infeasible to (1) generate the input string from its hash, (2) modify the input string without changing the hash, and (3) find two different input strings which produce the same hash.

More formally, the security properties of a hash function h with input message m and output $y = h(m)$ are defined as follows:

- preimage resistance: Given any hash y, it is computationally infeasible to find an m such that $h(m) = y$.
- second-preimage resistance: Given an input m, it is computationally infeasible to find a different input m' such that $h(m) = h(m')$
- collision resistance: It is computationally infeasible to find any two distinct inputs m and m' such that $h(m) = h(m')$.

Even stronger security properties are possible, for example it should be infeasible to find two inputs that produce *similar* hashes. Ideally, the hash function should behave like a random function, where each hash is equally probably, i.e., uniformly distributed.

There are two fundamental classes of hash functions: **unkeyed hash functions** and **keyed hash functions**. Keyless hash functions can be used to create *modification detection codes* (MDCs), whose main purpose is to confirm data integrity. There are two types of MDCs: *one-way hash functions* (OWHFs) which make it difficult to find an input string m that hashes to specific hash value, and *collision resistant hash functions* (CRHFs), which makes it difficult to find two input strings that map to the same hash. OWHFs are preimage and second-preimage resistant, and are considered **weak** one-way hash functions, while CRHFs typically have all three properties and are called **strong** one-way hash functions.

Keyed hash functions provide both message authentication and data integrity and are called *message authentication codes* (MACs) when used in symmetric-encryption protocols, and *digital signatures* when used in asymmetric encryption protocols. Both schemes hash the message and then sign it with a key. The receiver authenticates by applying the MAC or digital signature algorithm on the received message and verifies that the received hash matches the locally computed value. Hashing compresses the message and makes this data integrity check more efficient. Outside the scope of this expository, the chip area and computational complexity of cryptographic hash functions is much larger than that found in non-cryptographic hash functions [18, Ch. 9].

Similar to authentication protocols, secure hash algorithms continue to evolve, driving periodic changes, and additions to the public standards [23, 24]. The term *secure hash algorithm* (SHA) is used in reference to a set of public standards maintained by the National Institute of Standards and Technology (NIST). In particular, SHA-3 refers to subset of the cryptographic primitive family *Keccak*, a standard released in August of 2015 that is designed as an alternative to the SHA-2 family of secure hash functions [25].

6.3.3 Secure Sketches and Fuzzy Extractors

The introduction of PUFs as a primitive in authentication (and encryption) protocols made it necessary to enlist **error-correcting** and **randomness extraction** mechanisms into the suite of cryptographic primitives. The analog characteristics of the entropy source, as well as the embedded analog-to-digital instrumentation components of a PUF instantiation, combined with environmental (temperature, supply voltage), coupling and power supply noise sources make it difficult or impossible to precisely reproduce the bitstrings generated by PUFs from one run of the protocol to the next. When PUF bitstrings are used as input to traditional cryptographic primitives, such as hash functions or encryption algorithms, even a single bit-flip error in the bitstring causes a catastrophic failure in the protocol. Additionally, PUF-generated bitstrings, in many cases, are not ideal from the randomness perspective. Systematic bias effects and correlations inherent to the structure of the entropy source make it difficult for the PUF to produce a bitstring uniformly from the underlying distribution, i.e., such that all bitstrings of a given length are equally likely. Secure sketches, strong extractors, and fuzzy extractors are functions designed to deal with these deficiencies.

There are many types of error-correction algorithms that have been developed to fix errors that occur in bitstrings. The most popular algorithms used for PUFs produce **helper data** as a supplementary source of information during the initial bitstring generation (*Gen*) process, which is later used to fix bit-flip errors during reproduction (*Rep*). The helper data is typically transmitted and stored openly, in a *public* non-secure location, and therefore, it must reveal as little as possible about the bitstring it is designed to error correct.

The *Sketch* component of a **secure sketch** takes an input y and returns a helper data bitstring w [26, 27]. The *Recover* component takes a "noisy" input y' and a helper bitstring w and returns y'', which is guaranteed to match the original bitstring y as long as the number of bit flip errors is less than t (t is a parameter that can be selected based on the level of error correction that is needed). The algorithm is characterized by a *security property*, that guarantees that if y is selected from a distribution with **min-entropy** m, then an adversary can reverse-engineer y from m with probability no greater than $2^{-m'}$ (m' is defined below). Entropy is a measure of the disorder or randomness in a closed system, while min-entropy refers to the worst-case behavior of a random variable and is defined by Eq. 6.1. It is the negative $\log_2$ of the event with maximum probability.

$$H(X) = -\sum_{i=1}^{n} (p_i \log_2 p_i) \quad \text{The entropy of a random variable } X \text{ with probabilities } P_i, \ldots, p_n$$

$$H(X) = \frac{1}{\ln(2)} \log_2\left(\frac{1}{p}\right) \quad \text{When } \mathrm{P}_i = 1/n \text{(equal probabilities)}$$

$$H_\infty(X) = \min(-\log_2 P_i) = -\log_2(\max(P_i))\text{Min}-\text{entropy} \tag{6.1}$$

Dodis et al. [26, 27] proposed two algorithms for a secure sketch, both based on binary error correcting linear block codes. A linear block code is characterized with three parameters given as [n, k, t], which indicate that there are 2^k codewords of length n and each codeword is separated from all others by at least $2t - 1$ bits. The last parameter specifies the error correcting capability of the linear block code, in particular, that up to t bits can be corrected.

The **code-offset** construction is the simpler of the two linear block codes. The *Sketch*(y) procedure samples a uniform, random codeword c (which is independent of y), and produces an n-bit helper data bitstring w using Eq. 6.2 [19]. The bitstring w represents the binary offset between y and c.

$$w = y \oplus c \tag{6.2}$$

Recover (y', w) computes a noisy codeword c' using Eq. 6.3 and then applies an error-correcting procedure to correct c' as $c'' = \mathit{Correct}(c')$.

$$c' = y' \oplus w \quad => \quad c' = (y \oplus y') \oplus c \tag{6.3}$$

The error-corrected value of y' is computed as given by Eq. 6.4.

$$y'' = w \oplus c'' = y \oplus (c \oplus c'') \tag{6.4}$$

If the number of bits that are different between c and $c' < t$, where t represents the error-correcting capability of the code, then the algorithm guarantees $y = y''$. Also, w discloses at most n bits of y, of which k are independent of y (with k less than or equal to n). Therefore, the remaining min-entropy is $m - (n - k)$ (specified as m' above), where $(n - k)$ represents the min-entropy that is lost by exposing w to the adversary.

The second algorithm proposed in [26, 27] is referred to as the **syndrome** construction. The *Sketch(y)* procedure produces an $(n - k)$-bit helper data bitstring using the operation specified by Eq. 6.5, where H^T is a parity-check matrix dimensioned as $(n - k)\ x\ n$.

$$q = y \cdot H^T \tag{6.5}$$

The *Recover* procedure computes a syndrome s using Eq. 6.6.

$$s = y' \cdot H^t \oplus w \quad \Rightarrow \quad s = (y \oplus y') \cdot H^T \tag{6.6}$$

Error correction is carried out by finding a unique error word e such that the *hamming weight* (the number of '1's) in bitstring e is less than or equal to t (the error-correcting capability of the code). Also, the error word e satisfies Eq. 6.7.

$$s = e \cdot H^T \tag{6.7}$$

In both the code-offset and syndrome techniques, the *Recover* procedure is more computationally complex than the *Sketch* procedure. As discussed below, the first PUF-based authentication protocols implemented the *Recover* procedure on the resource-constrained hardware token. Subsequent work proposes a **reverse fuzzy extractor**, which implements *Sketch* on the hardware token and *Recover* on the resource-rich server, making the protocol more cost-effective and attractive for this type of application environment [28].

Similar to error-correction, there is a broad range of techniques for constructing a **randomness extractor**. Section 6.3.1 described the requirements for random number generation, and practical approaches for extracting randomness from non-ideal physical sources, e.g., those based on the use of *seeded cryptographic PRNGs*. Reference [19], Sect. 6.3.2 provides a survey of techniques proposed for extracting randomness.

Fuzzy extractors combine a secure sketch with a randomness extractor as shown in Fig. 6.1 (adapted from [19]). A PUF-based authentication protocol, with the *hardware token*, e.g., smart card, shown on the left and the *secure server*, e.g.,

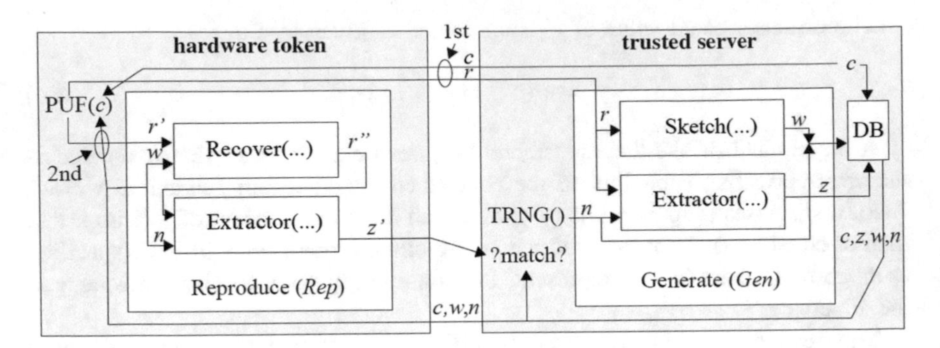

Fig. 6.1 Fuzzy extractor

bank, shown on the right is also shown to illustrate one possible usage scenario. The *Sketch*, as noted above, takes an input r, which, e.g., might be a PUF response to a server-generated challenge c, as input and produces helper data w (labeled 1st in the figure). The Extractor takes both r and a random number (seed) n and produces an entropy distilled version z, which can be stored as a tuple (c, z, w, n) in a secure database (DB) on the server. This component of the fuzzy extractor is called Generate or *Gen*.

Authentication in the field begins by selecting a tuple (c, z, w, n) from the DB and transmitting the challenge c, helper data w and the seed n to the hardware token. The PUF is challenged a second time with challenge c and produces a "noisy" response r' (labeled second in the figure). The Reproduce or *Rep* process of the fuzzy extractor uses the Recover procedure of the secure sketch to error correct r' using helper data w. The output r'' of Recover and the seed n are used by the Extractor to generate z'. As long as the number of bit flip errors in r' is less than t (the chosen error correction parameter), the z' produced by the token's Extractor will match the server-DB z and authentication succeeds. Note that the error corrected z' establishes a shared secret between the server and token, which can alternatively be used as input to traditional cryptographic primitives such as hash and block cipher functions (as opposed to being transmitted to the server as shown in the figure).

6.3.4 Statistical Metrics

PUF-generated bitstrings are often evaluated using techniques designed to measure the statistical quality of the bitstrings, which include characteristics such as uniqueness, reproducibility, and randomness. *Uniqueness* measures how different the bitstrings are from one device to another in the population. The probability mass function of the binomial distribution is the appropriate statistical characterization function for bitstrings and is given by Eq. 6.8, with mean and variance given by Eqs. 6.9 and 6.10, resp. [29]. Equation 6.8 gives the probability of getting exactly k successes in n trials.

$$f(k;n,p) = \frac{n!}{k!(n-k)!} p^k (1-p)^{n-k} \tag{6.8}$$

$$\mu_{\text{binomial}} = np \tag{6.9}$$

$$\rho_{\text{binomial}} = np(1-p) \tag{6.10}$$

Assuming the probability of a '1' in a bitstring of size n produced by a PUF is $p = 0.5$, then μ_{binomial} indicates that half of the bits will be '1' on average. The same characteristic holds across bitstrings from different devices if the probability of a '1' is 0.5 for any given bit position. It follows then that the average number of bits that are different from one bitstring to another, in this best case scenario, is 50 %. The metric used to measure uniqueness is **inter-chip hamming distance ($\mathbf{HD_{inter}}$)**. HD_{inter} counts the number of bits that are different. In a typical PUF application, the count is computed over all possible pairings of bitstrings produced by different devices in the population and then divided by the total number of bits and multiplied by 100 to yield a percentage. Note that the set of bits that differ between any two arbitrary bitstrings necessarily are distinct from one pairing to another.

Reproducibility measures the PUFs ability to regenerate its bitstring(s) over time and under different environmental conditions. The terms **enrollment** and **regeneration** are used in reference to the bitstring generation process. Enrollment is carried out when a new bitstring is required, while regeneration refers to the process of reproducing the same bitstring at some point later in time. The application determines whether precise regeneration is required, e.g., encryption requires exact replicas of the bitstring (when the bitstring is used as the key) while some authentication schemes have a built-in tolerance to allow some (small) fraction of bit flip errors to occur. Regeneration without errors is much more challenging when the process is carried out under different temperatures and/or supply voltages. The metric used to measure reproducibility is **intra-chip hamming distance ($\mathbf{HD_{intra}}$)**. Similar to HD_{inter}, HD_{intra} counts the number of bits that are different between pairings of bitstrings. For HD_{intra}, however, the pairings of bitstrings are composed from the set of bitstrings produced by a specific device, each regenerated possibly under different environmental conditions with respect to the enrollment conditions. The term TV corners is used in reference to the set of environmental conditions used to test the devices, e.g., 85 °C and +10 % V_{DD} is a TV corner. Similar to HD_{inter}, HD_{intra} is usually expressed as an average percentage over all devices tested in the experiment, by counting the total number of bit flip errors that occur, dividing by the total number of bits inspected and then multiplying by 100. The ideal case is an average HD_{intra} of 0 %, i.e., no devices produced any bit flip errors under any TV corner.

The NIST statistical test suite can be used to evaluate the *randomness* of PUF response bitstrings [30]. The NIST tests look for *patterns* in the bitstrings that are not likely to be found at all or above a given frequency in a "truly random" bitstring. For example, long or short strings of 0's and 1's, or specific patterns repeated in many places in the bitstring work against randomness. The output of the NIST statistical evaluation engine is the *number of chips* that pass the *null hypothesis* for a given test, when evaluated at a *significance level* α (α is set to the default value of 0.01 which reflects a confidence of 99 %). The null hypothesis is specified as the condition in which the bitstring-under-test is random. Therefore, a good result is obtained when the number of bitstrings that pass the null hypothesis is large.

The NIST test suite consists of 15 separate tests, all of which have constraints on the size of the bitstring. The following provides an intuitive overview of what the tests measure, with details regarding the bitstring size requirements and applied test statistics omitted (see [30]). The test is always conducted against what is expected in a truly random sequence of similar length.

- Frequency Test: Counts the number of '1' in a bitstring and assesses the closeness of the fraction of '1's to 0.5. All other tests assume this test is passed.
- Block Frequency Test: Same except bitstring is partitioned into *M blocks*. Ensures bitstring is "locally" random.
- Runs Test: Analyzes the total number of *runs*, i.e., uninterrupted sequences of identical bits, and tests whether the oscillation between '0's and '1's is too fast or too slow.
- Longest Run Test: Analyzes the longest run of '1's within *M-bit blocks*, and tests if it is consistent with the length of the longest run expected in a truly random sequence.
- Rank Test: Analyzes the linear dependence among fixed length substrings in the bitstring, and tests if the *number of ranks*, i.e., number of rows that are linearly independent, of size M, $M - 1$, etc., match the number expected in a truly random sequence.
- Fourier Transform Test: Analyzes the peak heights in the frequency spectrum of the bitstring, and tests if there are *periodic* features, i.e., repeating patterns close to each other.
- Non-overlapping and Overlapping Template Tests: Analyzes the bitstring for the number of times *pre-specified* target strings occur, to determine if too many occurrences of *non-periodic* patterns occur.
- Universal Test: Analyzes the bitstring to determine the *level of compression* that can be achieved without loss of information.
- Linear Complexity Test: Analyzes the bitstring to determine the length of the smallest set of LFSRs needed to reproduce the sequence.
- Serial and Approximate Entropy Tests: Analyzes the bitstring to test the frequency of all possible 2^m overlapping *m-bit* patterns, to determine if the number is uniform for all possible patterns.

- Cumulative Sums Test: Analyzes the bitstring to determine if the cumulative sum of incrementally increasing (decreasing) partial sequences is too large or too small.
- Random Excursions Test: Analyzes the total number of times that a particular state occurs in a cumulative sum random walk.

6.4 Traditional, Software-Oriented Authentication

Authentication refers to the process of "verifying the identity of the communicating principals to one another" [31]. It is usually subdivided into *entity authentication* and *message (or data origin) authentication* [18], with the former referring to authentication in "real-time" between two parties about ready to engage in communication while the latter refers to data such as email that may later need to be authenticated by the receiver as to the origin and time sent. Note that authentication of the origin of data also addresses data integrity, i.e., whether the message has been tampered with by unauthorized parties, because unauthorized changes imply the data has a new source.

Authentication is typically carried out between a **prover** (claimant) A, e.g., a hardware token such as a smart card, and a **verifier** B, e.g., a secure server operated by your bank. The verifier B either confirms or accepts the prover's identity as authentic or terminates without acceptance, i.e., *rejects*. The information exchanged with verifier B must be designed to prevent reuse by B, otherwise it could impersonate A to a third party C. Protocols should guarantee that the probability of *impersonation* is negligible, even when a polynomially large number of previous authentications occur between A and B.

Authentication can be used for security objectives including access control, entity authentication, message authentication, data integrity, non-repudiation and key authentication. Authentication can be carried out using symmetric encryption techniques, e.g., via *message authentication codes* or MACs, using public/private encryption schemes via *digital signatures* and through authenticated key establishment methods. The most common usage models include access control to a resource, e.g., to computer accounts, ATMs, to software, to a building, etc.

The capabilities provided in the authentication protocol depend on the security requirements. For example, an authentication protocol may be *unilateral*, i.e., from prover to verifier, or *mutual*. Some protocols may *preserve privacy*, to prevent malicious adversaries from tracking instances of authentications that occur between the prover and verifier over time. Others may be *symmetric* in nature, requiring the use of a shared secret between the prover and verifier provided by interactions, in real-time, with a *trusted third party* (*TTP*), or may be *asymmetric* with the prover and verifier maintaining their own private secrets. The computational and communication overheads associated with the protocols will depend on the type of protocol, its security requirements and the security properties that must be guaranteed.

6.4.1 Entity Authentication

Entity authentication techniques can be divided into three categories:

- *Something you know*: Passwords, PINs and secret or private keys whose knowledge is demonstrated in challenge-response protocols.
- *Something you possess*: Physical accessory, resembling a passport in function. Magnetic striped cards, smart-cards, and hand-held customized calculators (password generators) which provide time-variant passwords.
- *Something inherent*: Biometrics, e.g., human physical characteristics such as fingerprints, voice, retinal patterns, and signatures.

Passwords represent the most widely used form of authentication, but are considered **weak authentication** protocols. Passwords provide unilateral and time-invariant authentication, with the *userid* serving as the claim of identity and the *password* serving as evidence supporting the claim. Attacks include eavesdropping to enable *replay*, and password guessing such as *dictionary attacks*. On most systems, the passwords are encrypted using a *one-way function* (OWF) before being stored on disk (see Sect. 6.3.2). A technique called *salting* is also commonly used to make dictionary attacks more difficult by expanding the search space for the adversary.

Two-stage authentication and password-derived keys address the insufficient entropy issue associated with human chosen passwords. An n-digit PIN verifies the user to the token, e.g., smart card, in the first stage. The token typically embeds additional secrets for use in stage two between the token and the system. A variant uses *passkeys* to map a user password to a cryptographic key using a OWF. The most secure of the weak authentication schemes uses *one-time passwords*, which addresses eavesdropping and replay attacks on password schemes.

Challenge-Response protocols fall in the class of **strong authentication** protocols, whereby authentication requires the prover to demonstrate knowledge of a secret without revealing the secret itself to the verifier. Here, the prover provides a *response* to a *time-variant challenge*, with the response inseparably bound to both the secret and the challenge. The challenge can be a random number, called a *nonce* (for "used only once"), a sequence number or a timestamp. Time variant parameters are countermeasures to replay attacks and certain types of chosen-text attacks because the uniqueness and timeliness guarantees allow one protocol instance to be distinguished from another. Note that *challenge-response* protocols require some type of computing device and secure storage for long-term keying material.

Challenge-Response by Symmetric-key: Each pair of communicating parties share a secret key. In large communities, a trusted third party (TTP) can provide session keys in real time to circumvent the need to distribute n^2 key pairs. A common form of unilateral authentication uses random number(s) (RN) [18].

$$A \leftarrow B: r_B \qquad (B \text{ generates random nonce } r_B)$$
$$A \leftarrow B: E_K(r_B, B^*)$$

B generates random nonce r_B and transmits it to A (over an unsecured channel). A encrypts the nonce and the identifier B using a shared secret key K and transmits the encrypted message back to B. B then decrypts and (1) checks that the r_B received matches the r_B sent and (2) verifies $B*$ is equal to his own B. The shared secret K must be securely transmitted to A and B beforehand, typically using a mechanism involving a TTP, in order for this scheme to work.

Mutual authentication requires a second nonce r_A and a third message:

$$A \leftarrow B: r_B \qquad (B \text{ generates random nonce } r_B)$$
$$A \leftarrow B: E_K(r_A, r_B, B^*) \qquad (A \text{ generates random nonce } r_A)$$
$$A \leftarrow B: E_K(r_A, r_B)$$

Encryption ensures the nonce and identifiers are "inseparably" bound as discussed above.

Challenge-Response using Keyed One-Way Functions: Encryption is considered a "heavy weight" cryptographic primitive, and may be replaced by a one-way function (OWF) or a nonreversible function with shared key, and a challenge, for authentication in resource-constrained devices. The encryption algorithm E_K is replaced by a MAC algorithm h_K, i.e., a keyed hash function. The receiver also computes the MAC and compares it with the received MAC. These protocols require an additional cleartext field r_A to be transmitted [18].

$$A \leftarrow B: r_B \qquad (B \text{ generates random nonce } r_B)$$
$$A \leftarrow B: r_A, h_K(r_A, r_B, B) \qquad (A \text{ generates random nonce } r_A)$$
$$A \leftarrow B: h_K(r_A, r_B, A)$$

B confirms that the hash value received, designated as $h_k(r_A, r_B, B)$, is equal to the value he/she computes locally using the same hash function and shared secret K. A performs a similar validation using the transmitted hash $h_K(r_B, r_A, A)$ from B. As discussed in Sect. 6.3.2, the computational infeasibility of finding a second input to h_K that produces the same hash provides the security guarantee in this mutual authentication protocol.

Challenge-Response by Public-Key: Here, the prover decrypts a challenge using its secret key component of the public-private pair, which is encrypted by the verifier under its public key P_A. Alternatively, the prover can digitally sign a challenge.

$$A \leftarrow B: h(r), B, P_A(r, B)$$

$$A \leftarrow B: r$$

B chooses nonce r, computes the witness $x = h(r)$ (h is a OWF), where x demonstrates knowledge of r without disclosing it, and computes challenge $e = P_A(r, B)$. A decrypts e to recover r' and B', computes $x' = h(r')$ and rejects if x' does not equal x or if B' does not equal B, otherwise A sends $r = r'$ to B. B succeeds with unilateral entity authentication of A upon verifying the received r agrees with his r. The witness prevents chosen-text attacks.

6.5 Physical Unclonable Functions (PUFs)

Components needed for information security can be implemented using physical-layer security primitives. A long-standing assumption of software-based security systems has been that hardware implementations of security primitives are trustworthy "black boxes." In particular, for *keyed* security primitives such as block ciphers, key generation, and key storage are assumed to be trusted and secure, and operational state within black box implementations of security algorithms is assumed to be hidden and inaccessible. Unfortunately, models which assume a "hardware root of-trust" are becoming increasingly more vulnerable to attacks [32–34].

PUFs represent physical-layer security components that are designed to deal with threats to key generation and key storage. PUFs are circuit primitives that leverage within-die variations in ICs as a means of producing random bitstrings. Each IC is uniquely characterized by random manufacturing variations, and therefore, the bitstrings produced by PUFs are unique from one chip to the next. Cloning a PUF, i.e., making an exact copy, is nearly impossible because it would require control over the fabrication process that is well beyond our current capabilities. A PUF maps a set of digital "challenges" to a set of digital "responses" by exploiting these physical variations in the IC. The entropy in the responses is stored in the physical structures on the IC and can only be retrieved when the IC is powered up. The analog nature of the entropy source makes PUFs "tamper-evident," whereby invasive attacks by adversaries will, with high probability, change its characteristics.

PUFs have been proposed which leverage variations in transistor threshold voltages [35–37], speckle patterns [38, 39], delay chains and ROs [40–64], thin-film transistors [65], FPGAs [66, 67], SRAMs [68–74], leakage current [75, 76], metal resistance [77–81], transistor transconductance [82], the path delays of core logic macros [83–87], optics and phase change [88], sensors [89], switching variations [90], sub-threshold design [91], ROMs [92], buskeepers [93], microprocessors [94], using lithography effects [95, 96], optical proximity correction [97], aging [98], in subthreshold operation [99], memristors [100] and other non-volatile memories [101], in scan chains [102], phase change memory [103], and carbon-nanotubes [104]. Board-level authentication using PUFs has also recently been proposed [105] and for securing mobile system platforms [106, 107].

6.5.1 PUF-Based Authentication

As mentioned above, authentication is the process between a prover, e.g., a hardware token and a verifier, a secure server that confirms the identities, using corroborative evidence, of one or both parties. With the Internet-of-things (IoT), there are a growing number of applications in which the hardware token is resource-constrained, and therefore novel authentication techniques are required that are low in cost, energy and area overhead. Conventional methods of authentication which use area-heavy cryptographic primitives and non-volatile memory (NVM) are less attractive for these types of evolving embedded applications. PUFs, on the other hand, are hardware security and trust primitives that can address issues related to low cost because they eliminate (in many proposed authentication protocols) the need for NVM. Moreover, the special class of so-called "strong PUFs" can also reduce area and energy overheads by reducing the number and type of hardware-instantiated cryptographic primitives.

PUFs generate bitstrings that can serve the role of uniquely identifying the hardware tokens for authentication applications. The bitstrings are generated on-the-fly, thereby eliminating the need to store digital copies of them in NVM, and are (ideally) reproducible under a range of environmental variations. The ability to control the precise generation time of the secret bitstring and the sensitivity of the PUF entropy source to invasive probing attacks (which act to invalidate it) are additional attributes that make them attractive for authentication in resource-constrained hardware tokens.

PUF-based protocols have been proposed for applications including encryption, authentication, for detecting malicious alterations of design components and for activating vendor specific features on chips. Each of these applications has a unique set of requirements regarding the security properties of the PUF. For example, PUFs that produce secret keys for encryption are not subject to model building attacks (as is true for PUF-based authentication) which attempt to "machine learn" the components of the entropy source within the chip as a means of predicting the complete response space of the PUF. This is true for encryption because the responses to challenges are typically not "readable" from an interface on the chip. In general, the more access a given application provides to the PUF externally, the more resilience it needs to have to adversarial attack mechanisms. Authentication as an application for PUFs clearly falls in the category of extended access.

6.5.2 Strong Versus Weak PUFs

Weak PUFs are those whose challenge-response space is small while strong PUFs have very large, ideally exponential, challenge-response spaces [108, 109]. The distinction between strong and weak PUF is rooted in the amount of entropy that each class can access. The larger the entropy source, the more difficult it is for an adversary, who has access to the PUF, to collect and analyze challenge-response pairs (CRPs) until the complete behavior of the PUF can be predicted. The SRAM PUF is an early example of a weak PUF with only one CRP [68] while the arbiter PUF is traditionally considered a strong PUF because of its exponentially large challenge space [41]. However, if the size of the entropy source is considered a defining characteristic, then the arbiter PUF would fail to meet the definition of a strong PUF because its response space is derived from a relatively small entropy source, in particular, as small as a couple hundred gates. Given this latter consideration, very few of the proposed PUFs meet this expanded definition. Model-building resistance using machine learning techniques has emerged as an important criterion for determining whether a PUF is strong based only on the size of its CRP space or whether it is truly strong, i.e., attacks that attempt to learn and predict its behavior are infeasible [42, 110].

6.5.3 The Arbiter PUF

The most widely referenced strong PUF, the *arbiter* PUF, was the one of the first proposed, and is described in [41, 42]. However, it is also widely recognized that it is considered strong based only on the size of its input challenge space, and not on the amount of entropy it possesses.

The arbiter PUF measures path delays from a specialized test structure as its source of entropy as shown in Fig. 6.2. The test structure implements two paths, each of which can be individually configured using a set of challenge bits (stored in FFs along the top of the figure). Each of the challenge bits controls a "Switch box" that can be configured in either *pass mode* or *switch mode*. Pass mode connects the upper and lower path inputs to the corresponding upper and lower path outputs, while switch mode reverses the connections. A stimulus, represented as a rising edge on the left side of the figure, cause two edges to propagate along the two paths configured by the challenge bits. The faster path controls the value stored in the *arbiter* located on the right side of the figure. If the propagating rising edge on the upper input to the arbiter arrives first, the **response** bit output becomes a '0'. Otherwise, the response bit is a '1'. The switch boxes are designed identically as a means of avoiding any type of systematic bias in the delays of the two paths.[1] Within-die process variations cause uncontrollable delay variations to occur in the switch boxes, which in turn, makes each instance of the arbiter PUF unique in terms of its generated response bit(s). A bitstring can be obtained from the arbiter PUF by repeating the measurement process under a set of different challenges.

From this design, it is clear that the arbiter PUF has an exponential number of input challenges that can be applied, in particular, 2^n with n representing the number of switch boxes. However, the total amount of entropy is relatively small,

[1]Note that achieving an unbiased layout in an FPGA is a challenging and non-trivial process.

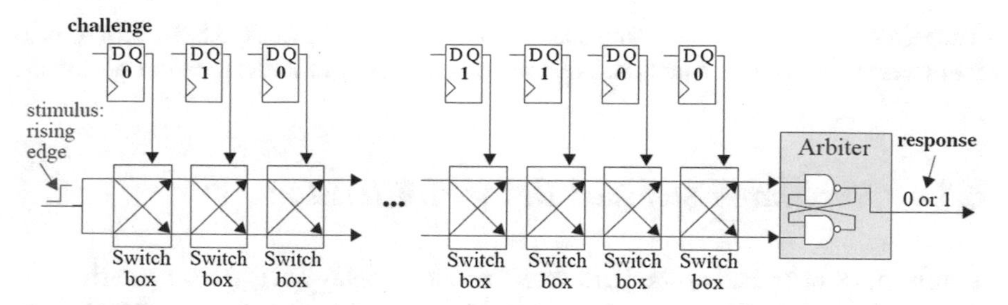

Fig. 6.2 Arbiter PUF [41]

and is represented by the four path segments in each of the switch boxes. For n equal to 128, the total number of path segments that can vary individually from one instance to another is 4 * 128 = 512. The exponential number of input challenges simply combines these individual sources of entropy in different ways. Model building attacks attempt to learn the delay relationships of the two configurations for each switch box [110]. Once known, the response under any challenge then becomes predictable (limited only by the noise margin of the arbiter measurement circuit).

The model-building weakness of the arbiter PUF is addressed in follow-on work, where the outputs of n arbiter PUFs are XOR'ed, to create a XOR-mixed arbiter PUF [44, 111, 112]. Figure 6.3 shows an example in which two arbiter PUF output bits are XOR'ed. The goal is to create an XOR network large enough to achieve the *avalanche criterion*. This criterion is commonly found in cryptographic hash and encryption functions where flipping one of the input bits (or a bit in the key for encryption) causes half of the output bits to flip. For the XOR-mixed PUF, the goal is to achieve the avalanche effect by flipping one of the challenge bits. Although this helps significantly with model building, particularly with networks of XORs greater than 4, larger XOR networks also reduce reliability by creating a *noise-based* avalanche effect, i.e., any odd number of bit flips that occur on the inputs of any given XOR network results in a response bit flip error. As reported in [111], if a single arbiter PUF has an HD_{intra} of 5 % (intra-chip HD measures the PUF's ability to reproduce the same bitstring over repeated applications of the challenge, usually under different environmental conditions), the HD_{intra} increases to 19 % for a 4-XOR-mixed arbiter PUF, i.e., nearly 1/5 of the response bits have bit flip errors.

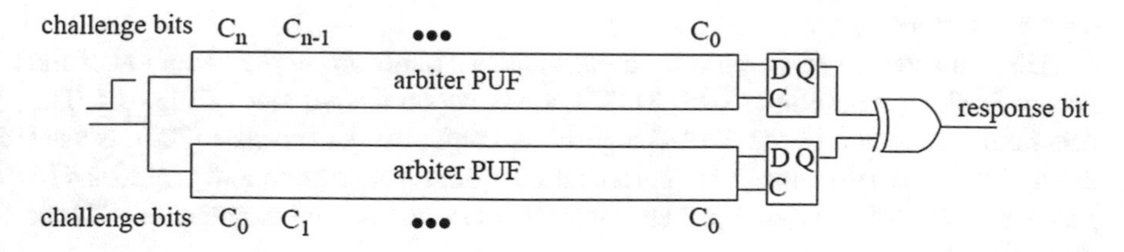

Fig. 6.3 XOR-mixed arbiter PUF [44, 111, 112]

Therefore, error-correction using techniques described in Sect. 6.3.3 become critical to ensuring proper functional operation when used in authentication protocols.

6.5.4 Hardware-Embedded Delay PUF (HELP)

Similar to arbiter PUFs, the hardware-embedded delay PUF (HELP) derives its entropy from variations in path delays. However, HELP measures delays from existing functional units. Therefore, no dedicated test structures are required. Another major benefit of using existing functional units is the amount of entropy that can be potentially leveraged. Cryptographic functional units are particularly attractive because of the complexity of their interconnection networks. On the down side, the lack of control over the configuration of paths in functional units creates issues related to systematic bias and reliability, as described in the following sections.

Interestingly, the authors of the first silicon-based PUF paper describe their notion of a "better PUF" in Ongoing and future work section, which turns out, based on our work, to be well founded [40]. The basic concept of measuring path delays from a core logic functional unit was implemented first by Li and Lach [83], but was not fully developed as a PUF primitive. In particular, the authors do not address the bias introduced by paths of different lengths nor do they deal with the reliability issues associated with paths that glitch.

Our development of HELP began in 2011 on a 90 nm ASIC implementation [86], but was fully developed as an intrinsic PUF (with full integration of the control logic, entropy source, and measurement components) on a 130 nm Xilinx V2Pro [84, 85], and more recently using a 28 nm Xilinx Zynq architecture [87]. We have developed solutions for path length bias and glitching that occur when core logic functional units are used as the source of entropy, as well as techniques that improve the attack resilience of HELP when used in low cost authentication applications. This section describes the characteristics of the most recent incarnation of HELP and presents new results.

The original version of HELP made use of an embedded test structure called REBEL [113] for measuring path delays and detecting glitches [84–86]. Recent implementations of HELP measure path delays in glitch-free functional units, which allow a simplified version of REBEL to be used [87]. The simplified version eliminates the delay chain component and instead samples the path delays at the capture FF directly.

HELP attaches to an on-chip module, such as a hardware implementation of the Secure Hashing Algorithm (SHA-3) [23], as shown on the left side of Fig. 6.4. The data path component of the SHA-3 algorithm, configured as *keccak-f* [200], is used in our FPGA experiments. This combinational data path component includes 416 primary inputs (PIs) and 400 primary outputs (POs) and is implemented on a Xilinx Zynq FPGA using 1936 LUTs.

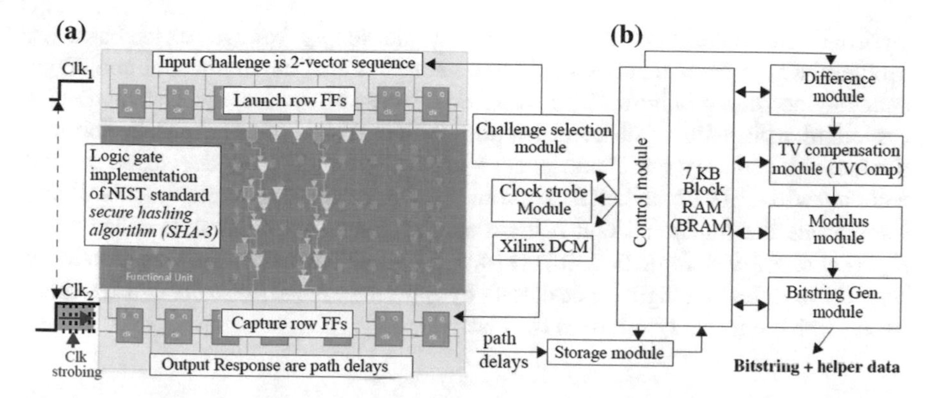

Fig. 6.4 HELP Block Diagram: **a** Instantiation of the HELP entropy source and **b** HELP processing engine

Similar to the arbiter PUF described in the previous section, within-die variations in path delays are the main source of entropy for HELP. Manufacturing variations change the relative path delays through the functional unit in different ways, and therefore each instance of the functional unit is uniquely characterized by these delays. However, the structure of the paths in the arbiter PUF is very different than those in a typical functional unit, i.e., the arbiter PUF paths are symmetric and regular (by design) while the paths within a typical functional unit exhibit no such regularity.

Functional unit paths exhibit *fan-out* and then *reconvergence* of fan-out at various points within the logic structure of the functional unit (called reconvergent-fanout), as shown on the right side of Fig. 6.5. Also, the lengths of the paths can vary widely, e.g., the *short paths* shown have 3 or fewer gates while the *long paths* are 5 or more gates in length. Both of these characteristics make it more difficult to build a PUF with good statistical characteristics. Reconvergent-fanout can cause ***glitching***, i.e., static and dynamic hazards, to occur on the primary outputs, whereby output signals transition more than once. Glitching creates ambiguity regarding the "correct" timing value to use for the path. Operating the functional unit under different

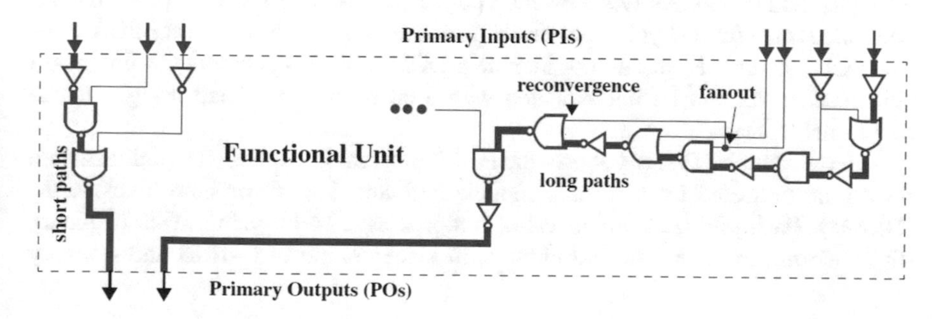

Fig. 6.5 Portion of a functional unit schematic, showing fan-out and reconvergence of paths

environmental conditions, e.g., temperature and supply voltage, exacerbates the problem, where paths that are glitch-free under nominal environmental conditions suddenly become glitchy under adverse conditions. Moreover, the systematic bias associated with paths of different lengths significantly degrades the statistical randomness and uniqueness characteristics of the PUF. We have developed several techniques to deal with both of these problems. Our most recent work, described here, implements the functional unit using a special *glitch-free* logic style called wave *differential dynamic logic* (WDDL) [114, 115], while the systematic bias introduced by paths of different lengths is dealt with by applying a *modulus* to the digitized path delay, which effectively removes the bias.

Clock Strobing

Path delay is defined as the amount of time (Δt) it takes for a set of 0-to-1 and 1-to-0 bit transitions introduced on the PIs of the functional unit (input challenge) to propagate through the logic gate network and emerge on a PO. HELP uses a clock-strobing technique to obtain high resolution measurements of path delays as shown on the left side of Fig. 6.4. A series of launch-capture operations are applied in which the vector sequence that defines the input challenge is applied repeatedly to the PIs using the Launch row flip-flops (FFs) and the output responses are measured on the POs using the Capture row FFs. On each application, the phase of the capture clock, Clk_2, is incremented forward with respect to Clk_1, by small Δts (on order of 20 ps), until the emerging signal transition on a PO is successfully captured in the Capture row FFs. A set of XOR gates connected to the Capture row FF inputs and outputs (not shown) provide a simple means of determining when this occurs. When an XOR gate value becomes 0, then the input and output of the FF are the same (indicating a successful capture). The first occurrence in which this occurs during the clock strobe sweep causes the current phase shift value to be recorded as the digitized delay value for this path. This operation is applied to all POs simultaneously.

The phase shifting module for Clk_2 is shown in the middle of Fig. 6.4. On-chip digital clock managers (DCMs) are commonly included in FPGA architectures. For example, Xilinx FPGAs typically incorporate at least one DCM with a digitally controlled *fine phase shift* control mechanism even on their lowest cost FPGAs. For low-cost components that do not include a DCM with this capability, a fine phase shift mechanism can be implemented with a small area overhead using a multi-tapped delay chain.

The right side of Fig. 6.4 shows the HELP processing engine. The digitized path delays are collected by a *storage* module and stored in an on-chip block RAM (BRAM). Each digitized timing value is stored as a 14-bit value, with 10 binary digits serving to cover the fine phase shift sweep range of 0–1023 and 4 binary digits of fixed point precision to enable up to 16 samples of each path delay to be measured and averaged. The 7 KByte BRAM allows 4096 path delays to be stored. We configure the applied challenges to test 2048 paths with rising transitions and 2048 paths with falling transitions. The 14-bit digitized path delays are referred to as PUFNums or **PN**.

PN Processing

Once the PN are collected, a sequence of mathematical operations is applied as shown on the right side of the Fig. 6.4 to produce the bitstring and helper data. The *difference* module creates unique, pseudo-random pairings between the rising and falling PN groups using two seeded linear feedback shift registers (LFSRs). The two 11-bit *LFSR seeds* are user-specified parameters. The PN differences, referred to as **PND**, are stored in the lower 2048 memory locations of the BRAM as values in the range ±511 with four binary digits of fixed point precision, overwriting the original set of rising-edge PN.

Figure 6.6a shows an example of this process using two groups of 38 curves, one curve for each Xilinx Zynq 7020 chip that was tested. The curves shown along the bottom depict the PN obtained from rising transition tests and those along the top are the PN from falling transition tests. The 13 line-connected points associated with each curve represent the chip's PN measured over a range of environmental conditions, called temperature-voltage (TV) corners. The PN at the x-axis position given by 0 are those measured under nominal conditions (referred to as **enrollment** values below), i.e., at 25 °C, 1.00 V. The PN at positions 1, 2 and 3 are also measured at 25 °C but at supply voltages of 0.95, 1.00, and 1.05 V. Similarly, the other groups of 3 consecutive points along the x-axis are measured at these supply voltages but at temperatures −40, 85, and 100 °C. The PN measured under TV corners numbered 1–12 are referred to as **regeneration** values. Figure 6.6b plots the **PND** defined by subtracting point-wise, each falling PN from the corresponding rising PN for the same chip.

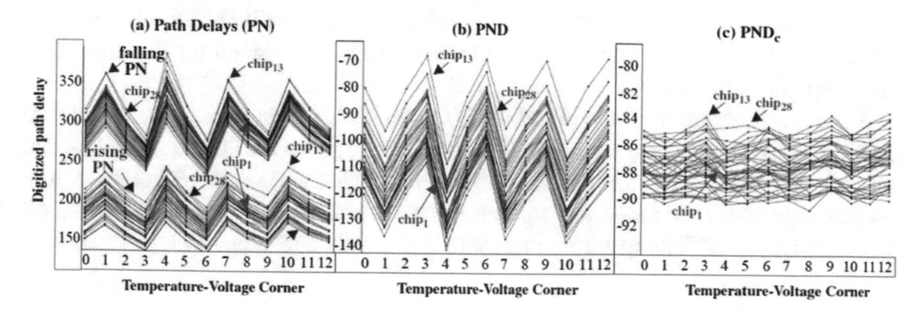

Fig. 6.6 **a** Example rising and falling path delays (PN), **b** PND and **c** PNDc

Temperature-Voltage (TV) Compensation

PUFs must be able to reproduce their bitstrings as precisely as possible, ideally without any bit flip errors, over a range of environmental conditions in which temperature and supply voltage are different from the conditions present during enrollment. No PUF construction to date is able to completely eliminate bit flip errors during regeneration, but some are more resilient to them than others. A method called temperature–voltage compensation (**TVComp** as shown on right side of Fig. 6.4) is proposed for the HELP PUF as a mechanism to improve its resilience to bit flip errors.

For HELP, bit flip errors occur because changes to the chip's ambient temperature and supply voltage change its path delays (called *TV noise*). TVComp applies a linear transformation to the path delay differences (PND) as a means of shifting and scaling them to a common reference. The goal is to define a transformation that eliminates the saw-tooth behavior in the curves shown in Fig. 6.6b, making them as flat and straight as possible.

TVComp is applied to the entire set of 2048 PND measured for each chip at each of the 13 TV corners separately (note, Fig. 6.6b shows only one of the PND from the larger set of 2048 *PND* that exist for each chip and TV corner). The TVComp procedure first converts the *PND* to "standardized" values. Equation (6.11) represents the first transformation, which makes use of two constants, μ_{TVx} and Rng_{TVx}, obtained from a histogram distribution of the measured PND. The second transformation is represented by Eq. (6.12), which translates the standardized z_{vals} to a new distribution with mean μ_{ref} and range Rng_{ref}. The reference mean and range values are user-selectable parameters of the HELP algorithm.

$$zval_i = \frac{(PND_i - \mu_{TV_X})}{Rng_{TV_X}} \tag{6.11}$$

$$TVCPNDiff_i = zval_i Rng_{ref} + \mu_{ref} \tag{6.12}$$

As an example, Fig. 6.7a shows the *PND* histogram distribution for chip C_1 at 25 °C, 1.00 V. The μ_{TVx} is shown as −40 while the Rng_{TVx} is computed between the 5 and 95 % as 136. Figure 6.7b superimposes the PND histograms for C_1 at 25 °C, 1.00 V and 100°C, 1.05 V. The TVComp process will shift (and scale) this distribution to the left to remove the adverse effects introduced by the change in environmental conditions.

A second illustration of the effect of TVComp is shown in Fig. 6.6b, c. The data in Fig. 6.6c is obtained by applying TVComp procedure to the 2048 PND measured under each of the 13 TV corners for each chip, i.e., 13 TV corners * 38 chips = 494 separate applications. Since the same reference mean and range are used for all transformations, TVComp eliminates both TV noise and chip-wide performance

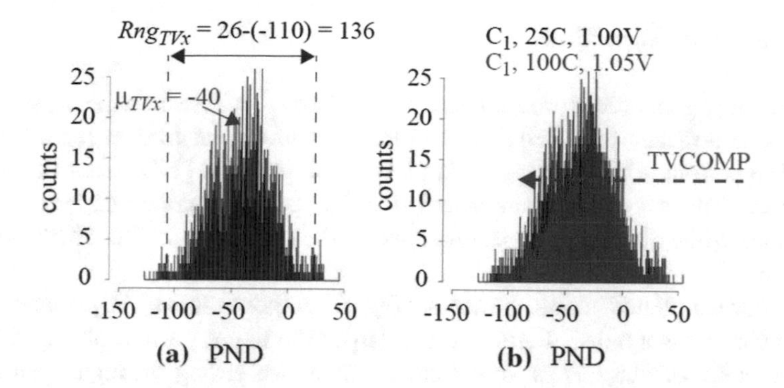

Fig. 6.7 **a** PND distribution for chip C_1 with μ_{TVx} and Rng_{TVx} depicted and **b** Chip C_1 PND distributions at 2 TV corners

differences between the chips. Note that the curves in Fig. 6.6c no longer exhibits the saw-tooth behavior introduced by TV noise.[2]

The differences that remain in the TVComp'ed PND (subsequently referred to as **PND$_c$**) shown in Fig. 6.6c are those introduced by *within-die process variations* (**WDV**) and *uncompensated* TV noise (**UC-TVNoise**). For this particular PND, the TVComp process is able to reduce TV noise to approx. 2 in the worst case, which translates to approx. 36 ps. In general, PNDc with larger levels of UC-TVNoise are more likely to introduce bit flip errors.

The implementation of the HELP algorithm shown in Fig. 6.4 constructs a histogram distribution in the upper 2048 memory locations of the BRAM using the 2048 PND stored in the lower portion and then parses the distribution to obtain μ_{TVx} and Rng_{TVx}. Once the distribution constants are available, the PND in the low portion of the BRAM are converted to PND$_c$.

The last operation applied to the PN is represented by the *Modulus* operation shown on the right side of Fig. 6.4. Modulus is a standard mathematical operation that computes the positive remainder after dividing by the modulus. The Modulus operation is required by HELP to eliminate the path length bias that exists in the PND$_c$, which acts to reduce randomness and uniqueness in the generated bitstrings. The value of the Modulus is also a user-selectable parameter, similar to the LFSR seed, mean and range parameters, and is discussed further in the following. The HELP engine shown in Fig. 6.4 overwrites the PND$_c$ after applying the Modulus. The final values, called **MPND$_c$**, are used in the bitstring generation process.

[2]TV compensation also serves as a countermeasure to prevent adversaries from manipulating temperature and supply voltage as a physical attack mechanism.

Bit Generation Algorithm

The bitstring generation process uses a fifth user-specified parameter, called the *Margin*, as a means of improving the reliability of the bitstring regeneration process. The bottom portion of Fig. 6.8a plots 18 of the 2048 PND_c from $Chip_1$ along the x-axis. The red curve line-connects the data points obtained under enrollment conditions while the black curves line-connect data points under the 12 regeneration TV corners.

The curves plotted along the top of Fig. 6.8a show the $MPND_c$ values after a modulus of 20 is applied. Figure 6.8b enlarges the upper portion of Fig. 6.8a and includes a set of margins of size 2 surrounding two strong bit regions of size 6. Designators along the top given as 's0', 's1', 'w0,' and 'w1' classify each of the enrollment data points as either a strong 0 or 1, or a weak 0 or 1, resp. Data points that fall on or within the hatched areas are classified as weak as a mechanism to avoid bit flip errors introduced by UC-TVNoise that occurs during regeneration.

The Margin method improves bitstring reproducibility by eliminating data points classified as "weak" in the bitstring generation process. For example, the data points at indexes 4, 6, 7, 8, 10, and 14 would introduce bit flip errors at one or more of the TV corners during regeneration because at least one of the regeneration data points is in the opposite bit value region from the corresponding enrollment value. We refer to this bitstring generation technique as the **Single Helper Data** (SHD) scheme since the classification of the MPNDc as strong or weak is determined solely by the enrollment data.

A second technique, referred to as the **Dual Helper Data** (DHD) scheme, requires that both the enrollment and regeneration $MPND_c$ be in strong bit regions before allowing the bit to be used in the bitstring during regeneration. The *helper data*, which represents the classification of the $MPND_c$ as strong or weak, is bitwise 'AND'ed, and then both the enrollment and regeneration bitstrings are generated (the enrollment data is assumed to be collected earlier in time and stored on a secure server). The DHD scheme doubles the protection provided by the margin against bit flip errors because the $MPND_c$ produced during regeneration must now change and

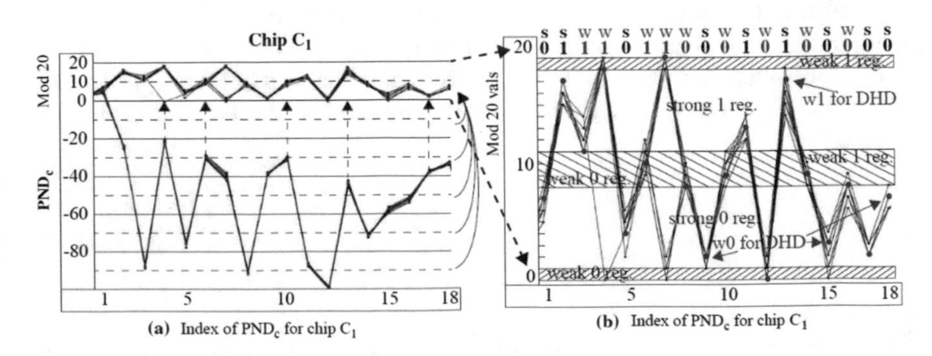

Fig. 6.8 Strong/Weak PND_c classification using margining

move across both a '0' and '1' margin before it can introduce a bit flip error. This is true because both the enrollment and regeneration $MPND_c$ must be classified as strong to be included in the bitstring and the strong bit regions are separated by 2 * margin.

Figure 6.8 highlights four cases where an enrollment-classified strong bit would be reclassified as weak in the DHD scheme because 1 or more of the regeneration PND_c falls within a weak region. This shows that in addition to doubling the protection against bit flip errors, the DHD scheme can potentially produce different bitstrings each time the chip regenerates it. Therefore, **DHD adds uncertainty** by leveraging UC-TVNoise (and sampling noise to a smaller degree). This feature is a benefit for authentication applications because only half of the helper data is revealed to the adversary while the other half is generated and kept on the chip or server. The missing helper data adds uncertainty for an adversary as to the final form of the bitstring. Encryption applications can leverage both of these DHD benefits as well by exchanging the chip and server helper data bitstrings while keeping the generated keys private. These benefits of DHD are expanded upon in the following sections.

Entropy Analysis

The Margin technique using either the SHD or DHD schemes adds uniqueness to the regenerated bitstring. This is true because weak bits are excluded from the bitstring based on the position of the PND_c and Margins and therefore, different chips utilize different bits in the constructed bitstring. Figure 6.9a, b depict several scenarios that show how the Margin and the position of the PND_c affect bitstring generation. The line-connected curves in Fig. 6.9 are analogous to those described earlier in reference to Fig. 6.6c. Figure 6.9a plots a set of 20 different PND_c to

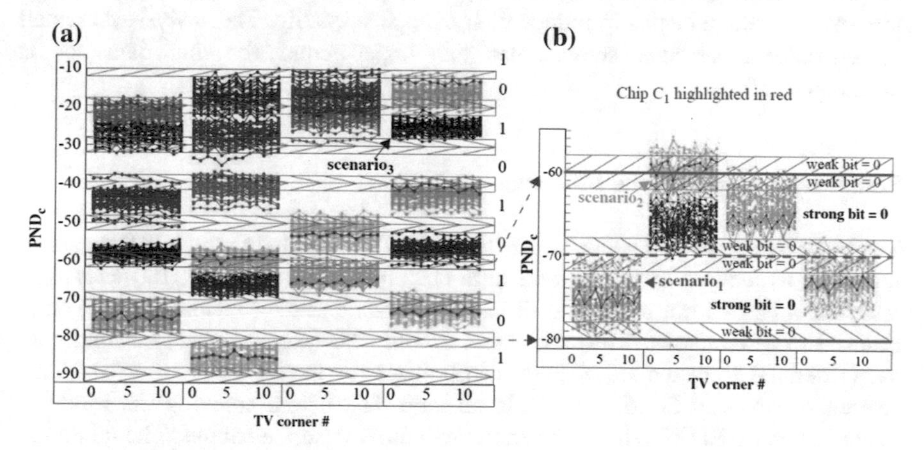

Fig. 6.9 **a** Example PND_c (20 groups) from 38 chips (y-axis) across 1 enrollment and 12 TV corners (x-axis), and **b** blow-up of −60 to −80 region

illustrate how PND_c distribute across the range defined by the Modulus, which is set to 20. Figure 6.9b is a blow-up of the bottom portion of Fig. 6.9a.

As indicated earlier, within-die process variations change path delays uniquely in different chips, which is reflected by the y-dimensional spread within each group of PNDc. For the data set labeled as $scenario_1$ in Fig. 6.9b, the range occupied by the PNDc is approx. 10. The y position of the overall data set is such that, except for a few points, the bit generated by this data will be 0 for all 38 chips.

However, the enrollment data points (left-most) for some chips fall within the weak bit regions and therefore, this bit is skipped for these chips using either the SHD or DHD schemes. Moreover, UC-TVNoise causes some of the regeneration data points to move from their strong bit positions in the enrollment data to weak bits during regeneration. The DHD scheme excludes this bit for these chips as well, creating differences in the generated bitstring for the same chip at different TV corners, while simultaneously providing a $2\times$ Margin to bit flip errors. Moreover, the relative position of the curve associated with each chip, with respect to the other chips, changes in each data set so it is unpredictable which data points are excluded during bitstring generation for any particular chip. The curve for chip C_1 is highlighted in red in each of the PNDc groups to illustrate the change in its relative position with respect to other chips in the group.

The data set labeled $scenario_2$ in Fig. 6.9b shows a second possibility, that is closest to the "ideal" case because the position and range of the curves spans the y-axis into both the strong 0 and strong 1 bit regions. The number of possible results regarding the status of the bit includes those described for $scenario_1$ plus an additional possibility that some chips generate a strong 1 bit and others a strong 0 bit. In contrast, $scenario_3$ labeled in Fig. 6.9a is closest to the "worst" case where nearly the entire data set is positioned with the strong 0 region. Note that this scenario is only possible when the Modulus is large enough to create strong bit regions that upper-bound the smallest range (WDV + UC-TVNoise) found among the $MPND_c$ groups. Generating bitstrings with Moduli larger than 4 * Margin + this smallest range begins to reduce their statistical quality. The analysis presented in subsequent sections shows that the upper-bound for this data set is Modulus = 28.

Statistical Analysis of the Bitstrings

The bitstrings generated using the DHD scheme is subjected to the NIST statistical test suite as well as Inter-chip and Intra-chip hamming distance (HD) tests. The analysis is carried out using two different *reference* scaling factors for TVComp, referred to as *minimum* (Min) and *mean* scaling. The μ_{ref} and Rng_{ref} scaling constants derived from the set of path distributions for the 38 chips are used as the reference values in Eq. 6.12 to scale all chip data before applying the Modulus operation and DHD bitstring generation procedures described above. The minimum scaling constants are derived from the chip with smallest distribution, i.e., smallest mean and range values. The mean scaling constants are computed from the average mean and range values across the distributions of all chips. We focus our analysis on these two scaling factors because they represent the extremes of the recommended range. We expect similar results to be produced for all scaling factors between these limits.

We use the acronym *SBS* to denote "strong bitstring." The DHD scheme requires two helper data bitstrings from the same chip as a means of constructing the two corresponding SBS's. The helper data bitstrings, which are derived from the 2048 $MPND_c$ using the Margin technique, are bitwise AND'ed and then used to select bits for use in the construction of the SBS's. The SBS's generated using enrollment data (TV_0) and the nominal regeneration TV corner data (TV_2) from the same chip are used in the NIST statistical tests and Interchip hamming distance (HD_{Inter}) calculations below. UC-TVNoise is smallest using this combination, and therefore it represents the worst case condition where the effect of the helper data AND'ing has the smallest impact on the additional entropy as discussed earlier. Only one of the SBS's from each chip is used in HD_{Inter} and NIST statistical tests, and the SBS's are truncated to the length of smallest bitstring among the 38 generated. The same criteria are used in the Intra-chip HD (HD_{Intra}) calculations except a much larger set of bits are processed by accumulating the results across a set of 256 different LFSR seeds (only one LFSR seed is used for NIST and HD_{Inter} tests because similar results are obtained using other seeds).

NIST Statistical Test Results The NIST statistical test results are shown in Fig. 6.10a, b for minimum and mean scaling, respectively. A test is considered "a pass" according to the NIST criteria if at least 35 of the 38 chips pass the test individually. The histogram bar heights indicate the number of chips that pass the test. The bitstrings generated using a Margin of 3 and a set of Moduli between 14 and 30 are subjected to 10 of the NIST tests. The size of the bitstring was too small for some values of the Modulus and therefore, the bar heights for these NIST test results are set to 0 (includes regions along back and left side of the 3-D histogram).

Under minimum scaling, all NIST tests are passed except for four associated with Modulus 30. These fails are related to $scenario_3$ discussed in reference to

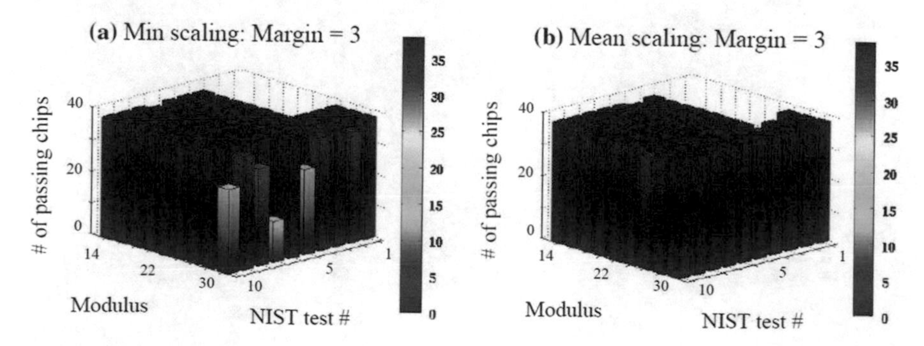

Fig. 6.10 NIST statistical test results using 38 chip bitstrings for each analysis and **a** Minimum scaled data and **b** Mean scaled data

Fig. 6.9, where the range of withindie variation fits entirely within the strong ‘0’ or ‘1’ regions defined by Modulus. This is supported by the results presented under the mean scaling, where the bitstrings for Modulus 30 pass all tests (only 1 test is failed under mean scaling, and with a value of 34 instead of 35). Mean scaling enlarges the y-dimensional spread of the data points over minimum scaling and reduces the probability that $scenario_3$ occurs. These results indicate that the bitstrings possess a high degree of randomness, which is a necessary condition for classifying the bitstrings as cryptographic quality. The results using Margins of 2 and 4 are very similar.

Interchip Hamming Distance (HD_{Inter})
HD_{Inter} is computed using Eq. 6.13. The symbols *NC*, *NB*, and *NCC* represent “number of chips,” “number of bits,” and “number of chip combinations,” respectively. This equation simply sums all the bitwise differences between each of the possible pairing of chip SBS’s (*NCC*), and then converts the sum into a percentage by dividing by the total number of bits that were examined. The XOR operator generates a 1 when the pair of bits in the SBS’s at the same position is different and 0 otherwise.

$$HD_{inter} = \left(\frac{1}{NCC \times NB} \sum_{i=0}^{NC} \sum_{j=i}^{NC} \left(\sum_{k=0}^{NB} (SBS_{i,k} \oplus SBS_{j,k}) \right) \right) \times 100 \qquad (6.13)$$

Figure 6.11a shows the HD_{Inter} results for a set of Moduli (x-axis) and Margins (y-axis). The ideal value for HD_{Inter} is 50 %, which indicates that half of the bits in any arbitrary pairing of bitstrings from the 38 chips have different values. The best values are produced for smaller Moduli, as expected. However, all values remain above 48.5 %, which indicates a high degree of uniqueness among the bitstrings from different chips.

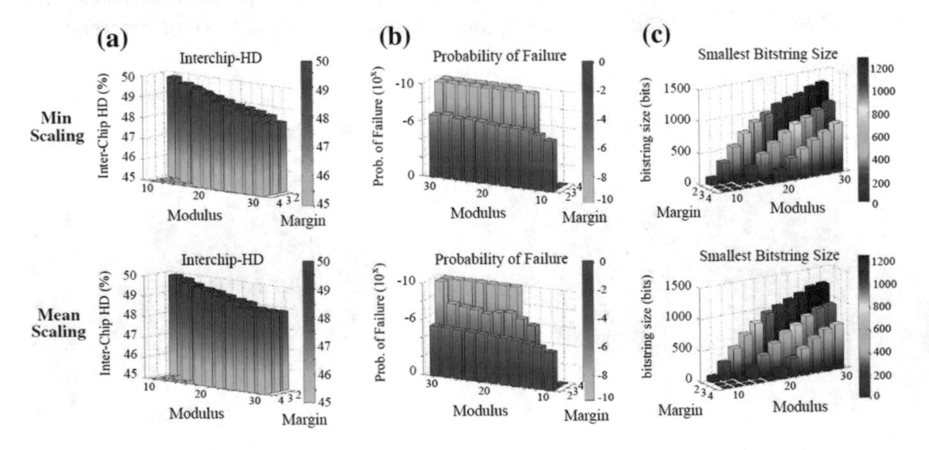

Fig. 6.11 **a** Interchip hamming distance (HD), **b** Probability of failure and **c** Smallest bitstring size statistics using 4096 PN

Intrachip Hamming Distance (HD_{Intra})
HD_{Inter} is computed using Eq. 6.14. The symbols *NS, NC, NB,* and *NT* represent "number of seeds," "number of chips," "number of bits," and "number of TV corners, respectively. As indicated earlier, we repeat the HD_{intra} analysis for 256 different LFSR seeds as a means of increasing the number of bits used in the analysis. *NT* is 12 to represent each of the TV corners used to compute the pair of chip SBS's under the DHD scheme. This equation sums all the bitwise differences between each of the enrollment SBS (TV_0) bitstrings and the 12 corresponding SBS bitstrings from the remaining TV corners for each chip and each LFSR seed, and then converts the sum into a percentage by dividing by the total number of bits that were examined. The value for N varied between approx. 12 million for Modulus 10 to more than 165 million for Modulus 30.

$$HD_{intra} = \left(\frac{1}{N} \sum_{z=0}^{NS} \sum_{i=0}^{NC} \sum_{j=0}^{NB} \sum_{k=1}^{NT} \left(SBS_{z,i,j,0} \oplus SBS_{z,i,j,k} \right) \right) \times 100 \tag{6.14}$$

Figure 6.11b reports HD_{Intra} as the probability of a bit flip failure for the same set of Moduli and Margins used in 11(a) (note the x-axis is reversed from that shown in Fig. 6.11a). The value of the exponent x is reported from the equation $1/10^{-x}$ so −6 indicates 1 chance in 1 million. Cases where no bit flips were detected as shown as −10. As expected, the larger Moduli produce lower probabilities of failure. The probability of failure for Margins 3 and 4 under minimum scaling are all set to 10^{-10} (no bit flip errors were detected), and are less than 10^{-6} for Margin 2 except for Modulus 10. The probability of failure under mean scaling is larger but remains below 10^{-6} for Margins 3 and 4.

Minimum Bitstring Size Figure 6.11c plots the smallest bitstring size for the same set of Moduli and Margins. Smaller Moduli have smaller strong bit regions for a given Margin and therefore, fewer bits quality as strong. However, the bitstring sizes grow quickly, with at least several hundred bits available for Moduli/Margin combinations with strong bit regions of size 2 and larger. Bitstring size can be increased as needed by increasing the number of tested paths beyond 4096.

Security Property Analysis

In this section, we investigate several important security properties of HELP that relate to its resistance to model building and to the number of bitstrings that each token can generate using the five user-defined parameters described earlier and a sixth parameter called the *Path-Selection-Mask* (which is discussed below and in Sect. 6.6 as it relates to proposed authentication protocol).

Parameter-Based Bitstring Diversity

Due to the interaction of the user-defined parameters, we present a conservative lower-bound estimate on the number of possible parameter combinations, i.e., those that ensure the generated bitstrings are random, reliable, and unique for each token. Note that the source of entropy is fixed in this analysis to a set of 4096 PN (in contrast to the analysis that includes the *Path-Selection-Mask* parameter as described in the next section). In other words, the set of five user-defined parameters, namely, μ, *Rng*, *Modulus*, *Margin,* and the *LFSR seeds*, apply different transformations to the same set of PN as a means of achieving bitstring diversity. As noted earlier, the two 11-bit LFSR *seed* parameters allow any of the 2048 rising edge PN to be paired with any of the 2048 falling edge PN, yielding 4,194,304 possible combinations. From the results shown in Fig. 6.11, the number of combinations of Margins and Moduli that yield high reliability ($<e^{-6}$) is 12 (using Moduli from 16–28 for Margin 3, and 20–28 for Margin 4, in steps of size 2). The number of different μ and *Rng* parameters is conservatively estimated to be 10 each. Therefore, a total of 4,194,304 * 12 * 10 * 10 ≅ 5 billion combinations of these five user-defined parameters are possible. This lower bounds the amount of effort required by an adversary in possession of the token to read out all the possible response bitstrings. The probability of achieving this lower bound is nearly zero in practice because, in the proposed protocol, the token and server generate nonces that are used to select values of the parameters and therefore, the adversary does not have direct control of the token's interface (details covered in Sect. 6.6).

Path-Selection-Mask-Based Bitstring Diversity

Unlike the parameter-based scheme, bitstring diversity introduced by the *Path-Selection-Mask* is based on changing the underlying entropy components. In other words, the 4096 PN are not fixed, but vary from one authentication to the next. In the protocol proposed in Sect. 6.6, path selection is performed by the server using a random number generator. Path selection involves choosing a subset x of y timing values from those produced simultaneously by the challenge. For example, assume that a challenge vector sequence produces 200 timing values and the server selects a random subset of 50. The number of ways of choosing 50 from 200 is a very large number and is given by Eq. 6.15. This number is then multiplied by the number of vectors required

$$Path-select-combos = \binom{200}{50} = 4.5e^{47} \tag{6.15}$$

to reach 4096 PN (as an example, we use 82 in our recent experimental evaluation). Therefore, the number of possible bitstrings using the path-selection-mask is exponentially related to the number of simultaneously sensitized paths produced by a challenge and the number of PN randomly selected. More importantly, the *Path-Selection-Mask* changes the characteristics of the PND distribution, which in turn impacts how each PND is transformed by the TVComp process. In other words, even with all 5 user-defined parameters held constant, the bit value generated by a

$MPND_c$ will vary because its value depends on the all of the 4096 PNs selected and used in the bitstring generation process. This complex relationship is leveraged as a security property in the HELP authentication protocol as a means of both preserving privacy and adding resilience to model-building attacks.

6.6 PUF-Based Authentication Protocols

The tamper-evident and unclonable characteristics of PUFs can be leveraged in authentication protocols to generate nonces and repeatable random bitstrings, to provide secure storage of secrets, to reduce costs and energy requirements and to simplify key management. Although weak PUFs have been proposed for authentication as described in the examples that follow, they increase the number and type of cryptographic primitives required on the token. Strong PUFs provide a distinct advantage by eliminating some of these cryptographic primitives while providing higher resistance to protocol attacks.

The cryptographic primitives required in an authentication protocol depend on the security requirements. For example, in the simplest form, the protocol can be designed to provide unilateral, e.g., server-based, authentication as discussed in Sect. 6.4. More advanced features such as mutual authentication and privacy-preserving protocols, i.e., those that prevent token tracking, require additional cryptographic primitives and message exchanges.

Entity authentication requires the prover (hardware token) to provide both an identifier and corroborative and timely evidence of its identity, e.g., a secret, that could only have been produced by the prover itself. From Sect. 6.4, PUFs carry out user authentication under the general model of "something you possess," e.g., a hardware token such as a smart card, which in turn, incorporate silicon-based fingerprint-like identities for authentication to a secure server, such as a bank. Bear in mind that PUFs do not address the task of identifying the user to the token. As discussed in Sect. 6.4, user-token authentication is layered on top of token-server authentication using passwords, PINs, actual human fingerprints, etc.

Although passwords, PINs, one-time passwords, etc. can be used for token-server authentication, they are considered weak authentication methods. The strong authentication methods described in Sect. 6.4 are based on a challenge-response mechanism but implicitly require the prover *A* to demonstrate knowledge of a secret known to be associated with prover *A* without revealing the secret itself to the verifier *B*. The challenge-response component provides a mechanism to enable the prover to maintain the secret while allowing, in the composition of exchanged messages, the prover to demonstrate its knowledge to the verifier. In order to ensure certain security properties, the random numbers (*nonces*) that are cryptographically bound to the secret and exchanged must have sufficient entropy. Cryptographic functions such as one-way hash functions, symmetric key encryption algorithms (for MACs), and public–private encryption algorithms (for digital signatures) may also be required. PUFs can certainly be used in these types of traditional authentication schemes, e.g., for generating nonces with sufficient entropy (which we discuss below), but the large number of CRPs available in strong PUF implementations also allow for simpler schemes with stronger security properties.

6.6.1 Protocol 1: Strong PUF with Unprotected Interface

The simplest mechanisms called *challenge-response entity authentication*, as proposed in [38, 40, 116, 117], exchange cleartext bitstrings directly, thereby eliminating area/energy-expensive cryptographic primitives associated with traditional schemes. A PUF whose inputs and outputs can be accessed directly, as in this scheme, is said to have *unprotected interfaces*. The protocol is shown graphically in Fig. 6.12 (referred to as *naive* in [117]), and consists of two phases:

- Enrollment: A process carried out in a secure environment between a token, *A* and verifier, *B*. Verifier *B* generates a sequence of randomly-chosen challenges, c_i, which are presented to token *A* and applied to the PUF, and the PUF responses, r_i are then recorded in a secure database as challenge-response pairs, crp_i, along with a unique identifier, ht_{ID} for the token.
- Authentication: Token *A* requests authentication by transmitting its ID, $\mathrm{ht_{ID}}$, to the verifier *B*. Verifier *B* selects one or more challenges from the database using the ht_{ID} and transmits them across an unsecured channel to the fielded token. Token *A* applies $\mathrm{c_i}$ to the PUF to generate $\mathrm{r_i}'$, which is transmitted to *B* for verification. *B* compares r_i with r_i' and accepts if the two bitstrings match with a tolerance, HD_{intra}, and rejects otherwise. Verifier *B* removes the crp_i from the database as a countermeasure to replay attacks.

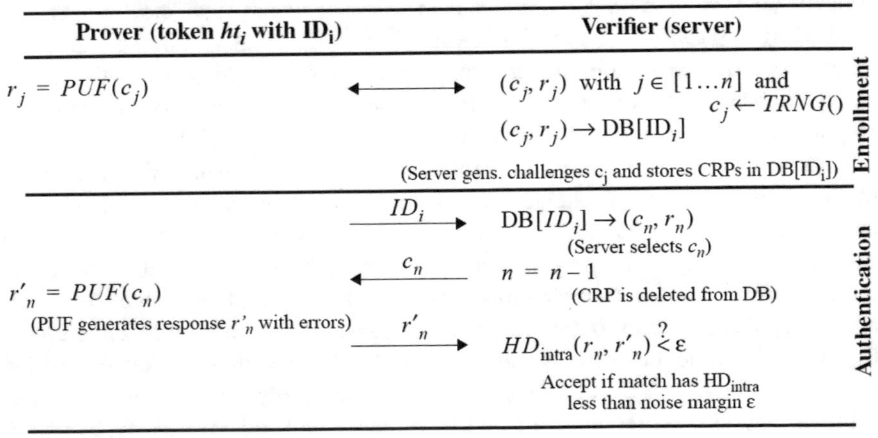

Prover (token ht_i with $\mathrm{ID_i}$)		Verifier (server)	
$r_j = PUF(c_j)$	⟷	(c_j, r_j) with $j \in [1 \ldots n]$ and $c_j \leftarrow TRNG()$ $(c_j, r_j) \rightarrow \mathrm{DB}[\mathrm{ID}_i]$ (Server gens. challenges $\mathrm{c_j}$ and stores CRPs in $\mathrm{DB[ID_i]}$)	**Enrollment**
	ID_i →	$\mathrm{DB}[ID_i] \rightarrow (c_n, r_n)$ (Server selects c_n)	**Authentication**
$r'_n = PUF(c_n)$	← c_n	$n = n - 1$ (CRP is deleted from DB)	
(PUF generates response r'_n with errors)	r'_n →	$HD_{\mathrm{intra}}(r_n, r'_n) \overset{?}{<} \varepsilon$ Accept if match has $\mathrm{HD_{intra}}$ less than noise margin ε	

Fig. 6.12 Naive strong PUF authentication [38, 117]

The protocol has the benefit of being simple to implement and is very lightweight for the token. The inability of the PUF to precisely reproduce the response r_i (in simple schemes that do not attempt error correction or error avoidance) makes it necessary to implement a error-tolerant matching scheme with $HD_{intra} > 0$. It should be noted, however, that large values of HD_{intra} increase the chance of impersonation, and act to reduce the strength of the authentication scheme.

A second drawback is the large number of challenge-response pairs that must be recorded during enrollment, as a means of ensuring that authentication can be carried out over a long period of time. This increases the storage requirements for the verifier, since the worst-case usage scenario must be accommodated, and/or creates inconveniences for users who exceed the stored CRP capacity. Other drawbacks include the lack of resistance to *denial of service* attacks, whereby adversaries purposely deplete the server database, the inability to carry out privacy-preserving or mutual authentication and the susceptibility of the scheme to model-building attacks [118]. The latter is the primary driver for the requirement that a *truly* strong PUF be used for authentication protocols with unprotected interfaces, of which this simple protocol is an example.

A growing list of proposed protocols address these short-coming by incorporating cryptographic primitives on the prover and verifier side [19, 21, 39, 40, 119]. The inclusion of cryptographic primitives enables significant improvements to the security properties of the protocols, and additionally allow for privacy-preserving and mutual authentication. However, their use, in many cases, requires error-free response bitstrings from the PUF, which in turn requires *helper data* to be stored with the CRPs on the server. Many recent protocols target low-cost, resource-constrained applications, e.g., RFID, and attempt to minimize the implementation footprint and energy profile on the token side. Error correction algorithms, such as *secure sketches* [26, 27], are asymmetric in terms of their computational cost, with helper data generation requiring fewer resources than the process of using the helper data to correct bit flip errors in the regenerated response. Recently proposed authentication protocols attempt to minimize the area and energy requirements for token-side operations by leveraging this asymmetrical relationship. We discuss several of these protocols below. An excellent review of these and other protocols [28, 38, 40, 120–133] is provided in [117, 134].

6.6.2 Protocol 2: Controlled PUF

The most straightforward countermeasure to model building attacks is to protect the challenge-response interface to the PUF using cryptographic hash function(s) [16, 117]. One possible implementation of the protocol proposed for a *Controlled PUF* is shown in Fig. 6.13. The hash of the challenge prevents *chosen-challenge* attacks. This is true because the hash is a one-way-function (OWF), which makes it computationally infeasible for the adversary to control the composition of the challenge applied to the PUF. Similarly, by hashing the output of the PUF,

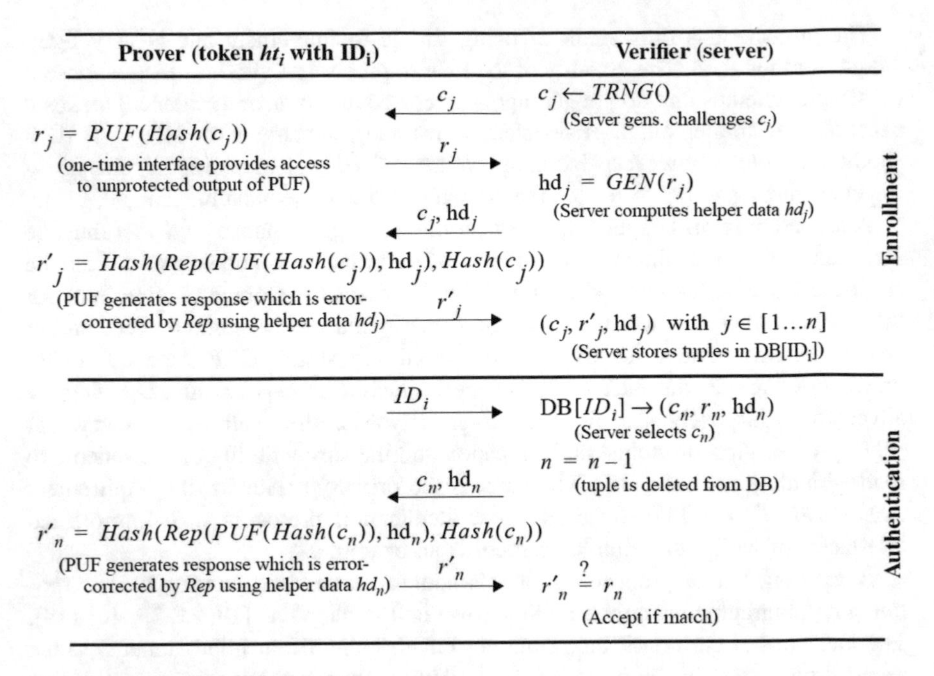

Fig. 6.13 Controlled PUF [16, 117]

correlations that may exist among different challenges are obfuscated, increasing the difficulty of model-building even further. The main drawback of using a OWF on the PUF responses as shown is a requirement that the responses from the PUF be error-free. This is true because even a single bit flip error in the PUF's response changes a large number of bits in the output of the OWF (avalanche effect). The functions *Gen* and *Rep* are responsible for error-correcting the response, using algorithms that were described earlier in Sect. 6.3.3.

The protocol works as follows. During enrollment in a secure environment, a one-time interface is used to allow the server to obtain PUF responses, r_j, produced from randomly generated, hashed challenges c_j. The Gen routine produces helper data hd_j for each r_j, which is sent to the token to produce a hashed version of the PUF response, r'_j. The 3-tuples $<c_j, r'_j, hd_j>$ produced by multiple iterations of this algorithm are stored in the database for token ht_{ID}. After enrollment, a fuse is blown to disable the one-time interface. Authentication is very similar except for the *Gen* operation. Note that the response r'_n must match the stored response r_n in order for the authentication to succeed, i.e., error-correction eliminates the need for the "fuzzy matching" component in Protocol 1. Otherwise, the benefits and drawbacks are similar as those described for Protocol 1 with additional drawbacks related to the need for a cryptographic hash function and the increased computational and energy cost associated with *Rep*.

6.6.3 Protocol 3: Reverse Fuzzy Extractor

Maes et al. proposes a protocol based on *reversed secure sketching* that is designed to address authentication in resource-constrained environments [19, 119]. Their protocol uses the syndrome technique proposed in [26] (see Sect. 6.3.3) for error correction but reverses the roles of the prover and verifier, i.e., the prover (resource-constrained token) performs the lighter-weight *Gen* procedure while the verifier (server) performs the compute-intensive *Rep* procedure. The same process is carried out during enrollment and regeneration. Given that the sketching procedure produces a unique bitstring with bits that are different every time it is executed on the token, in order to authenticate, the verificr is required to *correct the original bitstring* stored during enrollment to match each of the regenerated bitstrings. In order to accomplish this, the helper data produced by each run of *Gen* on the token is transmitted to the verifier.

The mutual authentication protocol proposed in [19] is graphically illustrated in Fig. 6.14. Similar to previous protocols, enrollment involves the verifier generating challenges and storing the PUF responses r_i for ht_i in a secure database (not shown). In the proposed protocol, *only a single CRP* is stored for each token, which is indexed by ID_i in the server's database, and then this interface is permanently disabled on the token. The authentication process begins with the token on the left generating the bitstring response again as r'_i and then multiplying it by the parity-check matrix $\mathbf{H}^{\mathrm{T}}$ of the syndrome-based linear block code to produce the helper data hd_i. A random number generator is used to produce nonce n_1 that is exchanged with the verifier as a mechanism to prevent replay attacks (see Sect. 6.4 for expository on traditional challenge-response authentication). The tuple $<ID_i$, hd_i and $n_1>$ is transmitted over an unsecured channel to the verifier.

Prover (token ht_i with ID_i)		Verifier (server)
$\mathrm{PUF}_i \rightarrow r'_i$ (PUF produces r'_i)		
$\mathrm{hd}_i = r'_i \bullet \mathrm{H}^{\mathrm{T}}$ (Helper data hd_i computed using *Gen*)		
$n_1 \leftarrow TRNG()$ (Nonce n_1 generated)	ID_i, hd_i, n_1 →	$\mathrm{DB}[ID_i] \rightarrow r_i$ (Server looks up r_i)
		$r''_i = \mathrm{Rep}(r_i, \mathrm{hd}_i)$ (And error corrects it to r''_i)
		$n_2 \leftarrow TRNG()$ (Nonce n_2 generated)
$h(ID_i, \mathrm{hd}_i, r'_i, n_1, n_2) \stackrel{?}{=} m_1$ (Accept if match, else abort)	← m_1, n_2	$m_1 = h(ID_i, \mathrm{hd}_i, r''_i, n_1, n_2)$ (Unkeyed hash of protocol vals)
$m_2 = h(ID_i, r'_i, n_2)$ (Unkeyed hash of protocol vals)	m_2 →	$h(ID_i, r''_i, n_2) \stackrel{?}{=} m_2$ (Accept if match, else abort)

Authentication

Fig. 6.14 "Reversed secure sketching" mutual authentication protocol proposed in [26]

The verifier looks up the response bitstring r_i generated by this token during enrollment in the secure database and invokes the *Rep* routine of the secure sketch error correction algorithm with r_i and the transmitted helper data hd_i. If the PUF response r'_i and corresponding helper data hd_i are within the error-correcting capabilities of the secure sketch algorithm, the output r''_i of *Rep* will match the r'_i generated by the token. A second nonce, n_2, is generated to enable secure mutual authentication (see Sect. 6.4) and a secure hash is applied to the ID_i, helper data hd_i, the regenerated response bitstring r''_i and both nonces n_1 and n_2 to produce m_1. The hash m_1 conveys to the token that the server has knowledge of the response r'_i, which allows the token to authenticate the server. This verification is carried out by the token by hashing the same values, except using its own version of r''_i and comparing the output to the transmitted m_1. If a match occurs, then r'_i must be equal to r''_i, and the token accepts, otherwise authentication of the server fails. The token then demonstrates knowledge of r'_i by hashing it with its ID_i and nonce n_2 and transmitting the result m_2 to the server. The server then authenticates the token using a similar process by comparing its result with m_2.

The helper data in this "*reverse*" implementation of the fuzzy extractor changes from one run of the protocol to the next, based on the number and position of the bits that flip during each regeneration. The main drawbacks of the proposed scheme are that it is not privacy-preserving and assumes that the helper data does not leak any information about the response r_i. Moreover, since most PUFs can reliably reproduce more than 80 % of the secret bitstring, any correlations that occur in the helper data bitstrings introduced by these "constant" secret bitstring components may reveal information that the adversary can use to increase the effectiveness of reverse-engineering attacks.

6.6.4 Protocol 4: Slender PUF Protocol

Majzoobi et al. proposed an authentication protocol [133] based on substring matching [112], again designed to address authentication in resource-constrained environments. Their protocol eliminates all types of cryptographic functions on the token, including hashing and error correction functions. The proposed protocol is demonstrated using a 4-XOR arbiter PUF, a variant of the arbiter PUF shown in Fig. 6.3, in which the output of four copies of the arbiter PUF are XOR'ed as a mechanism to increase its model-building resistance. The enrollment process involves building compact models of the arbiter PUFs using a one-time interface that allows access to the individual outputs and provides control over the input challenges. A compact model is a mathematical representation similar to what an adversary would construct when model-building the PUF.

The benefit of storing the compact models is the ability to estimate the response of the 4-XOR arbiter PUF for any arbitrary challenge. This capability is required in the proposed protocol because the challenge is composed of two parts, one part generated by the prover and one part generated by the verifier (using TRNGs).

This "on-the-fly" random challenge generation requires the verifier to generate a "simulated" PUF response from the compact model that closely matches that produced by the actual PUF on the token. The prover's contribution to the concatenated challenge makes it impossible for an adversary to carry out a chosen-challenge attack. A third feature of the protocol relates to the manner in which authentication is performed. A seeded LFSR is used to generate a sequence of challenges that are applied to the 4-XOR PUF to produce a response bitstring. The prover then selects a fixed length substring randomly from PUF-generated response bitstring and transmits it to the verifier. The verifier authenticates the token if it can find the substring (within a predefined noise tolerance) in the corresponding estimate of the response bitstring generated from the compact model. Revealing only part of the response bitstring adds again to the difficulty of model-building.

The protocol is graphically portrayed in Fig. 6.15. The compact model is built during enrollment in a secure environment using a sequence of CRPs applied to the individual arbiter PUFs. The access mechanism is then disabled by blowing fuses. Authentication begins with the generation of challenges c_V and c_P by the verifier and prover, resp., which are concatenated and applied to the PUF to produce response r. A random index i is then generated that serves as the staring index into bitstring r. A substring of r is extracted as r', and is returned to the verifier along with challenge c_P. The verifier uses the compact model to generate an estimate of the PUF response r'' using the same concatenated challenge $(c_V \mid c_P)$. Authentication succeeds if the verifier can locate the substring r' as a substring in r'' within an error

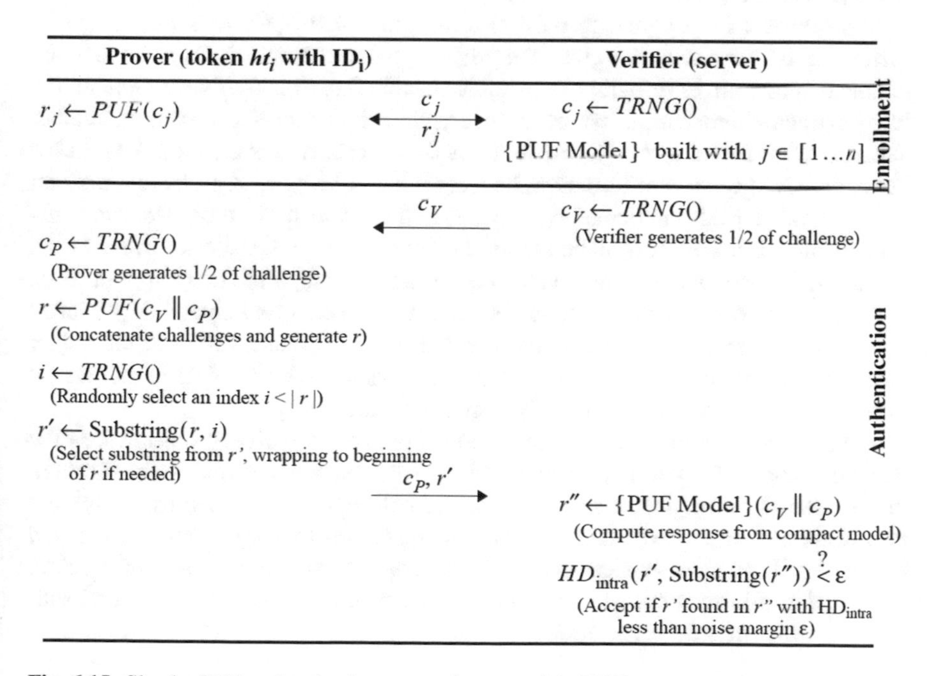

Fig. 6.15 Slender PUF authentication protocol proposed in [133]

tolerance of e. Although the protocol is very light weight for the token, and avoids NVM, the level of model to—hardware-correlation attained in the compact model must be very high and must be able to accommodate changes introduced by TVNoise, resulting in considerable time and effort at enrollment. PUFs that are easily modeled simplify the development of the compact model, but also represents somewhat of a contradiction to their required resilience to model-building attacks. Also, the proposed protocol does not preserve privacy.

6.6.5 Protocol 5: A Privacy-Preserving, Mutual Authentication Protocol

Aysu et al. recently proposed and implemented a PUF-based authentication protocol that provides both *privacy* and *mutual* authentication in resource-constrained environments [21]. They adapt the privacy protocol proposed by [135] to work as a reverse *fuzzy extractor*, as described in Sect. 6.3. The protocol ensures that an adversary is unable to identify or trace the tokens across multiple mutual authentications, despite the adversary having the ability to monitor and control communications and read out the contents of the token's non-volatile memory (NVM). The protocol assumes circuit-level countermeasures are implemented in the tokens to guard against other types of physical attacks, including fault injection and differential power analysis.

The protocol is designed to minimize the functional operations that are to be carried out by the token, but given the privacy goal, the protocol requires the token to implement four cryptographic primitives including the *Gen* operation of the fuzzy extractor algorithm, a symmetric encryption algorithm *Enc*, a random number generator *TRNG* and a pseudo-random function *PRF*. Moreover, the token makes use of an NVM to store information between authentications, in particular, a secret key sk_1 and a PUF challenge c_1. However, the protocol is designed such that leakage of this stored information cannot be used by an adversary to impersonate the token. In particular, the stored challenge is used to allow the token to reproduce a specific PUF response while the secret key is used to encrypt helper data produced by the fuzzy extractor's *Gen* operation on the token. The encryption of the helper data prevents the adversary from reverse engineering the helper data in an attempt to learn the PUF response to the NVM-stored challenge c_1.

Another key feature of the protocol, in support of the privacy objective, is the implementation of a key update mechanism. After each successful authentication, the key stored on the token and in the server's database is updated by applying a new challenge to the PUF and obtaining its response, thereby creating a *chained* sequence of keys across successive authentications. A copy of the state information to be replaced is maintained as a countermeasure to de-synchronization, and subsequent denial-of-service, attacks.

A graphical illustration of the protocol operation is shown in Fig. 6.16. The Enrollment operation is carried out in a secure environment. The server generates a secret key sk_1 and a challenge c_1 that is stored in NVM on the token. The token generates a response r_1 from the PUF and provides it to the server through a one-time interface. The server stores two copies of the sk_1 and r_1 in its secure database. The combination of sk_1 and r_1 is used to derive an ID for the token, as discussed below.

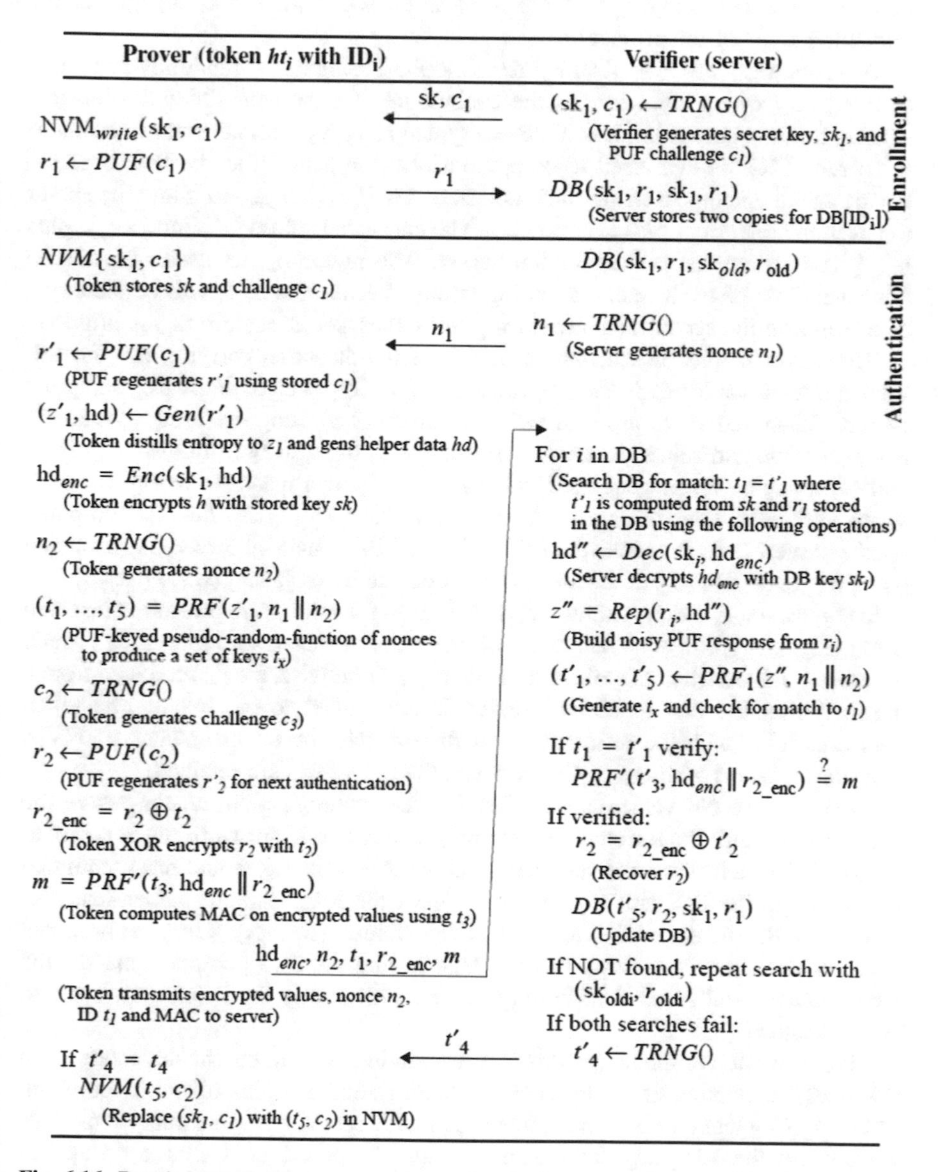

Fig. 6.16 Part 1: Mutual, Privacy-preserving authentication protocol proposed in [21, 135]

The server begins the authentication process by generating a nonce n_1, which is transmitted to the token. The token's challenge c_1 is read from the NVM and used to generate a noisy PUF response r'_1. The *Gen* component of the fuzzy extractor produces z'_1 (an entropy distilled version of r'_1) and helper data *hd*. Helper data *hd* is encrypted using the key sk_1 from the NVM to produce hd_{enc}. The token then generates a nonce n_2. The PUF-generated key z'_1 and the concatenated nonces $(n_1||n_2)$ are used as input to a pseudo-random function PRF to produce a set of unique values t_1 through t_5 that are used as an ID, keys, and challenges in the remaining steps of the protocol.

A second response r_2 is obtained from the PUF using a new randomly generated challenge c_2, which will serve as the chained key for the next authentication (assuming this one succeeds). It is XOR-encrypted as r_{2_enc} for secure transmission to the server. PRF' is then used to compute a MAC *m* using t_3 as the key, over the concatenated, encrypted helper data and new key $(hd_{enc}||r_{2\_enc})$ to allow the server to check the integrity of hd_{enc} and r_{2_enc}. The encrypted values hd_{enc} and r_{2_enc} plus n_2, t_1 and *m* are transmitted to the server. The nonce n_2, as usual, introduces "freshness" in the exchange, preventing replay attacks. The ID t_1 will be the target of a search in the server database during the server side execution of the protocol.

The server begins an exhaustive search of the database, carrying out the following operations for each entry in the DB: (1) decrypt helper data hd_{enc} using the current DB-stored sk_i to produce hd'', (2) construct z'' using the fuzzy extractor's *Rep* procedure and helper data hd'', (3) compute t'_1 through t'_5 from PRF(z'', $n_1||n_2$) and (4) compare token generated value t_1 with t'_1. If a match is found, then the server verifies that the token's MAC *m* matches the $PRF'(t'_3, h_{enc}||r_{2\_enc})$ computed by the server. If they match, then the token's PUF-generated key r_2 is recovered using (r_{2_enc} XOR t'_2), and the database is updated by replacing $(sk_1, r_1, sk_{old}, r_{old})$ with (t'_5, r_2, sk_1, r_1). If the exhaustive search fails, then the entire process is repeated using (sk_{oldi}, r_{oldi}). If both searches fail, the server generates a random t'_4 (which guarantees failure when the token authenticates). Otherwise, the t'_4 produced from a match during the first or second search is transmitted to the token. The token compares its t_4 with the received t'_4. If they match, the token updates its NVM replacing (sk_1, c_1) with (t_5, c_2). Otherwise, the old values are retained.

Note that the old values are needed for de-synchronization attacks where the adversary prevents the last step, i.e., the proper transmission of t'_4 from the server to the token. In such cases, the server has authenticated the token and has committed the update to the DB with (t'_5, r_2, sk_1, r_1) but the token fails to authenticate the server, so the token retains its old NVM values (sk_1, c_1). On a subsequent authentication, the first search process fails to find the t'_5, r_2 components but the second search will succeed in finding sk_1, r_1. This allows the token and server to re-synchronize.

The encryption of the helper data *hd*, as mentioned, prevents the adversary from repeatedly attempting authentication to obtain multiple copies of the helper data, and then using them to reverse engineer the PUF's secret. Note that encryption does not prevent the adversary from manipulating the helper data, and carrying out denial-of-service attacks, so the MAC operation is required to attain this security goal.

The weakest part of the algorithm is the very limited amount of PUF response information maintained by the server, i.e., effectively only one PUF response. Although the authors claim that circuit countermeasures can be used to prevent the PUF response from being extracted from the token using, e.g., differential power analysis, the entire security of the protocol is based on this premise. If, for example, the token's z'_1 is extracted, a clone that impersonates the token can be easily constructed (one that does not even need to embed a PUF), and once it authenticates successfully the first time, the authentic token is barred forever from succeeding (denial-of-service). The very limited amount of PUF response information stored on the server, although attractive from a storage overhead point-of-view, makes it vulnerable to this type of de-synchronization attack. Other issues relate to the requirement for NVM and the not-so-light-weight encryption function, which work against the low-cost, resource-constrained objective.

6.6.6 Protocol 6: The HELP Authentication Protocol

Similar to Protocol 5, the HELP authentication protocol is privacy-preserving and mutual, targets resource-constrained tokens and makes the same assumptions regarding adversarial threats to the token and server [136]. However, HELP does not make use of NVM, does not implement privacy using a *chained* key-update mechanism and requires only one cryptographic operation to be implemented on the token. The protocol is unique among those discussed in that it stores PUF *soft information* on the server instead of bitstrings or PUF models. Soft information refers to digitized path delay values, which from Sect. 6.5.4.2, can each be represented as a 14-bit value, depending on the digital clock manager parameters. When combined with the set of user-defined parameters described in Sect. 6.5.4, including *Modulus*, *Margin*, μ, and *Rng*, two 11-bit *LFSR Seeds* and a *Path-Selection-Mask*, this feature, i.e., storing path delay information, provides some distinct advantages over storing response bitstrings, as highlighted below.

The enrollment operation is graphically illustrated along the top of Fig. 6.17. The authentication protocol uses a common set of challenges $\{c_k\}$ for all tokens as a mechanism to preserve privacy, while establishing the token's identity on the server during the *ID Phase* of in-field authentication. The challenges $\{c_k\}$ are transmitted to the token in a secure environment during enrollment and applied as inputs to the PUF. A set of PN are produced and returned to the server as $\{PN_j\}$. The server generates an internal identifier ID_i for each token using *ServerGenID* and stores the set $\{PN_j\}$ under ID_i in the secure database.

A similar process is carried out during the *Authen Phase* of enrollment except that the challenges are selected from a large set using *SelectChallenges*(ID_i) for each token among those that have been generated using random vectors or automatic test pattern generation (ATPG). The server ensures that the selected set

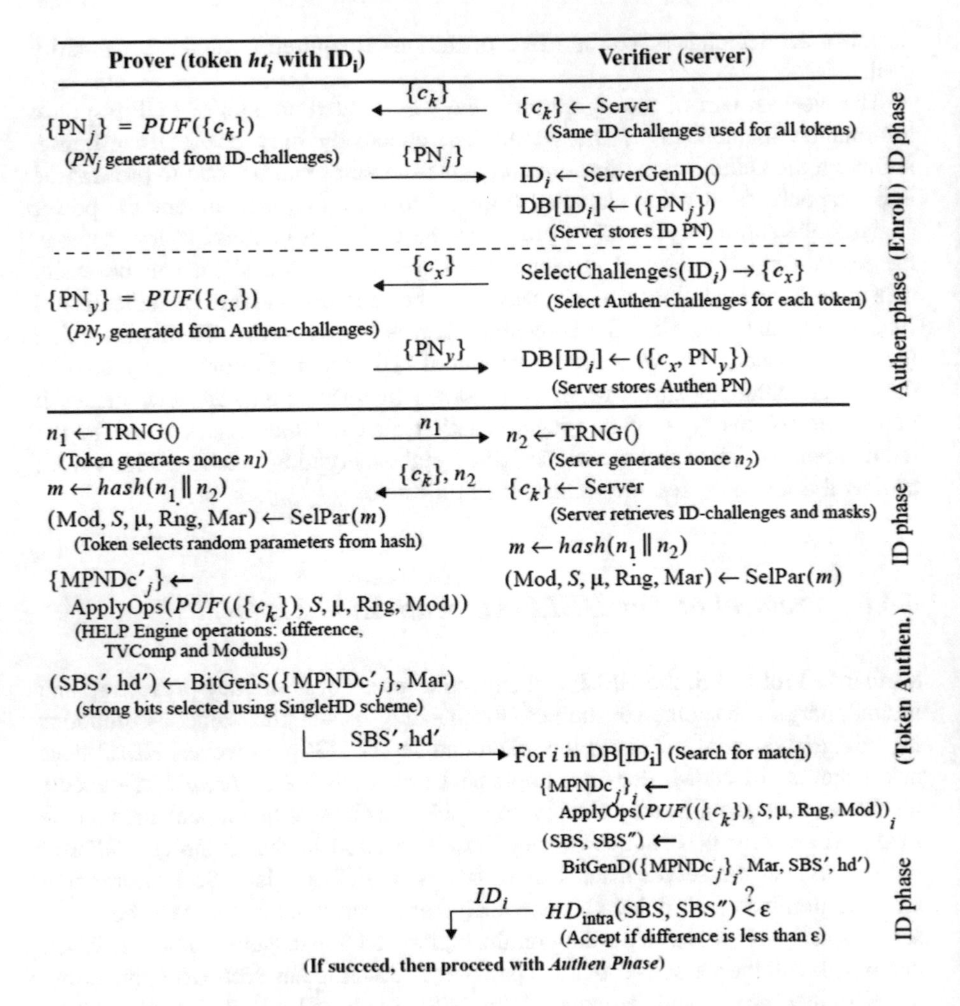

Fig. 6.17 The HELP authentication protocol

overlaps with those chosen for other tokens, but with no more than 50 % overlapping with any one token. This policy prevents the challenges used in the *Authen Phase* during in-field authentication from being used to track the token (explained further below). The set of PN $\{PN_y\}$ generated in the *Authen Phase* are also stored, along with the challenge vectors, in the secure database under ID_i. The number of structural paths for the data path component of SHA-3 is larger than 860,000, with more than 80 % testable, so the set of challenge vectors available is large. Note that the task of generating 2-vector tests for all paths is likely to be computationally infeasible for even moderately sized functional units. However, it is feasible and practical to use random vectors and ATPG to target random subsets of paths for the enrollment requirements.

The cardinality of $\{PN_y\}$ is approx. twice that of $\{PN_j\}$ at 8192 but both are relatively small because the parameters, particularly the *Path-Selection-Mask*, allow an exponential number of different combinations to be constructed over *successive* authentications. The example from Sect. 6.5.4.7 uses the *Path-Selection-Mask* to select 50 PN per challenge. In this case, the number of challenges that need to be applied in the *ID* and *Authen* Phases during enrollment is approx. 80 and 160, resp.

The protocol for token authentication is shown in the bottom portion of Fig. 6.17. The token initiates the process by generating and sending a nonce n_1 to the server. The server generates a nonce n_2 and transmits the fixed set of challenges $\{c_k\}$ and n_2 to the token. The concatenated nonce n_1 with n_2 is used as input to a hash function and a *SelPar* function is used to derive the *Mod*, *S*, μ, *Rng*, *Mar* parameters from the hash output *m*. The *SelPar* function selects bit fields in the hash output *m* for use in a table lookup operation to pseudo-randomly constrain the *Mod* and *Mar* parameters to a specific set of values (as given in Fig. 6.11). Other bit fields are used to define μ and *Rng*, constrained, in this case, to a range of fixed-point values. The same *SelPar* operation is carried out on the server. The hash function limits the amount of the control an adversary has over picking specific values for these parameters in an attack scenario in which the adversary has possession of the token. This component of the protocol is similar to the strategy proposed for the Slender PUF Protocol described in Sect. 6.4 [133] but is used there for challenge selection.

The set $\{c_k\}$ of challenges are applied to the PUF to generate the set $\{PN'_j\}$. The difference, TVComp and modulus operations shown on the right side of Fig. 6.4 are applied to $\{PN'_j\}$ to generate the set $\{MPNDc'_j\}$. Bitstring generation using the single helper data scheme, *BitGenS*, is then performed as described in Sect. 6.5.4.4 using the Mar parameter. *BitGenS* produces a strong bitstring *SBS′* and helper data string *hd′*, which are both transmitted to the server.

A search process is carried out on the server, where the $\{PN_j\}$i data for each token *i* in the database is processed in a similar fashion. However, bitstring generation is carried out using the dual helper data scheme (*BitGenD*). *BitGenD* returns an *SBS* computed using the server data and a modified bitstring *SBS″*, which is a reduced-in-size version of the token's SBS′ (see Sect. 6.5.4.4 for details). The search process terminates when the number of bits that differ in SBS and SBS″ is less than a tolerance ε (which may be zero) or the database is exhausted. In the former case, the token identifier ID_i is passed to the *Authen Phase*. Otherwise, authentication terminates with failure at the end of the *ID Phase*.

Note that token privacy is preserved in the ID Phase because, with high probability, the transmitted information *SBS′* and *hd′* will be different from one run of the protocol to the next, given the diversity of the parameter space provided by *Mod*, *S*, μ, *Rng*, *Mar*, and *Path-Select-Mask*. Also note that this is a compute-intensive operation for large databases because the difference, TVComp, modulus, and BitGenD operations must be applied to each server data set. However, the search operation can be carried out in parallel on multiple CPUs given the independence of the operations. Trial run experiments without any type of explicit parallelism yields runtimes of 200 us per database entry using a database of 10,000 elements when evaluated on an Intel i7-4702HQ CPU @ 2.2 GHz running Linux.

The *Authen Phase* is not shown but is identical to the *ID Phase* with the following exceptions. The subset of 80 token-specific challenges $\{c_1\}$ is randomly selected from the larger set of 160 in $\{c_x\}$ that were applied during enrollment. As indicated earlier, the 160 challenges selected for a token overlap with those selected for other tokens, making it impossible for adversaries to track specific tokens across multiple authentications. A second difference is that the *Authen Phase* represents the *mutual authentication* step, in which the server is authenticated to the token. Therefore, the server generates the SBS' and hd' using the Single Helper Data scheme, which is then transmitted to the token, and the token implements the Dual Helper Data scheme and *fuzzy* match operations (opposite to that shown in Fig. 6.17). This is possible in a resource-constrained environment because of the symmetry in energy requirements of the proposed error avoidance schemes, i.e., the work performed by the Single Helper Data and Dual Helper Data schemes are nearly the same. Note that an optional third phase can be implemented to carry out a second token authentication using the $\{c_x\}$ challenges if needed.

6.7 PUF-Based Authentication for SoC

System-on-chip (SoC) devices continue to proliferate as core components in IoT applications. Although not considered a resource-constrained device, the heterogeneous multi-core, multi-technology characteristics of SoCs, many of which integrate third party IP, make them easy targets for sabotage, reverse engineering, substitution, and cloning. The threat is exacerbated when the SoC integrates cryptographic IP blocks. PUF-based authentication mechanisms can be used to detect manipulation and substitution in the supply-chain and later as installed components in fielded systems.

Applications of PUF-based authentication in SoC are expanding. Recent work focuses on preventing scan chain attacks, carrying out entity authentication and providing authentication of bitstreams for FPGAs. For example, the authors of [137] propose a secure test wrapper which allows testing of multiple IP blocks using PUF-based authentication as a mechanism to improve the security of SoCs that embed IP cores. A low-cost PUF-based authentication architecture designed to secure code execution in IoT SoCs is proposed in [138]. The proposed architecture extracts a PUF-based key from the processor's cache to address threats against code and data authenticity and integrity. A scan chain PUF is proposed in [139] for authenticating SoCs as part of an Infrastructure IP designed to provide multiple security functions. An overview of traditional and modern-day bitstream authentication (and encryption) in FPGAs is provided in [140].

6.8 Conclusion

Authentication protocols, although proposed initially for digital systems over 40 years ago, continue to evolve as new cryptographic functions, such as the Physical Unclonable Function, become available as primitives for enabling physical layer security properties including secure key generation and storage. Adversarial attack surfaces are widening with the proliferation of low-cost and embedded devices for home automation, RFID, smart cards/cars/grids, embedded medical devices, and other types of Internet-of-Things applications. Adversarial attack mechanisms, including physical-layer information extraction techniques, model building and sophisticated network communication tracking algorithms, exacerbate the task of implementing secure unilateral, mutual, and privacy preserving authentication protocols. The introduction of PUFs as primitives can be leveraged to serve as significant countermeasures to adversarial attack mechanisms, particularly for authentication in resource-constrained environments.

This chapter covered both traditional and emerging PUF-based authentication protocols. The primary function of a PUF is to securely generate and store secrets, that can be converted, at any instance in time, into bitstrings for direct use in authentication functions and/or as keys for hashing and encryption functions within authentication protocols. The source of a PUF's entropy is based primarily on within-die variations that occur among circuit components of an integrated circuit. Within-die variations are uncontrollable and unique to each copy of the IC, which allows the PUF to produce unclonable and instance-specific bitstrings.

The integration of PUFs into commercial products is not yet wide-spread. However, published work on PUF constructions and their use in security and trust protocols is growing day-by-day. A wide variety of PUF primitives exist, each with distinctive characteristics related to the number of generated bits (weak vs. strong), robustness to on-chip noise sources, and the statistical quality of the generated bitstrings, e.g., randomness and uniqueness. Existing work shows how PUFs can address shortcoming and provide new capabilities to traditional software-based approaches to authentication but, as discussed in [117, 134], care must be taken to properly characterize the security properties of specific PUF constructions in order to ensure functional and/or practical implementations.

References

1. Goertzel, K.M.: Integrated circuit security threats and hardware assurance countermeasures. In: Real-Time Information Assurance, CrossTalk, Nov/Dec 2013
2. Pope, S., Cohen, B.S., Sharma, V., Wagner, R.R., Linholm, L.W., Gillespie, S.: Verifying Trust for Defense Use Commercial Semiconductors
3. Grand Challenges for Engineering. http://www.engineeringchallenges.org/cms/8996/9042.aspx

4. Defense Science Board Task Force On High Performance Microchip Supply, Office of the Under Secretary of Defense. http://www.acq.osd.mil/dsb/reports/2005-02-HPMS_Report_Final.pdf. Accessed Feb 2005
5. Dean Collins: TRUST, A Proposed Plan for Trusted Integrated Circuits. http://www.stormingmedia.us/95/9546/A954654.html
6. Senator Joe Lieberman: National Security Aspects of the Global Migration of the U.S. Semiconductor Industry. http://lieberman.senate.gov/documents/whitepapers/semiconductor.pdf. Accessed June 2003
7. TRUST in Integrated Circuits (TIC). http://www.darpa.mil/mto/solicitations/baa07-24/index.html
8. National Cyber Leap Year Summit 2009: Co-Chairs' Report. http://www.qinetiq-na.com/Collateral/Documents/English-US/InTheNews_docs/National_Cyber_Leap_Year_Summit_2009_Co-Chairs_Report.pdf. Accessed 16 Sept 2009
9. Integrity and Reliability of Integrated Circuits. DARPA-BAA-10-33 (2010)
10. Trusted Integrated Chips (TIC). IARPA-BAA-11-09 (2011)
11. Bureau of Industry and Security, U.S. Department of Commence. Defense Industrial Base Assessment: Counterfeit Electronics. http://www.bis.doc.gov/index.php/forms-documents/doc_download/37-defense-industrial-base-assessment-of-counterfeit-electronics-2010
12. Grow, B., Tschang, C.-C., Edwards, C., Burnsed, B.: Dangerous fakes. Businessweek. http://www.businessweek.com/stories/2008-10-01/dangerous-fakes (2008)
13. Kessler, L.W., Sharpe, T.: Faked parts detection. http://www.circuitsassembly.com/cms/component/content/article/159/9937-smt (2010)
14. Stradley, J., Karraker, D.: The electronic part supply chain and risks of counterfeit parts in defense applications. IEEE Trans. Compon. Packag. Technol. **29**(3), 703–705 (2006)
15. Ke, H., Carulli, J.M., Makris, Y.: Counterfeit electronics: a rising threat in the semiconductor manufacturing industry. In: International Test Conference (ITC), pp. 1–4 (2013)
16. Gassend, B., Clarke, D.E., van Dijk, M., Devadas, S.: Controlled physical random functions. In: Conference on Computer Security Applications, pp. 149–160 (2002)
17. https://en.wikipedia.org/wiki/Information_security
18. Menezes, A.J., van Oorschot, P.C., Vanstone, S.A.: Handbook of Applied Cryptography. CRC Press. ISBN 0-8493-8523-7. http://cacr.uwaterloo.ca/hac/. Accessed Oct 1996
19. Maes, R.: Physical Unclonable Functions, Constructions, Properties and Applications. Springer (2013). ISBN 978-3-642-41394-0
20. Barker, E., Kelsey, J.: Recommendation of random number generation using deterministic random bit generators. NIST SP800-90A. https://en.wikipedia.org/wiki/NIST_SP_800-90A
21. Aysul, A., Gulcan, E., Moriyama, D., Schaumont, P., Yung, M.: End-to-end design of a PUF-based privacy preserving authentication protocol. In: CHES (2015)
22. https://en.wikipedia.org/wiki/Cryptographic_hash_function
23. https://en.wikipedia.org/wiki/SHA-3
24. https://en.wikipedia.org/wiki/Secure_Hash_Algorithm
25. http://www.nist.gov/manuscript-publication-search.cfm?pub_id=919061
26. Dodis, Y., Reyzin, L., Smith, A.: Fuzzy extractors: how to generate strong keys from biometrics and other noisy data. In: Advances in Cryptology (EUROCRYPT), pp. 523–540 (2004)
27. Dodis, Y., Ostrovsky, R., Reyzin, L., Smith, A.: Fuzzy extractors: how to generate strong keys from biometrics and other noisy data. SIAM J. Comput. **38**(1), 97–139 (2008)
28. Van Herrewege, A., Katzenbeisser, S., Maes, R., Peeters, R., Sadeghi, A.-R., Verbauwhede, I., Wachsmann, C.: Reverse fuzzy extractors: enabling lightweight mutual authentication for PUF-enabled RFIDs. Lecture Notes in Computer Science, vol. 7397, pp. 374–389 (2012)
29. https://en.wikipedia.org/wiki/Binomial_distribution
30. NIST: Computer Security Division, Statistical Tests. http://csrc.nist.gov/groups/ST/toolkit/rng/stats_tests.html
31. Needham, R., Schroeder, M.: Using encryption for authentication in large networks of computers. Commun. ACM **21**(12), 993–999 (1978)

32. Lenstra, A.K., Hughes, J.P., Augier, M., Bos, J.W., Kleinjung, T., Wachter, C.: Ron was wrong, whit is right. Cryptology ePrint Archive, Report 2012/064 (2012)
33. Torrance, R., James, D.: The state-of-the-art in IC reverse engineering. In: Lecture Notes in Computer Science (LNCS), Workshop on Cryptographic Hardware and Embedded Systems, vol. 5747, pp. 363–381 (2009)
34. Kocher, P.C., Jaffe, J., Jun, B.: Differential power analysis. In: Lecture Notes in Computer Science (LNCS), Advances in Cryptology, vol. 1666, pp. 388–397 (1999)
35. Lofstrom, K., Daasch, W.R., Taylor, D.: IC identification circuits using device mismatch. In: International Solid State Circuits Conference, pp. 372–373 (2000)
36. Puntin, D., Stanzione, S., Iannaccone, G.: CMOS unclonable system for secure authentication based on device variability. In: Conference on Solid-State Circuits Conference, pp. 130–133 (2008)
37. Stanzione, S., Iannaccone, G.: Silicon physical unclonable function resistant to a 1025-trial Brute Force Attack in 90 nm CMOS. In: Symposium VLSI Circuits, pp. 116–117 (2009)
38. Pappu, R.: Physical one-way functions. Ph.D. thesis, MIT, ch. 9, 2001
39. Pappu, R.S., Recht, B., Taylor, J., Gershenfeld, N.: Physical one-way functions. Science **297** (6), 2026–2030 (2002)
40. Gassend, B., Clarke, D.E., van Dijk, M., Devadas, S.: Silicon physical random functions. In: Conference on Computer and Communications Security, 148–160 (2002)
41. Lee, J.W., Lim, D., Gassend, B., Suh, G.E., van Dijk, M., Devadas, S.: A technique to build a secret key in integrated circuits for identification and authentication applications. In: Symposium of VLSI Circuits, pp. 176–179 (2004)
42. Lim, D.: Extracting secret keys from integrated circuits. M.S. thesis, MIT, 2004
43. Lim, D., Lee, J.W., Gassend, B., Suh, G.E., van Dijk, M., Devadas, S.: Extracting secret keys from integrated circuits. Trans. Very Large Scale Integr. Syst. **13**(10), 1200–1205 (2005)
44. Suh, G.E., Devadas, S.: Physical unclonable functions for device authentication and secret key generation. In: Design Automation Conference, pp. 9–14 (2007)
45. Majzoobi, M., Koushanfar, F., Potkonjak, M.: Lightweight secure PUFs. In: Conference on Computer-Aided Design (2008)
46. Majzoobi, M., Koushanfar, F., Potkonjak, M.: Testing techniques for hardware security. In: International Test Conference, pp. 185–189 (2008)
47. Ozturk, E., Hammouri, G., Sunar, B.: Physical unclonable function with tristate buffers. In: Symposium on Circuits and Systems, pp. 3194–3197 (2008)
48. Ozturk, E., Hammouri, G., Sunar, B.: Towards robust low cost authentication for pervasive devices. In: Conference on Pervasive Computing and Communications, pp. 170–178 (2008)
49. Gassend, B., Van Dijk, M., Clarke, D., Torlak, E., Devadas, S., Tuyls, P.: Controlled physical random functions and applications. ACM Trans. Inf. Syst. Secur. **10**(4) (2008)
50. Devadas, S., Suh, E., Paral, S., Sowell, R., Ziola, T., Khandelwal, V.: Design and implementation of PUF-based 'Unclonable' RFID ICs for anti-counterfeiting and security applications. In: Conference on RFID, pp. 58–64 (2008)
51. Qu, G., Yin, C.: Temperature-aware cooperative ring oscillator PUF. In: Workshop on Hardware-Oriented Security and Trust, pp. 36–42 (2009)
52. Maiti, A., Schaumont, P.: Improving the quality of a physical unclonable function using configurable ring oscillators. In: Conference on Field Programmable Logic and Applications, pp. 703–707 (2009)
53. Maiti, A., Casarona, J., McHale, L., Schaumont, R.: A large scale characterization of ROPUF. In: Symposium on Hardware-Oriented Security and Trust, pp. 94–99 (2010)
54. Hori, Y., Yoshida, T., Katashita, T., Satoh, A.: Quantitative and statistical performance evaluation of arbiter physical unclonable functions on FPGAs. In: Conference on Reconfigurable Computing and FPGAs, pp. 298–303 (2010)
55. Yin, C.-E.D., Qu, G.: LISA: maximizing RO PUF's secret extraction. In: Symposium on Hardware-Oriented Security and Trust, pp. 100–105 (2010)

56. Costea, C., Bernard, F., Fischer, V., Fouquet, R.: Analysis and enhancement of ring oscillators based physical unclonable functions in FPGAs. In: Conference on Reconfigurable Computing and FPGAs, pp. 262–267 (2010)
57. Majzoobi, M., Koushanfar, F., Devadas, S.: FPGA PUF using programmable delay lines. In: Workshop on Information Forensics and Security, pp. 1–6 (2010)
58. Xin, X., Kaps, J., Gaj, K.: A configurable ring-oscillator-based PUF for Xilinx FPGAs. In: Conference on Digital System Design, pp. 651–657 (2011)
59. Qingqing, C., Csaba, G., Lugli, P., Schlichtmann, U., Ruhrmair, U.: The bistable ring PUF: a new architecture for strong physical unclonable functions. In: Symposium on Hardware-Oriented Security and Trust, pp. 134–141 (2011)
60. Qingqing, C., Csaba, G., Lugli, P., Schlichtmann, U., Ruhrmair, U.: Characterization of the bistable ring PUF. In: Design, Automation & Test in Europe Conference, pp. 459–1462 (2012)
61. Mansouri, S.S., Dubrova, E.: Ring oscillator physical unclonable function with multi level supply voltages. In International Conference on Computer Design, pp. 520–521 (2012)
62. Addabbo, T., Fort, A., Mugnaini, M., Rocchi, S., Vignoli, V.: Statistical characterization of a FPGA PUF module based on ring oscillators. In: Instrumentation and Measurement Technology Conference, pp. 1770–1773 (2012)
63. Maiti, A., Inyoung, K., Schaumont, P.: A robust physical unclonable function with enhanced challenge-response set. Trans. Inf. Forensics Secur **7**(1), Part: 2, pp. 333–345 (2012)
64. Meng-Day, Y., Sowell, R., Singh, A., M'Raihi, D., Devadas, S.: Performance metrics and empirical results of a PUF cryptographic key generation ASIC. In: Symposium on Hardware- Oriented Security and Trust, pp. 108–115 (2012)
65. Maeda, S., Kuriyama, H., Ipposhi, T., Maegawa, S., Inoue, Y., Inuishi, M., Kotani, N., Nishimura, T.: An artificial fingerprint device (AFD): a study of identification number applications utilizing characteristics variation of polycrystalline silicon TFTs. Trans. Electron Dev. **50**(6), 1451–1458 (2003)
66. Simpson, E., Schaumont, P.: Offline hardware/software authentication for reconfigurable platforms. In: Cryptographic Hardware and Embedded Systems, vol. 4249, Oct 2006, pp. 10–13
67. Habib, B., Gaj, K., Kaps, J.-P.: FPGA PUF based on programmable LUT delays. In: Euromicro Conference on Digital System Design (DSD), pp. 697–704 (2013)
68. Guajardo, J., Kumar, S.S., Schrijen, G.-J., Tuyls, P.: Physical unclonable functions and public key crypto for FPGA IP protection. In: Conference on Field Programmable Logic and Applications, 189–195 (2007)
69. Su, Y., Holleman, J., Otis, B.: A 1.6pJ/bit 96 % stable chip ID generating circuit using process variations. In: International Solid State Circuits Conference, pp. 406–407 (2007)
70. Guajardo, J., Kumar, S.S., Schrijen, G., Tuyls, P.: Brand and IP protection with physical unclonable functions. In: Symposium on Circuits and Systems, pp. 3186–3189 (2008)
71. Kumar, S.S., Guajardo, J., Maes, R., Schrijen, G.-J., Tuyls, P.: Extended abstract: the butterfly PUF protecting IP on every FPGA. In: Workshop on Hardware-Oriented Security and Trust, pp. 70–73 (2008)
72. Kassem, M., Mansour, M., Chehab, A., Kayssi, A.: A sub-threshold SRAM based PUF. In: Conference on Energy Aware Computing, pp. 1–4 (2010)
73. Bohm, C., Hofer, M., Pribyl, W.: A microcontroller SRAM-PUF. In: Conference on Network and System Security, pp. 25–30 (2011)
74. Bhargava, M., Cakir, C., Mai, K.: Reliability enhancement of bi-stable PUFs in 65 nm bulk CMOS. In: Workshop on Hardware-Oriented Security and Trust, pp. 79–83 (2012)
75. Alkabani, Y., Koushanfar, F., Kiyavash, N., Potkonjak, M.: Trusted integrated circuits: a nondestructive hidden characteristics extraction approach. In: Information Hiding (2008)
76. Ganta, D., Vivekraja, V., Priya, K., Nazhandali, L.: A highly stable leakage-based silicon physical unclonable functions. In: Conference on VLSI Design, pp. 135–140 (2011)

77. Helinski, R., Acharyya, D., Plusquellic, J.: Physical unclonable function defined using power distribution system equivalent resistance variations. In: Design Automation Conference, pp. 676–681 (2009)
78. Helinski, R., Acharyya, D., Plusquellic, J.: Quality metric evaluation of a physical unclonable function derived from an IC's power distribution system. In: Design Automation Conference, pp. 240–243 (2010)
79. Ju, J., Chakraborty, R., Rad, R., Plusquellic, J.: Bit string analysis of physical unclonable functions based on resistance variations in metals and transistors. In: Symposium on Hardware-Oriented Security and Trust, pp. 13–20 (2012)
80. Ju, J., Chakraborty, R., Lamech, C., Plusquellic, J.: Stability analysis of a physical unclonable function based on metal resistance variations. In: Symposium on Hardware-Oriented Security and Trust (HOST), pp. 143–150 (2013)
81. Ismari, D., Plusquellic, J.: IP-level implementation of a resistance-based physical unclonable function. In: Accepted to Symposium on Hardware-Oriented Security and Trust (HOST) (2014)
82. Chakraborty, R., Lamech, C., Acharyya, D., Plusquellic, J.: A transmission gate physical unclonable function and on-chip voltage-to-digital conversion technique. In: Design Automation Conference (DAC), pp. 1–10 (2013)
83. Li, J., Lach, J.: At-speed delay characterization for IC authentication and trojan horse detection. In: International Workshop on Hardware-Oriented Security and Trust (HOST), pp. 8–14 (2008)
84. Aarestad, J., Ortiz, P., Acharyya, D., Plusquellic, J.: HELP: a hardware-embedded delay—based PUF. IEEE Des. Test Comput. **30**(2), 17–25 (2013)
85. Aarestad, J., Acharyya, D., Plusquellic, J.: An error-tolerant bit generation technique for use with a hardware-embedded path delay PUF. In: Symposium on Hardware-Oriented Security and Trust (HOST), pp. 151–158 (2013)
86. Saqib, F., Areno, M., Aarestad, J., Plusquellic, J.: An ASIC implementation of a hardware-embedded physical unclonable function. IET Comput. Digit. Tech. **8**(6), 288–299 (2014)
87. Che, W., Saqib, F., Plusquellic, J.: PUF-based authentication, invited paper. In: International Conference on Computer Aided Design, Nov 2015
88. Kursawe, K., Sadeghi, A.-R., Schellekens, D., Skoric, B., Tuyls, P.: Reconfigurable physical unclonable functions—enabling technology for tamper-resistant storage. In: Workshop on Hardware-Oriented Security and Trust, pp. 22–29 (2009)
89. Rosenfeld, K., Gavas, E., Karri, R.: Sensor physical unclonable functions. In: Symposium on Hardware-Oriented Security and Trust, pp. 112–117 (2010)
90. Xiaoxiao, W., Tehranipoor, M.: Novel physical unclonable function with process and environmental variations. In: Conference on Design, Automation & Test in Europe, pp. 1065–1070 (2010)
91. Lin, L., Holcomb, D., Krishnappa, D.K., Shabadi, P., Burleson, W.: Low-power sub-threshold design of secure physical unclonable functions. In: Symposium on Low-Power Electronics and Design, pp. 43–48 (2010)
92. Ruhrmair, U., Jaeger, C., Bator, M., Stutzmann, M., Lugli, P., Csaba, G.: Applications of high-capacity crossbar memories in cryptography. Trans. Nanotechnol. **10**(3), 489–498 (2011)
93. Simons, P., van der Sluis, E., van der Leest, E.: Buskeeper PUFs, a promising alternative to D flip-flop PUFs. In: Symposium on Hardware-Oriented Security and Trust (HOST), pp. 7–12 (2012)
94. Maiti, A., Schaumont, P.: A novel microprocessor-intrinsic physical unclonable function. In: Field Programmable Logic and Applications, pp. 380–387 (2012)

95. Sreedhar, A., Kundu, S.: Physically unclonable functions for embedded security based on lithographic variation. In: Conference on Design, Automation & Test in Europe, pp. 96–105 (2012)
96. Kumar, R., Dhanuskodi, S.N., Kundu, S.: On manufacturing aware physical design to improve the uniqueness of silicon-based physically unclonable functions. In: International Conference on Embedded Systems, pp. 381–386 (2014)
97. Forte, D., Srivastava, A.: On improving the uniqueness of silicon-based physically unclonable functions via optical proximity correction. In: Design Automation Conference, pp. 7–12 (2012)
98. Meguerdichian, S., Potkonjak, M.: Device aging-based physically unclonable functions. In: Conference on Design Automation Conference, pp. 288–289 (2011)
99. Kalyanaraman, M., Orshansky, M.: Novel strong PUF based on nonlinearity of MOSFET subthreshold operation. In: Symposium on Hardware-Oriented Security and Trust (HOST), pp. 13–18 (2013)
100. Rose, G.S., McDonald, N., Lok-Kwong, Y., Wysocki, B., Xu, K.: Foundations of memristor based PUF architectures. In: IEEE/ACM International Symposium on Nanoscale Architectures (NANOARCH), pp. 52–57 (2013)
101. Che, W., Bhunia, S., Plusquellic, J.: A non-volatile memory-based physically unclonable function without helper data. In: International Conference on Computer-Aided Design (ICCAD) (2014)
102. Yu, Z., Krishna, A.R., Bhunia, S.: ScanPUF: robust ultralow-overhead PUF using scan chain. In: Asia and South Pacific Design Automation Conference (ASP-DAC), pp. 626– 631 (2013)
103. Zhang, L., Kong, Z.H., Chang, C-H.: PCKGen: a phase change memory based cryptographic key generator. In: International Symposium on Circuits and Systems (ISCAS), pp. 1444–1447 (2013)
104. Konigsmark, S.T.C., Hwang, L.K., Deming, C., Wong, M.D.F.: CNPUF: a carbon nanotube-based physically unclonable function for secure low-energy hardware design. In: Asia and South Pacific Design Automation Conference (ASP-DAC), pp. 73–78 (2014)
105. Zhang, F., Henessy, A., Bhunia, S.: Robust counterfeit PCB detection exploiting intrinsic trace impedance variations. In: VLSI Test Symposium, Apr 2015
106. Areno, M., Plusquellic, J.: Securing trusted execution environments with PUF generated secret keys. In: TrustCom (2012)
107. Areno, M., Plusquellic, J.: Secure mobile association and data protection with enhanced cryptographic engines. In: PRISMS (2013)
108. Guajardo, J., Kumar, S.S., Schrijen, G.T., Tuyls, P.: FPGA intrinsic PUFs and their use for IP protection. Cryptogr. Hardware Embedded Syst. **4727**, 63–80 (2007)
109. Rührmair, U., Busch, H., Katzenbeisser, S.: Strong PUFs: models, constructions, and security proofs. In: Sadeghi, A.-R., Naccache, D. (eds.) Towards Hardware-Intrinsic Security, pp. 79–95. Springer (2010)
110. Gassend, B., Lim, D., Clarke, D., van Dijk, M., Devadas, S.: Identification and authentication of integrated circuits. Concurr. Comput. **16**(11), 1077–1098 (2004)
111. Majzoobi, M., Koushanfar, F., Potkonjak, M.: Testing techniques for hardware security. In: International Test Conference, pp. 1–10 (2008)
112. Paral, Z., Devadas, S.: Reliable and efficient PUF-based key generation using pattern matching. In: Symposium on Hardware-Oriented Security and Trust, pp. 128–133 (2011)
113. Lamech, C., Aarestad, J., Plusquellic, J., Rad, R., Agarwal, K.: REBEL and TDC: two embedded test structures for on-chip measurements of within-die path delay variations. In: International Conference on Computer-Aided Design, pp. 170–177 (2011)
114. Tiri, K., Verbauwhede, I.: A logic level design methodology for a secure DPA resistant ASIC or FPGA implementation. In: DATE, pp. 246–251 (2004)

115. Tiri, K., Verbauwhede, I.: A digital design flow for secure integrated circuits. IEEE Trans. Comput.-Aided Des. Integr. Circ. Syst. **25**(7), 1197–1208 (2006)
116. Ranasinghe, D.C., Engels, C.W., Cole, P.H.: Security and privacy: modest proposals for low-cost RFID systems. In: Auto-ID Labs Research Workshop (2004)
117. Delvaux, J., Gu, D., Schellekens, D., Verbauwhede, I.: Secure lightweight entity authentication with strong PUFs: mission impossible? In: CHES, pp. 451–475 (2014)
118. Rührmair, U., Sehnke, F., Solter, J., Dror, G., Devadas, S., Schmidhuber, J.: Modeling attacks on physical unclonable functions. In: Conference on Computer and Communications Security, pp. 237–249 (2010)
119. Van Herrewege, A., Katzenbeisser, S., Maes, R., Peeters, R., Sadeghi, A.-R., Verbauwhede, I., Wachsmann, C.: Reverse fuzzy extractors: enabling lightweight mutual authentication for PUF-enabled RFIDs. In: International Conference on Financial Cryptography and Data Security (2012)
120. Bolotny, L., Robins, G.: Physically unclonable function-based security and privacy in RFID systems. In: PerCom, pp. 211–220 (2007)
121. Ozturk, E., Hammouri, G., Sunar, B.: Towards robust low cost authentication for pervasive devices. In: PerCom, pp. 170–178 (2008)
122. Hammouri, G., Ozturk, E., Sunar, B.: A tamper-proof and lightweight authentication scheme. Pervasive Mobile Comput. 807–818 (2008)
123. Kulseng, L.,. Yu, Z., Wei, Y., Guan, Y.: Lightweight mutual authentication and ownership transfer for RFID systems. In: INFOCOM, pp. 251–255 (2010)
124. Sadeghi, A.-R., Visconti, I., Wachsmann, C.: Enhancing RFID security and privacy by physically unclonable functions. In: Information Security and Cryptography, pp. 281–305 (2010)
125. Katzenbeisser, S., Unal Kocabas, Van Der Leest, V., Sadeghi, A., Schrijen, G.J., Schroder, H., Wachsmann, C.: Recyclable PUFs: logically reconfigurable PUFs. In: CHES, pp. 374–389 (2011)
126. Kocabas, U., Peter, A., Katzenbeisser, S., Sadeghi, A.: Converse PUF-based authentication. In: TRUST, pp. 142–158 (2012)
127. Lee, Y.S., Kim, T.Y., Lee, H.J.: Mutual authentication protocol for enhanced RFID security and anticounterfeiting. In: WAINA, pp. 558–563 (2012)
128. Jin, Y., Xin, W., Sun, H., Chen, Z.: PUF-based RFID authentication protocol against secret key leakage. Lect. Notes Comput. Sci. **7235**, 318–329 (2012)
129. Xu, Y., He, Z.: Design of a security protocol for low-cost RFID. In: WiCOM, pp. 1–3 (2012)
130. Lee, Y.S., Lee, H.J., Alasaarela, E.: Mutual authentication in wireless body sensor networks based on physical unclonable function. In: IWCMC, pp. 1314–1318 (2013)
131. Yu, M.-D.M., M'Rahi, D., Verbauwhede, I., Devadas, S.: A noise bifurcation architecture for linear additive physical functions. In: HOST, pp. 124–129 (2014)
132. Konigsmark, S.T.C., Hwang, L.K., Chen, D., Wong, M.D.F.: System-of-PUFs: multilevel security for embedded systems. In: CODES, pp. 27:1–27:10 (2014)
133. Majzoobi, M., Rostami, M., Koushanfar, F., Wallach, D.S., Devadas, S.: Slender PUF protocol: a lightweight, robust, and secure authentication by substring matching. In: Symposium on Security and Privacy Workshop, pp. 33–44 (2012)
134. Delvaux, J., Gu, D., Peeters, R., Verbauwhede, I.: A survey on lightweight entity authentication with strong PUFs. Cryptology ePrint Archive: Report 2014/977
135. Moriyama, D., Matsuo, S., Yung, M.: PUF-based RFID authentication secure and private under complete memory leakage. IACR Cryptology ePrint Archive 2013, 712 (2013). http://eprint.iacr.org/2013/712
136. Che, W., Saqib, F., Plusquellic, J.: A privacy-preserving, mutual PUF-based authentication protocol. Submitted to special issue "Physical Security in Cryptography Environment", Cryptogr. J. http://www.mdpi.com/journal/cryptography. Accessed Aug 2016

137. Das, A., Kocabas, U., Sadeghi, A.-R., Verbauwhede, I.: PUF-based secure test wrapper design for cryptographic SoC testing. In: Design, Automation and Test in Europe, pp. 866–869 (2012)
138. Hoffman, C., Cortes, M., Aranha, D.F., Araujo, G.: Computer security by hardware-intrinsic authentication. In: Hardware/Software Codesign and System Synthesis, pp. 143–152 (2015)
139. Wang, X., Zheng, Y., Basak, A., Bhunia, S.: IIPS: infrastructure IP for secure SoC design. Trans. on Comput. **64**(8), 2226–2238 (2015)
140. Trimberger, S.M., Moore, J.J.: FPGA security: motivations, features, and applications. Proc. IEEE 1248–1265 (2014)

Chapter 7
FPGA-Based IP and SoC Security

Debasri Saha and Susmita Sur-Kolay

7.1 Introduction

Field-programmable gate arrays (FPGAs) have become almost indispensable in embedded reprogrammable systems for a plethora of applications in recent times. An application-specific design can be transformed into a configuration bitstream to program an already fabricated architecture so that the specific application is realized on this targeted hardware. FPGA-based IPs are used in communication infrastructure, digital camera, high-performance signal and image processing applications, automotive electronics, industrial control, and distributed database applications. A modern FPGA is becoming increasingly complex, typically comprising embedded multi-core processors, gigabit serial transceivers, clock managers, analog-to-digital converters, digital signal processing blocks, ethernet controllers, megabytes of memory, and other functional blocks in addition to the arrays of basic logic elements.

This introductory section presents a typical FPGA architecture and a design flow. It continues with brief description of the various associated IPs, typical threats, and security aspects. Section 7.2 discusses cryptographic primitives because a bitfile or a partial bitstream is loaded on an FPGA architecture in encrypted form to prevent an unauthorized access of the IP. This encryption of bitfile may be cracked through side-channel attacks. In Sect. 7.3, methods for authentication of genuine IP vendor and the authorized IP user by including their binary signatures in the FPGA bitstream are described. Section 7.4 deals with various effective techniques to combat the threat of Hardware Trojan Horse (HTH) [1] in FPGAs. Finally, as SoCs are also

D. Saha (✉)
A. K. Choudhury School of Information Technology,
University of Calcutta, Kolkata, India
e-mail: sahadebasri@gmail.com

S. Sur-Kolay
Advanced Computing & Microelectronics Unit,
Indian Statistical Institute, Kolkata, India
e-mail: ssk@isical.ac.in

S. Bhunia et al. (eds.), *Fundamentals of IP and SoC Security*,
DOI 10.1007/978-3-319-50057-7_7

being implemented with FPGAs, security issues in IP distribution, IP management, and inter-communication which are more complex and challenging are discussed in Sect. 7.5. Section 7.6 represents the summary.

7.1.1 Brief Description of an FPGA Architecture and Its Design Flow

An FPGA vendor, such as Xilinx, Altera (acquired by Intel in 2015), MicroSemi, is a proprietor of its FPGA chips available in several device families. A design software tool is used to design, optimize, and load an FPGA design meant for any member of its device family. For example, Xilinx provides a number of architecture families such as Spartan, Virtex, Kintex, and Zync. A standard FPGA architecture consists of various types of basic components—configurable logic blocks (CLBs), memories (block select RAMs), programmable routing matrix, input/output blocks (IOB), and clock distributions. Modern architectures are becoming more nonhomogeneous as those contain block RAM (BRAM), multipliers, and even DSP blocks. Each CLB contains a large number of slices (logic cells). Each logic cell contains a few lookup tables (LUTs) to implement fixed capacity Boolean functions, and flip-flops as storage elements. SRAM controlled MUXes and fast carry logic connect outputs of LUTs (Fig. 7.1a). BRAMs are used for caching and accessing common data. Programmable routing matrix consists of routing resources for local routing and for general-purpose routing [2].

The design tool of each vendor broadly follows an FPGA design flow (Fig. 7.1b), which is somewhat different from that of ASICs. In the design entry phase, the design description in a hardware description language (HDL) such as Verilog or VHDL is simulated, and then synthesized to generate the netlist in *.ngc* or *.edif*

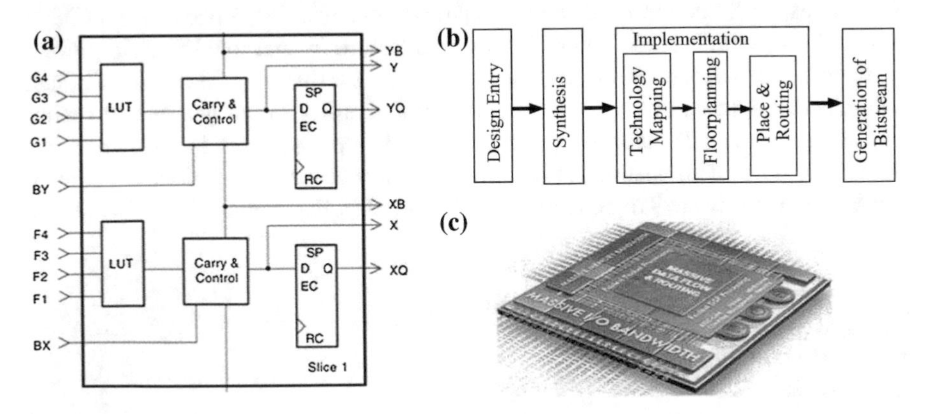

Fig. 7.1 FPGA architecture: **a** a slice of a CLB in Xilinx Virtex family, **b** FPGA design flow, and **c** Xilinx FPGA chip with ultrascale architecture (Courtesy: Xilinx)

format. The next phase is implementation. In map subphase of implementation, i.e., in technology mapping of the FPGA design flow, the netlist is split into small parts such that each part can be mapped onto a physical component of the already fabricated architecture, namely slices and IOBs. In place and route (P&R) subphase, the netlist components are placed in slices and interconnected using the routing resources to realize the netlist so that the timing report ascertains specified timing closure. Once the design is implemented, the configuration bitstream is generated in .bit format and is downloaded in encrypted form either directly onto an FPGA chip, or in external memory, e.g., flash PROM in Xilinx platform. The bitfile on decryption configures the FPGA to realize the design.

Modern FPGAs support partial reconfiguration (PR), which facilitates reconfiguration of onc or more part(s) of the design with different bitstreams. PR enhances flexibility and the speed of applications within limited hardware resources, thereby rendering the design to be suitable for adaptive computing. Module-based PR is widely used to reconfigure distinct modular parts of the design, whereas in difference-based PR, a partial bitstream contains information only about the minute differences between the present and the next design. PR may be dynamic or static. Dynamic (active) PR permits reconfiguring some parts of an FPGA while the rest of the device is running, whereas static PR does not. Dynamic PR is very effective in real-time applications but requires fast decryption and other primitive security components for the partial bitstream. A Microblaze soft processor as microcontroller may be used to initiate reconfiguration in a reconfigurable system. Other ways to initiate reconfiguration are JTAG, RS232 commands, timer- based, and event-based reconfiguration. The new partial configuration bitstream is then loaded into FPGA using an internal configuration access port (ICAP).

PR is supported in the Xilinx architectures Virtex-4 to Virtex-7, Kintex-7, Artix-7 and the Zynq-7000 programmable SoC family. Xilinx ultrascale architecture (Fig. 7.1c) enables PR of almost all FPGA resources including I/O and clock network. Xilinx PlanAhead tool is used to specify the static partition and the modules to be reconfigured. It also defines and manages the buses across the partition interface between the reconfigurable modules and the other parts of the design.

7.1.2 IPs in FPGA

Application-specific ICs (ASICs) are reused as intellectual property (IP) cores in larger circuits, but ASICs need enormous design effort and long design time before their fabrication and thereby entailing high cost. In order to cope up with the requirements of high productivity, low cost, and low time-to-market, nowadays application-specific designs are also being configured in FPGA. Therefore, in addition to hard-coded ASICs, FPGA-based designs have been well-accepted as IP cores. Support to partial reconfigurability enhances the flexibility, and high parallelism increases efficiency of FPGA-based IPs. Often, in a system-on-chip (SoC), a piece of FPGA architecture is included along with several ICs on the SoC to enhance flexibility of its usage. Alternatively, an SoC with multiple IP cores from different IP core vendors may entirely be configured on an FPGA. An increasing number of companies are providing IP support to FPGA.

In an FPGA, an application-specific design is optimized in performance, power, and in size for being loaded in the smallest possible FPGA chip and is taken in the form of a configuration bitstream file, also known as *bitfile*. IP for an FPGA may be in various forms, namely HDL design, FPGA-based design after place-and-route, stored bitfile before loading, or even bitfile running on an FPGA. The soft bitfile is the most valuable and vulnerable IP.

7.1.3 Typical Threats for an FPGA-Based IP

In the context of FPGA-based IP security, the following attacks create concern:

T1 unauthorized access of a bitstream IP through hacking a bitfile in transmission, or through cloning, i.e., copying the bitstream by intercepting it from the FPGA;
T2 reverse engineering of a bitstream IP to extract information about the lower level design or any embedded secret information, such as a signature used for authentication or a secret key for decryption;
T3 attempt to extract any secret information from side-channel information of an FPGA chip, i.e., from measurable manifestation of a circuit in operation, such as delay, power, and electromagnetic emissions of the FPGA chip;
T4 counterfeiting, i.e., selling a low-quality FPGA IP at the price of a branded product after extraction of secret signature of the IP core designer from the branded IP followed by its insertion into a low-quality IP;
T5 malicious modification of an IP through tampering of embedded secret information, or through spoofing, i.e., replacing the authentication portion of an FPGA bitstream by that of the attacker;
T6 modification by inclusion of extraneous logic known as Hardware Trojan Horse (HTH) in order to affect the performance or lifetime of an IP or to extract secret design information;
T7 untrustworthy design tool;
T8 fabrication of an FPGA chip with intentional structural defect, or may be a lower generation device with lower performance.

Some of these threats such as threats T1, T2, and T3 are more alarming for FPGAs compared to ASICs. As the design tool and the FPGA chip are proprietary to the FPGA vendor, the last two threats T7 and T8 can cause serious concerns only if the design team and fabrication facility of the FPGA vendor are not trustworthy.

Nowadays, support to reconfiguration provides enhanced flexibility, but at the cost of additional security threats such as intrusion through the reconfiguration controller, damage of the base array by corrupted bitstream during reconfiguration.

Widespread usages of FPGA IPs on an FPGA-based SoC, and support for partial reconfiguration, create causes of concerns, specifically in (i) IP exchange, (ii) identification and partial decoding of bitstream, (iii) IP management on FPGA-based SoCs, and also (iv) detection of Trojan sources active across the IPs. Discussion on these threats on FPGA-based SoCs appears in Sect. 7.5.

7.1.4 Security Aspects: Countermeasures and Vulnerabilities

In order to counter the above-mentioned attacks on FPGA IP cores, a number of security aspects have been implemented in FPGAs, namely,

(a) encryption of the bitstream to prevent cloning, reverse engineering, and integrity check of the bitstream to detect malicious modification;
(b) appropriate design of cryptoprocessor unit on an FPGA to protect secret information from side-channel attacks;
(c) embedding in the design a signature that is tamper resistant as well as resilient against reverse engineering to protect the IP against counterfeiting;
(d) applying techniques for detection of Trojans.

These mechanisms ensure trusted use of an IP. On one hand, only the genuine IP core vendor gets the patent by correctly proving his ownership and also the desired royalty fees for each legal IP instance as only an authorized user can access his IP core. On the other, an IP core purchased by a legal buyer is of desired IP value ensuring protection of buyer's right [3]. Some of the above-mentioned security aspects are usually included in standard FPGA products, whereas the rest arc in research domain. The FPGA products are still vulnerable to the threats which have not been covered by the security aspects in them, along with other new types of attacks.

Several surveys on FPGA security, such as [4–6], have on a number of security challenges and their countermcasures. In the current perspective of partial reconfigurability of FPGAs and FPGA-based SoCs, this chapter highlights the present state-of-the-art of FPGA security and existing vulnerabilities.

In commercial FPGAs, the following techniques as countermeasures have been introduced to enhance FPGA security [7].

- The bitfile (i.e., the .bit file) generated in the final stage of a design tool for FPGA is difficult to read, and the bitstream in the bitfile for Xilinx starting from the old Virtex-II family to the recent Virtex-7 series remains encrypted, and therefore cloning of the design is prevented.
- In Virtex-6, Spartan-6, and other recent families from Xilinx, a unique but public 57-bit device identifier, known as Device DNA, is programmed into a one-time programmable (OTP) e-fuse to uniquely identify an FPGA device. It attempts to make the encryption device-specific so that the encrypted bitfile cannot be decrypted and utilized in other chips.

- The format of configuration bitstream is in general proprietary and kept confidential to the vendor, otherwise an adversary with the knowledge of the configuration format may reverse-engineer the bitstream to obtain the logic design.
- *Readback* reveals the present state of an operating FPGA. Therefore, readback of any configuration file, even encrypted, is not permitted in Xilinx, or in Altera Stratix II and Stratix II GX devices. The Altera device families of Arria II GX, Arria V, Cyclone III LS, Cyclone V, and Stratix V have user mode anti-tamper features such as controlling interfaces with JTAG. For Xilinx, Readback via JTAG and other external interfaces is disabled after the device is loaded with an encrypted design. This helps in prevention of cloning as well as reverse engineering.
- In reconfigurable FPGAs, temporary data storage is cleared and the current communication is terminated before reconfiguration, otherwise an adversary may take advantage of using the data for reverse engineering.
- Hashed message authentication code (HMAC) is incorporated in Xilinx Virtex-6 to ensure authentication of an FPGA design. It also ensures its integrity by detecting spoofing. Cyclic redundancy check (CRC) as integrity check is applied on the bitstream to detect malicious modification /tampering of the same.
- Xilinx includes error-correcting codes for each configuration data frame (the smallest addressable segments of the device configuration memory space) and a bitstream scrubbing hardware to monitor configuration data and to correct altered bits.
- Internal monitoring on voltage and temperature in Xilinx has been employed to identify possible environmental attacks on an operating design. Any attempt to alter the key or configuration bitstream causes the key and bitstream content to be cleared.
- Further, verification, cross-checking, and constraint checking are done at various stages of the design tool to check integrity and correctness of the design with possible detection of any tamper or extraneous logic present.
- Mixing of encrypted and non-encrypted data for a single application is prevented, as the latter may contain Trojans.

In spite of several security measures adopted in the (i) manufacturing flow, (ii) design tools, (iii) during configuration, and (iv) operation of FPGAs, all the threats described in Sect. 7.1.3 cannot be tackled. While countermeasures to the threats T1, T2, and T5 have been implemented, several research proposals are there for T3 and T4. Although the security against the risk in threat T6 for HTH has only been enhanced, the risk is still alive. An unbiased verification team within an FPGA vendor for verifying their design tool and testing their device can help to tackle threats T7 and T8.

Moreover, few new attacks have been introduced into the security mechanisms incorporated. For example, the entire secret key for decryption has already been recovered by analyzing side-channel information. After measuring the power consumption of a single power-up of Xilinx Virtex-II Pro, all the three different keys used by its triple DES encryption module could be retrieved. The full 128-bit AES key [8] of an Altera Stratix II has been discovered by applying side-channel analysis with 30,000 measurements in less than three hours. Some internals of hardware crypto engines of the corresponding Xilinx and Altera devices have also been revealed through these attacks. A keyed test mechanism has already been discovered for enabling readback for Microsemi FPGAs.

A large number of possible Trojan sources in a design tool as well as in the hardware render complete assurance of a trusted environment to be difficult. The bitfile with high IP value is more vulnerable for such Trojan intrusion. Of late, attackers can even interpret the format of a bitfile and insert Trojans very effectively.

A rich adversary may employ costlier attacks like tampering of configuration data by applying radiation or physical stress. Nowadays, high-profit applications also utilize FPGA IPs. Therefore, these costlier attacks become feasible as an adversary can gain monetarily notwithstanding the cost of the attacks. Techniques with low performance overhead to counter costlier attacks are on demand.

Sometimes, certain security measures to counter various types of attacks may be conflicting with each other—such as encryption and trustworthy signature verification, encryption and reconfigurability.

In the platform of partially reconfigurable FPGA-based SoCs and embedded systems, some security mechanisms have been adopted to cope up with the increased vulnerability to attacks mostly targeting the content of memory, the operations of processors and configuration controller, interfaces and data transmission on a system. But, more effective and efficient measures are on demand. Further, during distribution of multiple IPs to several IP tool vendors, key management is quite complex, and the possibility of partial interception of IPs remains to be handled properly.

The above-mentioned vulnerabilities to attacks are still alive and new security holes are being introduced. Therefore, IP security in FPGA domain has interesting challenges and needs special attention.

7.2 Cryptographic Primitives on FPGA Bitstream

For an FPGA design, the corresponding bitfile core is to be kept encrypted when it is shipped or transmitted to its buyer and also at buyers' site to prevent cloning as well as reverse engineering of the bitfile. The encryption algorithm to be used must be fast, robust against cryptanalysis and the decryption hardware implementing the decryption algorithm must consume low area and low power, is of high speed and robust against side-channel attack. Besides encryption, this section discusses other crypto primitives for authentication, integrity check and freshness of bitstream, generation of chip-specific encryption keys and partial encryption.

7.2.1 Symmetric Key Encryption of Bitstream on FPGAs for Confidentiality

For bitstream encryption, symmetric key encryption is used. The configuration bitfile is encrypted using a secret key at the vendors' end. At the users' end, the encrypted configuration bitstream from some external nonvolatile memory is loaded into the FPGA at each system power-up. The same secret key stored on-chip is used to decrypt the configuration file. Both encryption of bitstream and storing of the key for decryption take place at the IP vendors' end and the legitimate user cannot access the private key.

Xilinx FPGAs apply either triple-data encryption standard (3DES) or 256-bit advanced encryption standard (AES) [8] in cipher block chaining (CBC) mode. Altera Stratix II and Stratix II GX devices use 128-bit AES in counter (CTR) mode for configuration bitstream encryption. Key length in this range provides desired strength against attacks. CBC or CTR mode prevents the propagation of errors.

Configuration bitstream on decryption is typically stored in SRAM memory, which facilitates higher performance, greater logic density, improved power efficiency, reduced manufacturing cost, and higher flexibility of self-test. But, SRAM is volatile, i.e., loses data at power-off. So, battery-backed SRAM is used in Xilinx to support throughout the application life. In addition, SRAM-based memory facilitates fast in-site partial reconfiguration. However, the possibility of data interception due to the external memory persists in SRAM-based memory.

Actel FPGAs of Microsemi and some other FPGAs use nonvolatile on-chip flash resource for configuration bitstream to eliminate the risk of using external memory. But, integration of flash memory on SRAM-based FPGA is costlier as it requires complex fabrication steps. Use of flash memory for PR is also technologically possible and several research directions for partially reconfigurable flash memory are available. But, the presence of configuration data permanently on flash memory-based chip and reconfigurability of flash may cause similar security threats as in SRAM.

For key storage, Xilinx uses either battery-backed SRAM with a key clear property as volatile storage, or OTP e-fuse as nonvolatile storage. For enhanced secrecy and persistence of the key in the buyers' site, OTP nonvolatile memory (flash or e-fuse) is preferred. The key may be programmed on-chip, or off-chip during the regular manufacturing flow.

Encryption of bitstream has an overhead due to the additional bitstream storage and the decryption unit on FPGA. Instead of a built-in decryptor, the decryption unit may be configured in the configuration logic of the FPGA also.

FPGA Implementations of AES Processors

An AES decryption unit in an FPGA must incur very low overhead in terms of area and power. Table 7.1 shows the implementation details of a few fast and compact

Table 7.1 FPGA implementations of AES and AES-enhanced processors

Processors	Device	Throughput	Resource	Efficiency
AES (fastest) [9]	Spartan-III	25 Gbps	–	1.441 Mbps/slice
AES (fast) [10]	Virtex-II	24.922 Gbps	–	6.97 Mbps/slice
AES (compact w/o mem) [11]	Spartan-6	58.13 Mbps	80 slices	–
AES-SHA [12]	Virtex-5	575 Mbps	1938 slices	–
AES-GCM [12]	Virtex-5	913 Mbps	1538 slices	–
Pl AES-GCM [13]	Virtex-5	27.7 Gbps	3211 slices	8.62 Mbps/slice
4-Pl AES-GCM [13]	Virtex-5	102.4Gbps	12152 slices	8.42 Mbps/slice

AES processors in FPGAs. Two FPGA designs for AES are presented by Good and Benaissa [9]—while one is the fastest design, the second one based on 8-bit datapath and only 124 slices and 2 BRAMs on Spartan-II is believed to be the smallest compared to the other designs using 32-bit datapath.

AES or any other cryptoprocessor is designed for FPGA implementation in a way to defend leakage of the secret key through side-channel information. In order to defend against differential power analysis (DPA)-based side-channel attacks, which are measured in terms normalized energy deviation (NED) and normalized standard deviation (NSD), randomization of computations and equalization of consumed power are applied in general. These techniques will be discussed in detail in another chapter. For FPGAs, ROM-based substitution box (S-box) for AES is proposed [14] which outperforms logic S-box in area, power, performance, and also in power-analysis resistance. For power-analysis resistance, they propose modification of traditional ROM to create matched bitline and wordline capacitances across the memory. One tricky approach to resist any power analysis-based attack is to place a sensor or detector circuit, which can detect whether any device is attached with the power pin or not. In order to prevent leakage of information from electromagnetic field measured by an antenna, the computations are distributed across the FPGA.

7.2.2 Primitive Crypto Units for Authentication and Integrity Check

Encryption of bitstream cannot ensure authentication of the IP vendor as well as integrity of the bitstream. For integrity check, cyclic redundancy check (CRC) is applied to recover from a bit flip in the bitstream, caused due to remote transmission of bitfile or application of electromagnetic field on an FPGA. For malicious and more complex bitstream alteration, integrity check using Secure Hash Algorithm (SHA-2) [8] is effective. Use of two different algorithms—AES for confidentiality and SHA-2 for integrity—may cause speed mismatch and significant area overhead. Furthermore, such an attempt cannot ensure authentication. For confidentiality, authentication, and integrity check, message authentication code (MAC) function is to be used over encryption [8]. Alternatively, AES in Galois/Counter mode (GCM) is preferred as fast authenticated encryption (AE) algorithm with integrity check. It facilitates area efficiency and high-speed implementation using dynamic PR. For encryption/decryption, counter (CTR) mode is used which is highly parallel. For MAC-based authentication, hashing based on product–sum operation in Galois field $GF(2^w)$ (GHASH) [12] is used which enables faster and more compact hardware implementation. Interleaving of CTR and GHASH in a single function improves performance. Several implementation details of AES-GCM are given in Table 7.1. Analyzing parallel implementations, 8-block parallel implementation is found as a sweet point.

7.2.3 Confidentiality, Authentication, Integrity and Freshness of Bitstream in Remote Dynamic Partial Reconfiguration

In recent times, for real-time applications, remote (online) configuration of FPGA has been proposed, where complete or partial configuration bitstream is sent over the internet in a compressed form. The compressed bitstream is obtained through lossless compression technique such as run-length encoding, Lampev–Ziv algorithm or Huffman coding having high compression ratio, high-speed decompression, and low resource cost of decompressor. Such remote transmission has the risk of man-in-the-middle attack and bitstream spoofing [8], so on-the-fly reconfiguration without integrity check is not allowed in general as it may damage the FPGA hardware with corrupted bitstream. There may be possibilities of replay attack and even denial of service in remote transmission. Drimer et al. in [15] proposed secure remote configuration of FPGAs facilitating authentication, decryption, integrity cheek, freshness of bitstream against replay attack, and protection against denial of service.

Finally, authors in [16] propose a single-chip solution (without using any external memory) for secure remote *re*configuration of the partial bitstream while the current configuration is active. They integrate a fast decompression unit with the cryptographic unit offering all the objectives mentioned in [15]. Station-to-station (STS) protocol is used to obtain a session key, from which one key for encryption and another key for message authentication via HMAC are obtained. A microblaze microcontroller is used to receive a partial bitstream, which is temporarily stored in BRAM and cryptographic operations are performed through a cryptoprocessor consisting of AES, SHA-256, random number generator and elliptic curve processor, and finally this encrypted version is sent to ICAP.

7.2.4 Generation of FPGA Chip-Specific Encryption Key

If the embedded secret encryption key is specific to an FPGA architecture, i.e., device family, the user may re-sell the encrypted bitfile, which can be decrypted in any FPGA chip of the same architecture family. In order to prevent such an attempt, the secret encryption key is chip-specific, i.e., for each fabricated FPGA chip, a unique key is used for encryption of the bitfiles to be loaded into that specific chip. The authors in [17] propose to encrypt the bitfile IP core based on secure device identification. Using both public-key and secret-key cryptography, the system and the IP exchange protocol are designed in this work. Several techniques have been developed to generate the FPGA chip-specific secret information—either secret keys, or some random number used as initialization vector or a seed for cryptographic primitives.

PUF Design in FPGA for Generation of Encryption Key

FPGA chip-specific secret information is obtained using a circuit for physically unclonable function (PUF), where for a set of challenge (input) vectors, the responses are unpredictable. In addition to the challenge vectors, manufacturing variability and other conditions determine PUF responses. The PUF circuit is unclonable in nature, therefore PUF response and hence PUF-based encryption key remains unpredictable to an adversary. Among several types of PUFs, arbiter PUFs, memory-based PUFs, and ring oscillator (RO) PUFs [18] may easily be implemented in FPGAs. The simplest is the delay-based arbiter PUF, which permits minute variation in delay by a single lookup element. RO PUF is more stable and reliable under a wide range of temperature variations, and memory-based PUF provides longer secret information, therefore enhanced security.

In PUF-based key generation, such as [19], first the uniquely distinguishable responses from PUFs are collected, then the responses are post-processed by applying error correction using either artificial neural network (ANN), or Bose–Chaudhuri–Hocquenghem (BCH) code or Von-Neumann corrector, or sometimes more complicated one-way hash function such as SHA-1 or SHA-2 to prevent prediction of challenges.

PR facilitates more condition-specific key generation. In [20], the authors measure the delay of a linear PUF structure by setting the clock frequency at the frequency of PUF using digital clock management (DCM) and a phase-locked loop (PLL) of Xilinx Virtex architecture. For different challenge configurations, a set of linear equations are solved to obtain each switch delay of the PUF circuit. Next, using PR, a secure and robust PUF structure consisting of PUFs connected in parallel rows, the logic circuits for I/O and wire interconnects are configured. The input network is attached to parallel PUFs to satisfy strict avalanche criteria (SAC) (i.e., each output bit changes with a probability of 0.5 whenever a single input bit is complemented) for the PUF circuit. A nonlinear transformation applied on the responses of PUFs to generate the output ensures robustness against reverse engineering. The mixing property of XOR logic or in general parity generator provides the resilience against emulation and statistical guessing.

Random Number Generation in FPGA

Partial reconfiguration is effectively used in random number generation (RNG) techniques. In RNG technique [21], the unpredictability of a random number is enhanced by driving the bi-stable flip-flop in a logic component to its metastable state using an at-speed monitor and a control mechanism that establishes a closed-loop feedback system. The monitor system keeps track of the probabilities of the output bit over repeated time intervals. Using this information, the control unit decides to add/subtract the delay to/from the top/bottom paths. The length of the propagation path is incremented or decremented minutely only by a single LUT in a CLB of an FPGA architecture so that the delay difference is close to zero. The design of a true random number generator (TRNG) for FPGA implementation is shown in Fig. 7.2.

Simultaneous generation of random bits and PUF responses has been attempted in Actel Fusion FPGA using some special hardware circuit known as universal transition effect ring oscillator (UTERO) [22].

7.2.5 *Partial Encryption and Security Issues*

Encryption of the entire bitfile discussed so far, without any provision of encrypting a partial bitstream, causes security concern in the case of partial reconfiguration (PR). Earlier, authors in [23] suggested encryption of only judiciously selected portions

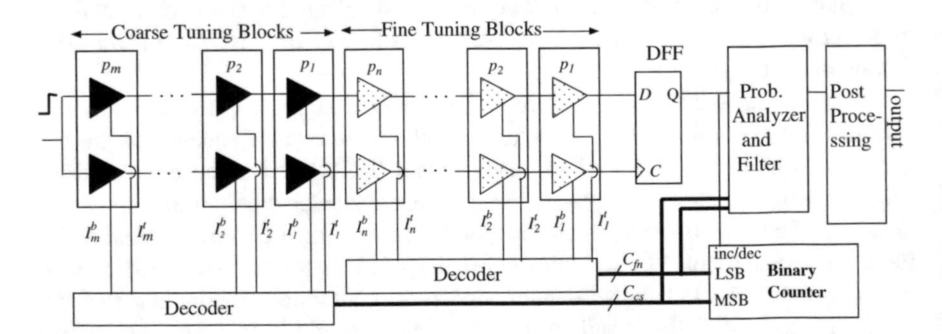

Fig. 7.2 True Random Number Generator [21]: coarse blocks create delay differences over a wide range; decoder block maps the values of the counter to the number of 1's in the input to the programmable delay lines. The bit rate is 16 Mbit/sec and the propagation delay is 61.06 ns. This TRNG core uses 128 LUTs packed into 16 Virtex-5 CLBs

Table 7.2 Effect of partial reconfiguration on bitstream encryption

	Encryption of bitstream	Design of cryptoprocessor	Generation of PUF-based key
Pros	A decryption unit in static partition, decrypts partial bitstreams for all reconfigurable modules	Different cryptoprocessors may be configured for different applications	Reconfiguarable PUF generates more condition-specific keys; reconfigurable RNG enhances randomness
Cons	Requires support for encryption of partial bitstream	Reconfiguring different cryptoprocessors causes identification of its bitstream & guessing its logic through reverse engg.	Requires binding of PUF design with h/w before reconfiguring to reproduce the same behavior and the same key

of the configuration bitstream. Later on, various FPGA tools have been updated so that those facilitate partial encryption, i.e., encryption of partial bitstreams for the reconfigurable modules to be loaded thereafter. For Xilinx, Virtex-6 supports partial encryption but Virtex-5 does not. BitGen pads the FPGA partial bitstream with NOP commands so that the entire bitstream is evenly divided into AES-256 encryption blocks and encrypted. Then, encrypted partial reconfiguration (EPRC) system is used to perform a frame by frame CRC check before loading it.

PR has both positive and negative security aspects (Table 7.2). Loading different cryptoprocessors using module-based PR facilitates the use of different encryption algorithms for different applications. Difference-based PR may be used if the design for security is adaptive in nature, requiring minute changes at consecutive times based on some controls. Both types of PR introduce the security threat of identifying the bitstream for the cryptoprocessors or other crypto designs, such as PUF circuit to regenerate the secret key. Therefore, the encryption algorithm should be strong enough to prevent reverse engineering of the bitstream. There is another concern in generation of PUF-based key. For partially reconfiguring an FPGA with a design module including a PUF circuit, it is mandatory to bind the PUF design to a proper location of the hardware at the time of reconfiguration; otherwise the metrics of the PUF circuit are likely to be changed, so will its response. Thus, the PUF-based secret key cannot be reproduced.

However, all these techniques cannot prevent a legal buyer from intentional reselling of his encrypted bitfile core along with the corresponding FPGA hardware to an unauthorized user, who can download the bitfile core into that particular FPGA hardware. In order to prevent such events, authentication of legitimate buyers is required by the IP vendor.

7.3 Authentication of an IP in FPGA

Techniques for embedding signatures of the IP vendor and the legitimate buyer, and for verification of those signatures for purpose of authentication of an FPGA-based IPs, are discussed in this section.

7.3.1 *Embedding Signatures of IP Vendor and Legitimate User*

The techniques for embedding signatures of the IP vendor and the legitimate buyer are termed as *watermarking* and *fingerprinting*, respectively. Sometimes, design tool-specific information is embedded instead of the signature of the IP vendor. A signature embedding technique must be fast, robust against tampering, and incur low overhead. In a design tool for FPGA, the constraint file, which is used to incorporate constraints on objectives such as timing or on technology mapping or placement of I/Os and logic, may be used to embed signatures.

Embedding Signature During Technology Mapping

Authors in [24] propose two protocols for embedding user- and tool- specific information into a logic circuit while performing multilevel logic minimization and technology mapping. It embeds additional constraints, derived uniquely from the authors' signature, into the problem specification, such that the final solution can be retrieved only within a subset of the set of all solutions for multilevel logic minimization and technology mapping.

Watermark may be embedded during logic synthesis through incremental technology mapping of selective disjoint closed cones [25]. A closed cone is a portion of a logic network, which contains no outgoing edge to other logic nodes. In order to minimize and isolate perturbations of the topology, disjoint closed cones are used as watermark hosts. After logic synthesis, some disjoint closed cones in the optimized circuit are selected, and re-mapped based on signature bits. In order to retrieve the signature, the watermarked copy is needed to be compared against the original master copy.

Embedding Signature in Timing Constraints

The technique proposed in [26] embeds the bits of the watermark in a constraint file as the least-significant bit of the timing constraints on signal delays. It has practically zero overhead on delay. However, the watermark bits can easily be tampered. Moreover, the watermark embedded through this technique lacks in verification possibilities.

Embedding Signature in Bitstream

A widely used approach is to embed the signature of an IP vendor in the form of a watermark in the LUTs of unused CLBs in the CLB array of an FPGA. Thus the signature remains hidden in the configuration bitstream of the FPGA design. Further, the netlist level modification is performed to connect those marked LUTs with the other parts of the design to protect the watermark against reverse engineering. This technique is then extended in [27] for embedding a fingerprint along with the watermark in the LUTs of distinct CLBs of distinct design instances for each buyer. For these techniques, the signatures can be verified from the bitfile core.

In [28], the authors propose to use a master key secret to both the IP vendor and the buyer, to select the LUTs for embedding watermark from the unused LUTs once the design is implemented, thereby requiring negligible change in the original design. The same key is used to extract the watermark, provided the design is verified not to be tampered by an additive attack.

Embedding Signature in HDL Design

Besides the above-mentioned techniques, a signature included in an HDL design is propagated through the design flow irrespective of the architecture. Therefore, such a technique also remains effective for FPGAs. One such technique is signature hosting [29], where unused locations in a memory implemented using a direct lookup are used to store the signature. For example, a memory structure, which is a part of modulo-11 arithmetic, has five (11–15) input patterns, i.e., memory locations unused. The signature embedded through this technique may be identified at the output using some additional extraction logic.

Extracting Feature Vector from an FPGA Device Running an IP

For protection of a bitstream IP which is running on an identifiable device against counterfeiting, a low-cost soft physical hash (SPH) function has been proposed in [30]. A feature vector is extracted from any physical emanation of the target device running the IP to form an SPH. The SPH should be robust against IP preserving transformations such as re-synthesis of design under different sets of constraints or addition of parasitic IP running in parallel. It should also be sensitive to more significant variations of an IP.

7.3.2 Techniques for Verification of Signatures in FPGA

There are several works which emphasize verification issues of the watermark embedded in FPGA design. Centralized verification team may be biased and untrustworthy. Distributed verification with a centralized database requires secure site and incurs communication overhead to access centralized database. The most convincing trustworthy way of verifying an embedded signature is public verification of the signature of the IP vendor by an authorized buyer from the IP. However, for the widely used techniques of embedding signature in bitstream, the signature does not remain secure in case of public verification as the locations of the signature and the signature pattern are revealed during such verification. Therefore, there is a trade-off between security and trustworthiness for verification of a signature in FPGA design.

Trustworthy Verification of Signatures

The two-level verification method discussed in [31] works by embedding a separate public watermark for public detection. The public mark consists of a short header (company name) and a long message body generated by applying a public one-way hash function on the header. During public verification which is applied on the public mark, the locations of the public mark are revealed. Since the mark is embedded as a bitstream of LUTs of a programmable FPGA, the attacker can easily tamper the mark or override the header and the message body in his favor to result in wrong identification of the IP vendor. Therefore, although the technique facilitates convincing public verification, it is not secure.

Secure Verification of Signatures

Recent watermark verification techniques collect watermark information either from a port, or in the form of power or electromagnetic emanations from an FPGA chip. The work in [32] generates a signature-based power consumption pattern using a shift register, measures the corresponding voltage at power supply pins of the FPGA using an oscilloscope, and quantizes this voltage pattern into the signature. In order to keep the extracted signature encrypted at the user end and thereby secure, the authors in [33] suggest encoding a security tag with a spread-spectrum code followed by its transmission as temperature signal which is then measured using a thermocouple. Both the techniques use a centralized signature database for verification of a signature.

Trustworthy Yet Secure Verification of Signatures

In order to find an amicable solution to facilitate trustworthy yet secure verification of a signature hidden in an FPGA design, the concept of the zero-knowledge protocol is applied in [34]. Zero-knowledge protocol (ZKP) is a an interactive method between two parties or a two-person game, involving several rounds so that the prover can prove to the verifier that a "statement is true," without revealing any knowledge other than the veracity of the statement. The entire proof is split into two parts, say, P_1 and P_2. There are several rounds, in each of which the verifier randomly picks one of the two parts and asks the prover to prove that part. The prover is declared successful if the part provided by the verifier in each round is proven correctly. Failure in any one round is taken as failure in verification of the signature.

In the ZKP-based signature verification technique *Verify_ZKP*, the watermark is obtained from the public information (say, company name) I_{Pb} of the IP vendor, encoded with a function h_s representing s shifts of a nonlinear feedback shift register (NLFSR). It is embedded as the configuration bitstring of some of the unused CLBs of an FPGA based on a secret key K_C.

The encoding key and the locations of the watermark remain private to the IP vendor. *Verify_ZKP* proves with trust that the "desired watermark is present" in the bitfile core D of the FPGA design without revealing the locations or the bitstring in the mark. The watermark is verified in a mapped version of D to keep the mark secure. For verifying a watermark in a mapped design, the proof part P_1 is verification of the mapping function, whereas the part P_2 is verification of the presence of the public information I_{Pb} in the watermark from the mapped design. In each round r,

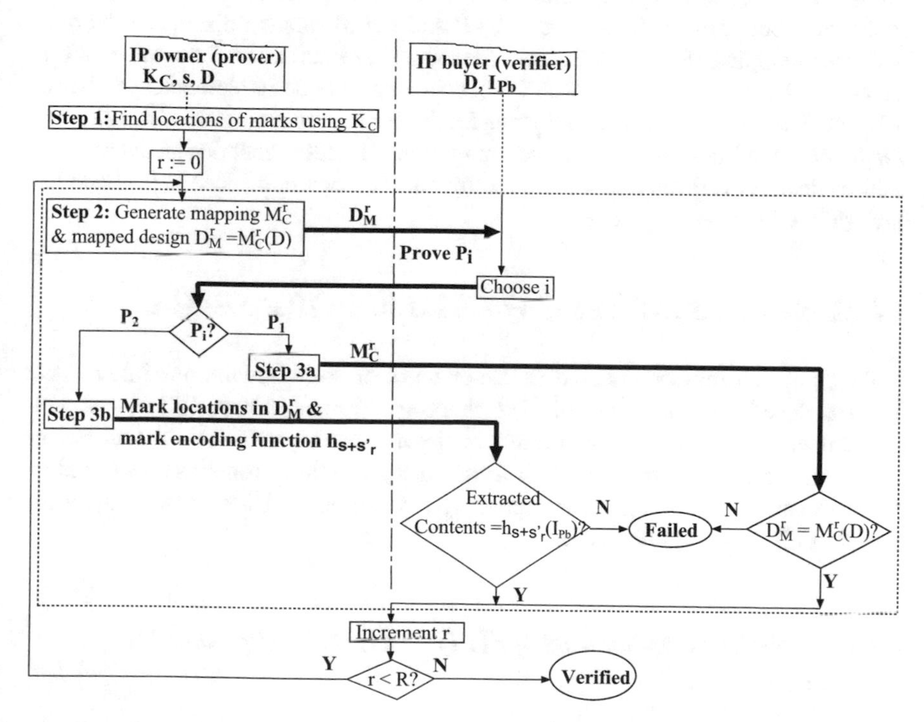

Fig. 7.3 Overview of interactive zero-knowledge protocol *Verify_ZKP* [34] for secure yet trustworthy signature verification

the prover (IP vendor) generates a distinct mapped design D_M^r from D and commits it to the verifier (buyer). A genuine IP vendor succeeds to prove P_1 or P_2 as demanded by the verifier in that round. If the core D does not have the desired watermark, or the committed (marked) design is not a mapped version of D, the probability of success in a round is $1/2$ in either case. Therefore, after sufficiently large number of rounds R, the cheating prover can succeed with a very low probability $1/2^R$. The main steps of the ZKP-based verification protocol *Verify_ZKP* are illustrated in Fig. 7.3.

In order to enforce zero-knowledge property, the mapping function M_C^r in each round r should not be self-mapping, should be distinct, and have low correlation with the mappings in the other rounds. Mapping in each round consists of two main steps: (i) location mapping which generates a different assignment of the CLBs to the CLB locations, and (ii) content encoding which encodes the configuration bitstring of each CLB separately. For location mapping, space-filling curves and Latin rectangles are used to achieve the desired properties enlisted above. For content encoding, the shifts of the NLFSR used differ for each round. The time complexity for *Verify_ZKP* is linear in the size of the design. The embedded watermark is robust against typical attacks like tampering or deletion, finding ghost signature and additive attack. Using partial reconfiguration, different mappings can be configured for verification. The strength of *Verify_ZKP* is determined by the Pearson's product–moment correlation coefficients between the location of a CLB and the Manhattan distance to its location after mapping, for all CLBs over 20 rounds of interaction. The values are of the order of 10^{-2} for FPGA IWLS'05 benchmark designs implemented by Xilinx ISE tool. Similar correlation coefficients for the content encoding are also quite low. *Verify_ZKP* facilitates public verification without any additional design overhead. It reduces design overhead due to marking by 56.2 % when compared to [31] and has negligible CPU time requirement.

7.4 Insertion and Detection of HTH in an IP for FPGAs

A Trojan circuit may be inserted in one or more of the following possible modes: (i) HDL description of a design, (ii) technology mapped design, (iii) placed-and-routed design, or (iv) the configuration bitstream for a target FPGA. An adversary aims at inserting a Trojan in an FPGA-based design so that it remains undetected by the design and validation tool corresponding to the product. Different techniques for detecting HTH are highlighted below.

7.4.1 *RO-Based Detection of HTH in HDL Design and Its Challenges*

A widely used technique for detection of a Trojan inserted in the HDL design is to use ring oscillators (RO) as a locking mechanism for binding an FPGA design to a specific area of FPGA hardware. This results in a specific physical placement of the design on the hardware. A ring oscillator is a delay loop circuit, typically composed of wires and inverters, that oscillates at a particular frequency, which is very sensitive to wire length, gate delay, and process variation. When malicious modifications (hardware Trojans) are inserted in the design, the placement gets altered. Thus the altered wire length of the ring oscillator leads to a discrepancy in its frequency which can be detected by ModelSim HDL simulator. However, the authors in [35] were able to circumvent the ring oscillator-based protection mechanism by (i) design Lockdown approach that keeps the locations of the ring oscillators fixed by applying additional P&R constraints using Xilinx PlanAhead, and more successfully by (ii) a ring oscillator emulation approach that reproduces the functionality of the ring oscillators with a lookup table, in spite of the presence of small Trojans in the design. The Trojans remain undetected for a limited challenge–response set. The impact of the ring oscillator emulator module on area is minimal, requiring 1 % more F/Fs, 5 % more LUTs, 4 additional BRAMs and it causes 15 % increase in delay.

7.4.2 Parity-Based Detection of HTH Inserted in HDL, Mapped or P&Red Design

The authors of [1] propose an IP protectin (IPP) technique for detection of tampers such as changes, deletion of existing logic, and addition of extraneous logic such as Trojans, inserted in FPGA design files. The technique is parity-based and uses an error-correcting code structure for this purpose. For each test vector, the parity of outputs of the CLBs in a parity group (PG) produces one parity bit; For a test set, a parity vector (PV) is generated for each PG. During a trust-checking phase, a test-pattern generator (TPG) and an output response analyzer (ORA) are configured in FPGA. The TPG is connected to the inputs of each PG of CLBs, one at a time, and it feeds identical input/test vectors to each CLB in a parity group, while the output vector produced by the ORA is checked against the expected PV for this PG (Fig. 7.4). Failing to detect a desired parity relation signals the possible existence of additional circuitry, i.e., Trojan in the FPGA design. The technique uses two-level randomization: (a) randomization of the mapping of the parity groups to the CLB array, and (b) randomization within each parity group of odd and even parities for different input combinations. The two-level randomization is meant to counter the attacks by an adversary who tries to either detect the parity groups and inject tampers to mask each other, or tamper with the TPG and the ORA in an undetectable manner.

This method using an underlying error-correcting code and its 2-level randomization was validated by inserting 1–10 circuit CLB tampers and 1–5 extraneous logic CLBs in two medium-size circuits and a RISC processor circuit implemented on a Xilinx Spartan-3 FPGA. The results of 100 % tamper detection and 0 % false alarms, obtained at a hardware overhead of only 7–10 %, were promising. This technique can detect extraneous logic implemented completely in some unused CLBs, as it maps the error-correcting code in all the CLBs, irrespective of functional or not. Support to partial reconfiguration in modern FPGAs facilitates the TPG to be connected to the inputs of each PG of CLBs.

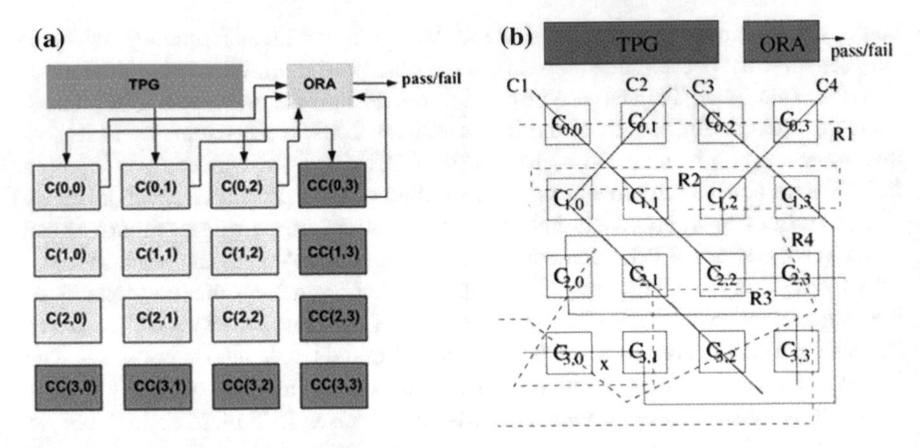

Fig. 7.4 Detection of Trojans [1] in FPGA using error-correcting code: **a** test-pattern generator (TPG) for sending test vectors to each CLB of a parity group and output response analyzer (ORA) for checking responses against desired parity vector, **b** column and row parity groups in solid and dashed line, respectively

7.4.3 *Trojans Inserted in Bitstream and the Challenge of Detection*

The authors in [36] propose inclusion of Trojans directly in the unencrypted bitstream for an FPGA. The process of inclusion has the difficulty of understanding the structure or format of the bitfile, but has the advantage of bypassing all the checks in the FPGA design tool except CRC, which can also be disabled. The Trojan circuits are based on ring oscillator, which have the effect of increasing the operating temperature, and hence causes increased device aging. The bitstrings corresponding to these Trojan circuits are inserted into appropriate locations in the configuration bitstream corresponding to some unused CLBs. Two different types of Trojans have been inserted, namely (i) isolated Trojans, whose insertion is quite easy and (ii) Trojans connected with the original design, which requires appropriate modification at several locations in the bitstream for their insertion.

7.4.4 *Tamper or Fault Injection in FPGAs and Overhead of Detection*

Applying radiation on an FPGA may cause bit flips in its memory blocks resulting in a change in its functionality. If the supply voltage or the external clock is altered intentionally, it may induce glitch and introduce faulty operations. These attacks may change or leak secret information embedded in the FPGA. Constant monitoring, or applying tamper detection for clock pulse or voltage entails additional overhead on the process.

7.5 Security in SoCs and Other Advanced Architectures on FPGAs

Let us discuss now the major security issues in FPGA-based advanced architectures such as system-on-chips, embedded systems, and cloud architectures.

7.5.1 Security in SoC

An SoC usually contains reusable IPs, based on either ASIC or FPGA, embedded processor(s) (a general-purpose processor and multiple special-purpose processors depending on the requirements) or controller(s), memory elements such as SRAM, ROM, bus architecture for interfacing IPs and other components on SoC, programmable blocks (FPGA). Figure 7.5 shows the components of an SoC, an SoC configured on FPGA, and the way of IP reuse in SoC environment.

There are two possible ways of using FPGAs in an SoC. The first is inclusion of FPGA blocks on a system along with other ASIC components, where the

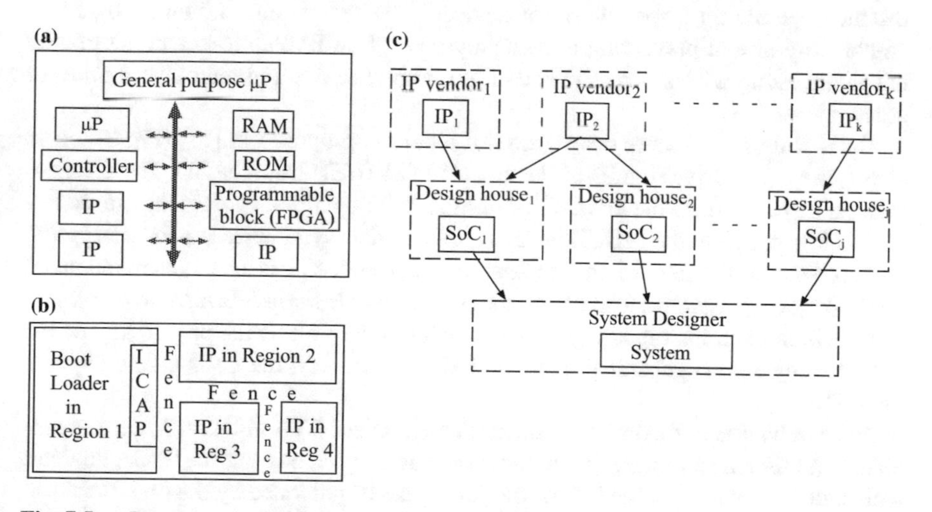

Fig. 7.5 **a** Programmable FPGA with other components on an SoC, **b** entire SoC configured on an FPGA chip [6], **c** reuse of IPs in SoCs and in systems

reprogrammable block facilitates reprogrammable interfacing between the IPs, and reconfiguring of different cryptoprocessors or various FPGA IPs based on applications. The second one is to configure an entire SoC with all FPGA-based IPs on a single FPGA chip. As FPGA-based systems are growing increasingly complex, modular, IP-based approach is becoming popular.

In the context of FPGA-based IP security on an SoC, several parties are involved:

- IP core vendor (CV),
- FPGA vendor (FV),
- system developer (SD),
- user of the system and sometimes a trusted third party (TTP).

The prospective adversaries of an FPGA IP core are the system developer and the user of the system.

For an FPGA-based SoC, an SD receives multiple IPs from various CVs, therefore the IP cores may be intercepted during their distribution. Further, multiple IPs from different CVs are simultaneously in operation. Hence, the presence of a Trojan infected adversary IP may (i) affect the other IPs, (ii) intrude in the controller zone, or (iii) extract valuable information from the communications among the IPs.

Secure Distribution of FPGA IP Cores

The first requirement in SoC design is to ensure secure distribution of FPGA IP cores to the system developer. The objectives are to (i) assure confidentiality and authenticity of the core, (ii) limit the number of instances of the core, and (iii) make every instance operate on a specific set of devices. The second and the third objectives combinedly aim at preventing over-deployment of an IP. Public-key cryptography is applied along with symmetric key cryptography to ensure secure distribution of cores.

One solution for secure distribution of a core is proposed in [37] (Fig. 7.7a). In step 1, the CV receives the ID of the target FPGA (FID) from the SD. The CV generates a key K_{CV} from the private key of his private–public-key pair, the public key of FVs key pair, and the FID. This key K_{CV} is used as symmetric key for encrypting the bitstream of the IP core. In step 2, at the SD's end, K_{CV} is again generated using public-key cryptography. The FV provides a *personalization bitstream* in encrypted form, which contains the key generation function as well as the private key of the FV. The key K_{CV} is generated from the FV's private key, the CV's public key and the FID.
In order to handle different IP cores from independent CVs, different K_{CV_i}s are generated and used accordingly. When the same core is used on multiple FPGA devices, a distinct $K_{CV_{ij}}$ based on the FID of the jth device is generated by the ith CV.

Another solution is symmetric key-based pay-per-use licensing scheme [38] (Fig. 7.6b). A TTP acts as a metering authority (MA) to generate a license to a SD to use an IP core only once at a small fee paid by the SD to the CV. The SD does not

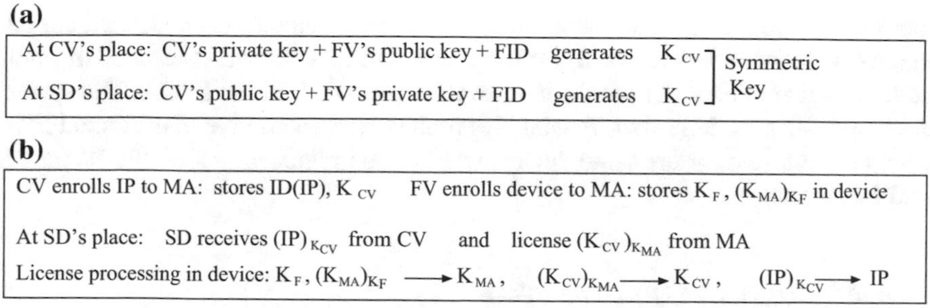

(a)

At CV's place: CV's private key + FV's public key + FID generates K_{CV}] Symmetric Key

At SD's place: CV's public key + FV's private key + FID generates K_{CV}]

(b)

CV enrolls IP to MA: stores ID(IP), K_{CV} FV enrolls device to MA: stores K_F, $(K_{MA})_{KF}$ in device

At SD's place: SD receives $(IP)_{K_{CV}}$ from CV and license $(K_{CV})_{K_{MA}}$ from MA

License processing in device: K_F, $(K_{MA})_{KF} \longrightarrow K_{MA}$, $(K_{CV})_{K_{MA}} \longrightarrow K_{CV}$, $(IP)_{K_{CV}} \longrightarrow$ IP

Fig. 7.6 Secure core distribution protocols **a** described in [37], **b** described in [38]

need to make a large payment for indefinite use of the IP, so his IP remains protected from over-use (overbuilding).

In step 1, each CV enrolls each of his IP cores with the MA, when the MA stores the ID(IP) and the secret key K_{CV} of the CV. Similarly, each FV enrolls each of his devices with the MA, when the MA programs a unique device secure key K_F into the nonvolatile memory (NVM) of the device. The secret key K_{MA} of the MA encrypted with K_F is also incorporated so that K_{MA} can be generated within the device at the time of license processing. In step 2, in order to build an IP on a particular FPGA device, the SD receives the bitstream IP encrypted with K_{CV} from CV, but he needs a license from the MA. So, the MA sends the license containing K_{CV} encrypted with K_{MA}. In step 3, during license processing, first K_{MA} is generated as mentioned above, and then K_{CV} is extracted from the license for loading the bitstream IP on the device.

If the FPGA IPs are in HDL form and the system integrator needs to apply *multiple* EDA tools to synthesize (and P&R) the IPs into a single chip, the encryption scenario is more complex. One possible way is to encrypt each IP using the secret key K_{CV} of its vendor, then encrypt K_{CV} using the public key of the vendor of the EDA tool, and send both the encrypted data to the EDA tool.

The output of an EDA tool for a design phase is sent to another EDA tool for the next design phase. This scenario has the following security risks:

(i) The secret keys of the CVs of many IPs involved, may be extracted by an untrusted EDA tool.
(ii) One IP vendor may recover one part of the synthesized output netlist.

In order to overcome the risk in (ii), either *All-or-Nothing* principle is used, or a separate set of keys other than the K_{CV}s is used to send the output netlist to the next EDA tool. However, the risk in (i) can only be tackled if end-to-end security can be established from core to device using PUF response of the device. For example, a pre-routed IP core is encrypted using the public part of a key, obtained from the IP core and the PUF response of the device [39].

The work in [40] proposes a method for relocation of partial bitstream IPs on a device. It ensures enhanced security in IP exchange as it does not disclose to the IP vendor the information about the FPGA design on which an IP is to be deployed. The method receives information about the resource requirements of the IP and the bus macros at the boundary of the IP from the IP vendor. It calculates the value of frame address register (FAR) of the desired location for the IP in the FPGA. The system developer SD may deploy an IP using this protocol onto a number of different FPGA designs reconfiguring the same device, without communicating with the IP vendor multiple times.

Spatial Separation of IPs on an FPGA

When an entire SoC is configured on an FPGA, various IPs obtained from different IP vendors may not be trustworthy. The system developer needs to partition the FPGA to allocate space to each of these FPGA-based IPs. The partitions for various cores should be hard (not flexible). By controlling the mapping onto the device at the floorplanning stage, spatial isolation of IP cores can be maintained in FPGA. Fences containing buffers or unused logic are placed between the IPs to isolate their regions. Fences are wide enough so that a single-bit failure in configuration cannot connect the neighboring partitions. Continuous monitoring on the configuration data, known as bitstream scrubbing, is applied particularly in the isolation fences to detect any bit flip. Restricted use of FPGA interconnects is allowed through the fence for communication between the IPs. Increased modularization reduces the possibility of interference and enhances the ease of checking correctness.

Configuration Controller in a Trusted Zone

A cryptoprocessor, which is a special-purpose processor to execute a cryptographic algorithm, ensures secure communication of data between the IPs. In addition to secure communication, for the purpose of authentication of IPs and protection of its firmware, configuration bitstream IPs and data from external attacks as well as from attacks across the fence, a cryptographic coprocessor (acting as cryptographic kernel) is designed and placed in an isolated partition. This partition implements the idea of a trusted platform module (TPM) available in hardcoded SoCs. This module provides a trusted environment to initiate the system.

In the case of FPGAs, this module primarily acts as a configuration controller and performs the above-mentioned tasks. For Xilinx, it is termed as "boot loader." Authenticated encryption (AE) mechanism ensures trust for it. This trusted controller then applies the similar concept of trust for loading other FPGA IPs.

An FPGA-based coprocessor is more flexible as well as efficient. This trusted controller is often realized in the FPGA fabric itself, rather than using FPGA functions to provide more flexibility in choosing the decryption algorithm and handling keys. Any malicious modification of this trusted module containing a coprocessor and the firmware, by an external source, is prohibited. For modern reconfigurable FPGAs, the trusted controller clears the present application in case of any external attempt to configure a device. But, access to the configuration controller from the internal configuration bitstream is allowed through ICAP for self-reconfiguration. As a consequence, proper security mechanisms are enforced at this level.

Secure Communication Between the IPs

IP cores may communicate through either shared memory, or direct/RF communications, or a shared bus. In the case of shared memory, a reference monitor and static analysis for direct connection enforce security. A reference monitor is a control mechanism that possesses three properties: it is self protecting, its enforcement mechanism cannot be bypassed, and it should be correct and complete. It enforces legal sharing of memory among the cores.

In order to guarantee secure communication of data through the shared bus between the IPs on an FPGA-based SoC or by RF across the SoCs, encrypted transmission of data is required. FPGA-based designs for symmetric key encryption AES have been discussed earlier in Sect. 7.2.1. Here, we focus on the FPGA implementations of some recent cryptoprocessors for asymmetric key algorithms, such as RSA, elliptic curve cryptography (ECC), and a few other special types of curves. ECC provides an equally strong security level with a smaller key length, in comparison to RSA. ECC processors on a prime field are important from the aspect of security.

The characteristics of a few fast and/or compact FPGA-based ECC processors are given in Table 7.3. In [41], ECC implementation over $GF(2^{163})$ takes only 5.1 μs in Virtex-5 and 3.5 μs in Virtex-7 for an ECC point multiplication (pm). Roy et al. in [42] implement ECC over $GF(2^{163})$ which consumes only 3513 slices in Virtex-5 to perform an ECC pm in 9.5 μs. The design of an FPGA-based ECC processor [43]

Table 7.3 FPGA implementations of ECC and ARX processors

Processors	Device	Throughput	Resource
ECC (fastest, binary field) [41]	Virtex-7	286×10^3 pm/s	8736 slices
ECC (fast, binary field) [42]	Virtex-5	105×10^3 pm/s	3513 slices
ECC (fast, prime field) [43]	Virtex-4	37000 pm/s	24452 slices, 468 DSPs
ECC (smallest) [44]	Virtex-5	90.1 pm/s	81 slices, 22 BRAMs, 8 DSP multipliers
ECC (compact) [45]	Virtex-II	63 pm/s	2085 slices, 9 BRAMs, 7 multipliers
Salsa20-sr	Virtex-II	38 Mbps	193 CLBs
ChaCha_config2	Virtex-6	595 Mbps	77 CLBs, 2 BRAMs
BLAKE-64	Virtex-5	314 Mbps	108 CLBs, 3 BRAMs
Skein	Virtex-6	179 Mbps	240 CLBs

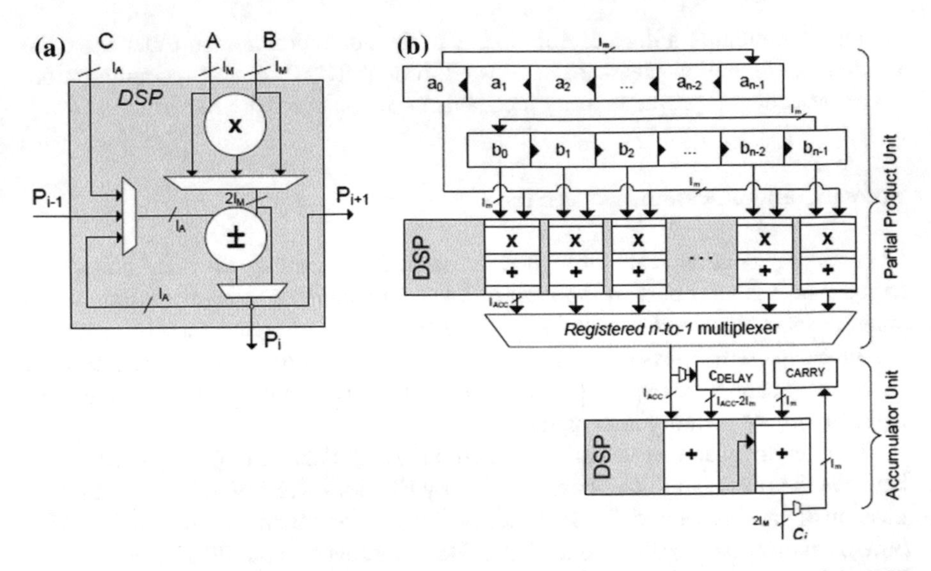

Fig. 7.7 FPGA-based design of an ECC cryptoprocessor [43]: **a** simplified structure of the DSP blocks in advanced FPGAs, **b** l-bit multiplication circuit with a cascade of parallel DSP blocks

for P-224 (i.e., the key length is 224) using DSP cores with other FPGA resources (Fig. 7.7) provides the highest speed among the ECC processors over prime field. Among the area efficient compact implementations, the details of the designs in [44] and in [45] have been provided in Table 7.3. The ECC design [44] with less than 100 slices may be considered as the smallest or most compact one implemented in modern Virtex-5 device. Benaissa et al. in [46] achieved throughput/slice figures of 19.65, 65.30, and 64.48 (10^6/(sec $\times$ slices)), respectively on Virtex-4, Virtex-5, and Virtex-7 FPGAs.

The implementations discussed so far are made resistant against simple power analysis at the algorithmic level, using the Montgomery ladder for modular multiplication [47], and by making the number of integer additions and subtractions independent of the input values in the modular addition/subtraction component. In order to ensure resistance against DPA, several models for elliptic curve, such as Edward curve, binary Huff curve, have been employed. The details of DPA-resistant FPGA implementations of Edward curve are given in several works, e.g., [48], and those of Huff curves are in the works like [49]. These details appear in another chapter.

Cryptographic algorithms based on addition (A), rotation (R), and exclusive-or (X) operations are classified as ARX algorithms. Many FPGA implementations of the cryptoprocessors supporting ARX-based cryptographic primitives are present in the literature [50] as given in Table 7.3.

7.5.2 Security in Embedded Systems

The major application of an SoC is in embedded systems, where the hardware components along with the software components are embedded in a larger electrical or mechanical system, termed as cyber-physical systems. In an embedded system, an entire OS and several security-sensitive software are loaded in a processor which is kept in a trusted zone to keep it separated from other application-specific coprocessors, certain reconfigurable and/or hardcoded chips, and software processes. The processor controls access to the application-specific hardware IPs by the other software components. In an FPGA-based embedded system, the processor and the application-specific coprocessors are configured in the FPGA. In addition to reverse engineering and cloning, code injection is an additional important attack in an FPGA-based embedded system.

In an FPGA-based embedded system, the configuration controller or the boot loader in the processor decrypts and authenticates not only the configuration data but also the software using session keys. The processor along with cache, memory, and peripherals may remain hardcoded on a system, and connected with one another as well as the FPGA programmable logic using AMBA AXI buses.

For example, a multi-core ARM processor exists in the Zynq family of Xilinx. Then, the cryptoprocessor module of the processor is generally configured in the FPGA programmable logic. Even the decryption and authentication units are built in the FPGA fabric to allow an user to choose his decryption and authentication algorithms, as well as to perform authentication before decryption and reduce the risk of side-channel attack.

Extraneous Circuits to Enforce Security in Embedded Systems

The work in [51] assumes that the entire hardware component of an embedded system is in FPGA, and a software executes on an FPGA hardware only if the hardware manifests a uniqueness based on process variation and device aging. A benign hardware Trojan (BHT) is designed as delay-logic arbiters and is implemented in an FPGA platform supporting reconfigurability, in order to realize all the required device aging. The output of the BHT exploiting process variation and device aging of the hardware either enables or disables writing to particular general-purpose registers (GPR). Thus the BHT embedded in the FPGA guarantees that the software is authorized to execute on the hardware and vice versa. It is shown in [51] that the worst-case performance penalty is 8 % for zlib benchmark with 22 GPRs.

Security in Embedded Systems Having FPGA-Based MPSoCs

McIntyre et al. in IOLTS 2010 first proposed the use of a distributed software scheduling algorithm to avoid low trust cores in a *hardcoded* multi-core processor. Later on, trustworthy systems in FPGAs with multi-core processor or multi-processors were adopted.

Nowadays, FPGAs are sufficiently large to host one or more multi-processor system-on-a-chip (MPSoCs). For the sake of security, several processors are segregated into a number of domains. Each software on an embedded system is allowed to access only the processors of some specific domains, not all. Software solution, such as real-time operating system for time and space partitioning, is the state-of-the-art technique for domain segregation. The authors in [52] propose a robust, reliable, and efficient architecture, which targets reconfigurable platform to offer hardware-enforced segregation. This domain segregation is achieved with a hierarchical connection of memory buses by secure bus bridges.

7.5.3 Security in Cloud Architecture

Among the modern architectures in FPGAs, the cloud architecture is growing popular, after SoCs and embedded systems. In secure yet fast cloud computing, computation logic targeting a secure hardware is separated from the code for I/O and coordination, which may run on an untrusted hardware. A tamper-proof FPGA with small feature size is suitable for a computation-specific processing chip. FPGA architecture, due to its partial reconfigurability, provides better power-performance ratio and does not have the security threats caused due to cache sharing [53]. The FPGA-based computation circuit has a strong secure guarantee. Further, the communication channel between a processor chip and a state chip for storing a state across power cycles, such as in a smart card, is also made secure. In order to prevent side-channel attacks between the circuits sharing an FPGA, a supervisor module such as a TPM is used.

7.6 Summary

In the domain of FPGAs, several IP threats are still alarming. Most of the crypto units in FPGA are side-channel cracked. FPGA vendor-specific configuration bitstream format, support to partial reconfigurability in modern FPGA, enhanced demand for FPGAs on system-on-chip, increase in remote computing are the sources of open challenges for IP and SoC security in FPGAs.

References

1. Dutt, S., Li, L.: Trust-based design and check of FPGA circuits using two-level randomized ECC structures. ACM Trans. Reconfig. Technol. Syst. **2**(1) (2009)
2. http://homepages.cae.wisc.edu/~ece554/website/Lectures/Xilinx_Vertex_Tech_s03.pdf

3. Qu, G., Potkonjak, M.: Intellectual Property Protection in VLSI Designs: Theory and Practice. Springer, Heidelberg (2003)
4. Drimer, S.: Volatile FPGA design security–a survey. http://www.cl.cam.ac.uk/techreports/UCAM-CL-TR-763.pdf (2008)
5. Majzoobi, M., Koushanfar, F., Potkonjak, M.: FPGA-oriented security. In: Introduction to Hardware Security and Trust, Chapter 1. Springer (2011)
6. Trimberger, S.M., Moore, J.J.: FPGA security: motivations, features, and applications, invited paper. Proc. IEEE **102**(8) (2014)
7. McNeil, S.: Solving Today's Design Security Concerns, Xilinx White paper FPGAs, WP365 (v1.2) July 30 (2012)
8. Menezes, A., Oorschot, P., Vanstone, S.: Handbook of Applied Cryptography. CRC Press (1996)
9. Good, T., Benaissa, M.: AES on FPGA from the fastest to the smallest. In: CHES 2005: Proceedings of International Conference on Cryptographic Hardware and Embedded Systems, LNCS 3659, pp. 427-440. Springer (2005)
10. Granado-Criado, J.M., Vega-Rodríguez, M.A., Sánchez-Pérez, J.M., Gómez-Pulido, J.A.: A new methodology to implement the AES algorithm using partial and dynamic reconfiguration. Integr. VLSI J. **43**(1), 72–80 (2010)
11. Chu, J., Benaissa, M.: Low area memory-free FPGA implementation of the AES algorithm. In: FPL 2012: Proceedings of International Conference on Field Programmable Logic and Applications, pp. 623–626 (2012)
12. Hori, Y., Katashita, T., Sakane, H., et al.: Bitstream protection in dynamic partial reconfiguration systems using authenticated encryption. IEICE Trans. Inf. Syst. **E96-D**(11), 2333–2343 (2013)
13. Abdellatif, K.M., Chotin-Avot, R., Mehrez, H.: Improved method for parallel AES-GCM cores using FPGAs. In: Proceedings of International Conference on Reconfigurable Computing and FPGAs, pp. 1–4 (2013)
14. Teegarden, C., Bhargava, M., Mai, K.: Side-channel attack resistant ROM-based AES S-box. In: HOST 2010: Proceedings of IEEE International Symposium on Hardware-Oriented Security and Trust, pp. 124–129 (2010)
15. Drimer, S., Kuhn, M.G.: A protocol for secure remote updates of FPGA configurations. In: Proceedings of International Workshop on Applied Reconfigurable Computing, Reconfigurable Computing: Architectures, Tools and Applications, pp. 50–61. Springer, Berlin (2009)
16. Vliege, J., Mentens, N., Verbauwhede, I.: A single-chip solution for the secure remote configuration of FPGA using bitstream compression. In: Proceedings of International Conference on Reconfigurable Computing and FPGAs, pp. 1–6 (2013)
17. Adi, W., Ernst, R., Soudan, B., Hanoun, A.: VLSI design exchange with intellectual property protection in FPGA environment using both secret and public-key cryptography. In: ISVLSI 2006: Proceedings of IEEE Computer Society Annual Symposium on VLSI, pp. 24–29 (2006)
18. Morozov, S., Maiti, A., Schaumont, P.: An analysis of delay based PUF implementations on FPGA. In: Proceedings of International Conference on Reconfigurable Computing: Architectures, Tools and Applications, pp. 382–387 (2010)
19. Pappala, S., Niamat, M., Sun, W.: FPGA based trustworthy authentication technique using physically unclonable functions and artificial intelligence. In: HOST 2012: Proceedings of IEEE International Symposium on Hardware-Oriented Security and Trust, pp. 59–62 (2012)
20. Majzoobi, M., Koushanfar, F., Potkonjak, M.: Techniques for design and implementation of secure reconfigurable PUFs. ACM Trans. Reconfig. Technol. Syst. **2**(1) (2009)
21. Majzoobi, M., Koushanfar, F., Devadas, S.: FPGA-based true random number generation using circuit metastability with adaptive feedback Control. In: CHES 2011: Proceedings of Cryptographic Hardware and Embedded Systems, LNCS 6917, pp. 17–32. Springer (2011)
22. Varchola, M., Drutarovsky, M., Fischer, V.: New universal element with integrated PUF and TRNG capability. In Proceedings of International Conference on Reconfigurable Computing and FPGAs, pp. 1–6 (2013)

23. Yip, K., Ng, T.: Partial-encryption technique for intellectual property protection of FPGA-based products. IEEE Trans. Consum. Electr. **46**(1), 183–190 (2000)
24. Kirovski, D., Hwang, Y., Potkonjak, M., Cong, J.: Protecting combinational logic synthesis solutions. IEEE Trans. Comput.-Aided Des. Integr. Circuit Syst. **25**(12), 2687–2696 (2006)
25. Cui, A., Chang, C.H., Tahar, S.: IP watermarking using incremental technology mapping. IEEE Trans. Comput.-Aided Des. Integr. Circuit Syst. **27**(9), 1565–1570 (2008)
26. Jain, A., Yuan, L., Puri, P., Qu, G.: Zero overhead watermarking technique for FPGA designs. In: GLSVLSI 2003: Proceedings of ACM Great Lakes symposium on VLSI, pp. 147–152 (2003)
27. Lach J., Mangione-Smith, W.H., Potkonjak, M.: Fingerprinting techniques for field-programmable gate array intellectual property protection. IEEE Trans. Comput.-Aided Des. Integr. Circuit Syst. **20**(10), 1253–1261 (2001)
28. Saha, D., Sur-Kolay, S.: Robust intellectual property protection of VLSI physical design. J. IET Comput. Dig. Tech. **4**(5), 388–399 (2010)
29. Castillo, E., Meyer-Baese, U., Garcia, A., Parrilla, L., Lloris, A.: IPP@HDL: efficient intellectual property protection scheme for IP cores. IEEE Trans. Very Large Scale Integr. (VLSI) Syst. **15**(5), 578–591 (2007)
30. Kerckhof, S., Durvaux, F., Standaert, F., Gerard, B.: Intellectual property protection for FPGA designs with soft physical hash functions: first experimental results. In: HOST 2013: Proceedings of IEEE International Symposium on Hardware-Oriented Security and Trust, pp. 7–12 (2013)
31. Qu, G.: Publicly detectable techniques for the protection of virtual components. In: Proceedings of Design Automation Conference, pp. 474–479 (2001)
32. Ziener, D., Teich, J.: Power signature watermarking of IP cores for FPGAs. J. Signal Process. Syst. **51**(1), 123–136 (2008)
33. Kean, T., McLaren, D., Marsh, C.: Verifying the authenticity of chip designs with the design tag system. In: HOST 2008: Proceedings of IEEE International Workshop on Hardware-Oriented Security and Trust, pp. 59–64 (2008)
34. Saha, D., Sur-Kolay, S.: Secure public verification of IP marks in FPGA design through a zero-knowledge protocol. IEEE Trans. VLSI (VLSI) Syst. **20**(10), 1749–1757 (2012)
35. Rilling, J., Graziano, D., Hitchcock, J., et al.: Circumventing a ring oscillator approach to FPGA-based hardware Trojan detection. In: ICCD 2011: IEEE International Conference on Computer Design, pp. 289–292 (2011)
36. Chakraborty, R.S., Saha, I., Palchaudhuri, A., Naik, G.K.: Hardware Trojan insertion by direct modification of FPGA configuration bitstream. IEEE Des. Test Comput. **30**(2), 45–54 (2013)
37. Drimer, S., Guneysu, T., Kuhn, M.G., Paar, C.: Protecting multiple cores in a single FPGA design. http://www.saardrimer.com/sd410/papers/protect_many_cores.pdf (2007)
38. Maes, R., Schellekens, D., Verbauwhede, I.: A Pay-per-use licensing scheme for hardware IP cores in recent SRAM-based FPGAs. IEEE Trans. Inf. Forensics Secur. **7**(1), 98–108 (2012)
39. Guajardo, J., Guneysu, T., Kumar, S.S., Paar, C.: Secure IP-block distribution for hardware devices. In: HOST 2009: IEEE International Workshop on Hardware-Oriented Security and Trust, pp. 82–89 (2009)
40. Ebrahim, A., Benkrid, K., Khalifat, J., Hong, C.: A platform for secure IP integration in Xilinx Virtex FPGAs. In: International Conference on Reconfigurable Computing and FPGAs, pp. 1–6 (2013)
41. Khan, Z.U.A., Benaissa, M.: High speed ECC implementation on FPGA over $GF(2^m)$. In: FPL 2015: Proceedings of International Conference on Field Programmable Logic and Applications, pp. 1–6 (2015)
42. Roy, S.S., Rebeiro, C., Mukhopadhyay, D.: Theoretical modeling of elliptic curve scalar multiplier on LUT-based FPGAs for area and speed. IEEE Trans. Very Large Scale Integr. (VLSI) Syst. **21**(5), 901–909 (2013)
43. Guneysu, T., Paar, C.: Ultra high performance ECC over NIST primes on commercial FPGAs, In: CHES 2008: Proceedings International Workshop on Cryptographic Hardware and Embedded Systems, LNCS 5154, pp. 62–78. Springer (2008)

44. Basu Roy, D., Das, P., Mukhopadhyay, D.: ECC on your fingertips: a single instruction approach for lightweight ECC design in GF(p). IACR Cryptology ePrint Archive 2015: 1225
45. Vliegen, J., Mentens, N., Genoe, J., Braeken, A., Kubera, S., Touhafi, A., Verbauwhede, I.: A compact FPGA-based architecture for elliptic curve cryptography over prime fields. In: ASAP 2010: Proceedings of IEEE International Conference on Application-specific Systems Architectures and Processors, pp. 313–316 (2010)
46. Khan, Z.U.A., Benaissa, M.: Throughput/area-efficient ECC processor using Montgomery point multiplication on FPGA. IEEE Trans. Circuits Syst. **62-II**(11), 1078–1082 (2015)
47. Cho, S.M., Seo, S.C., Kim, T.H., Park, Y.-H., Hong, S.: Extended elliptic curve Montgomery ladder algorithm over binary fields with resistance to simple power analysis. J. Inf. Sci. **245**, 304–312 (2013)
48. Azarderakhsh, R., Reyhani-Masoleh, A.: Efficient FPGA implementations of point multiplication on binary Edwards and generalized Hessian curves using Gaussian normal basis. IEEE Trans. Very Large Scale Integr. (VLSI) Syst. **20**(8), 1453–1466 (2012)
49. Chatterjee, A., Sengupta, I.: High-speed unified elliptic curve cryptosystem on FPGAs using binary Huff curves. In: VDAT 2012: Proceedings of VISI Design and Test, LNCS 7373, pp. 243–251. Springer (2012)
50. Shahzad, K., Khalid, A., Rkossy, Z.E., Paul, G., Chattopadhyay, A: CoARX: a coprocessor for ARX-based cryptographic algorithms. In: Proceedings of Annual Design Automation Conference, Article No. 133 (2013)
51. Zheng, J.X., Chen, E., Potkonjak, M.: A benign hardware Trojan on FPGA-based embedded systems. In: FPL 2012: Proceedings of International Conference on Field Programmable Logic and Applications, pp. 464–470 (2012)
52. Kliem, D., Voigt, S.-O.: Scalability evaluation of an FPGA-based multi-core architecture with hardware-enforced domain partitioning. Microprocess. Microsyst. (2014)
53. Costan, V., Devadas, S.: Security challenges and opportunities in adaptive and reconfigurable hardware. In: HOST 2011: Proceedings of IEEE International Symposium on Hardware-Oriented Security and Trust (2011)

Chapter 8
Physical Unclonable Functions and Intellectual Property Protection Techniques

Ramesh Karri, Ozgur Sinanoglu and Jeyavijayan Rajendran

8.1 Introduction

Mathematically strong cryptographic primitives and protocols assume that the underlying hardware is trustworthy and rely on them to store secrets. However, because of the vulnerabilities in the hardware, an attacker can retrieve these secret keys [1]. Thus, one needs to prevent an attacker from extracting secret keys from the hardware. Additionally, semiconductor companies invest billions of dollars in designing a chip. Such designs become their intellectual property (IP), and hence, they are called as IP designs or IP cores. However, because of the vulnerabilities in the hardware design flow, one needs to prevent an attacker from stealing IP designs [2]. In this book chapter, we will explore hardware design techniques that can thwart these attacks.

8.1.1 Storing Secret Keys on a Chip

Mobile and embedded devices, which are becoming more ubiquitous day by day, often handle sensitive private information. Such devices need to authenticate the user and the data, and protect against attackers who have physical access to those devices. Furthermore, several security protocols that run on these devices require cryptographic applications such as encryption, which require secret keys. However,

R. Karri (✉) · J. Rajendran
Department of Electrical and Computer Engineering, Polytechnic School of Engineering, New York University, 5, MetroTech Center, Brooklyn 11201, USA
e-mail: rkarri@nyu.edu

J. Rajendran
e-mail: jv.ece@nyu.edu

O. Sinanoglu
Department of Electrical and Computer Engineering, New York University Abu Dhabi (NYUAD), Building C1, Office: 166, NYUAD Campus Saadiyat Island, Abu Dhabi 129188, United Arab Emirates
e-mail: os22@nyu.edu

S. Bhunia et al. (eds.), *Fundamentals of IP and SoC Security*,
DOI 10.1007/978-3-319-50057-7_8

when secret keys are stored in an IC, an attacker can easily retrieve them by performing side channel attacks and/or tampering with the chip. Traditionally, secret keys are stored in a nonvolatile electrically erasable programmable read-only memory (EEPROM) or battery-backed static random access memory (SRAM). Unfortunately, such techniques are not only prone to tampering attacks but also result in tremendous area, power, and delay overhead, thereby increasing the cost of the chip.

To thwart such attacks, researchers have developed a security primitive referred to as physical unclonable functions (PUFs) [3]. PUFs use random, process variations inherent in chip manufacturing to produce keys unique to the chip. PUFs are attractive because one can use them to store secret keys in an efficient and secure way. Section 8.2 details about the two main types of PUFs (weak- and strong-PUFs), security metrics to evaluate their capabilities, their applications and protocols for using PUFs, and the challenges in implementing PUF circuits.

8.1.2 The Need for IP Protection Techniques

Due to the ever increasing complexity and cost of constructing and/or maintaining a foundry with advanced fabrication capabilities, many semiconductor companies are becoming fabless. Such fabless companies design ICs and send them to an advanced foundry, which is usually off-shore, for manufacturing. Also, the criticality of time-to-market has forced companies to buy several IC/IP blocks to use them in their systems-on-chips (SoCs). The buyers and sellers of these IP blocks are distributed worldwide.

As the IC design flow is distributed worldwide today, hardware is susceptible to new kinds of attacks such as counterfeiting, reverse engineering, and IP piracy [4–7]. ICs may be recycled or remarked and sold illegally. An attacker, anywhere in this design flow, can reverse engineer the functionality of an IC/IP. One can then steal and claim ownership of the IP. An untrusted IC foundry may overbuild ICs and sell them illegally. Finally, rogue elements in the foundry may insert malicious circuits (hardware Trojans) into the design without the designer's knowledge [8]. Because of the IP right violations alone, the semiconductor industry loses up to $4 billion annually [2]. The annual losses due to counterfeit ICs, which include recycled, remarked, tampered, and overproduced ICs, are estimated to be about $169 billion [9]. Such attacks have led IP and IC designers to reevaluate trust in hardware [10].

To thwart such attacks, researchers have developed IP protection techniques: watermarking [11], fingerprinting [12], metering [13], logic locking [6, 14], and split manufacturing [15]. Along with these techniques, Sect. 8.3 explains the different classes of IP protection techniques in detail and security metrics used to evaluate their effectiveness.

8.2 Physical Unclonable Functions (PUFs)

8.2.1 Motivation

Physical unclonable functions (PUF) are a cost-effective alternative to storing keys on EEPROM and SRAM that can be used for authentication and key generation [3]. A PUF is an unclonable circuit that uses inherent random variations to produce unique, stable responses (output) for a given challenge (input). Instead of storing the secret, PUFs generate the secret on the fly. PUFs are attractive because: (i) unlike conventional memories, tampering attacks are made difficult, (ii) unlike battery-powered SRAMs, one does not need to keep the PUF circuit on all the time, and (iii) unlike EEPROMs, the PUF circuit does not require an additional layer of mask, thereby reducing manufacturing cost.

8.2.2 Types of PUFs

PUFs are classified based on the number of challenge–response pairs (CRPs) that they generate. The two main types of PUFs are weak-PUFs and strong-PUFs, which are described below. Apart from these two types of PUFs, one can also consider **unique objects** as a type of PUF [16]. Unique objects use random unclonable properties which require an external equipment to measure their responses.

Weak-PUFs

In this type of PUF, the number of CRPs generated is polynomial in the number of components in the PUF circuit. The responses generated by weak-PUFs are used as a "fingerprint" of the chip and/or as the physical keys of the chip. Thus, they are also called physically obfuscated keys. Since the number of CRPs is only polynomial in the number of components in the PUF, these PUFs may not be useful in certain applications such as authentication, which requires a large number of CRPs (in the order of millions).

Example—SRAM-based PUF. Static random access memory (SRAM) cells can be used as a weak-PUF [17]. An SRAM cell, shown in Fig. 8.1, consists of two cross-coupled inverters and access transistors. "A" and "B" are the two nodes in this cell, and the voltages at these two nodes determine the output (response) of the SRAM. When an SRAM cell is powered on, its output transitions to either 0 (AB = 01) or 1 (AB = 10). This transition is usually driven by the strength of cross-coupled inverters, which is, in turn, determined by process variations. The strength of the cross-coupled inverters is dictated by different transistor parameters such as length of the channel (L_{eff}), threshold voltage (V_{th}), dopant concentration, etc.

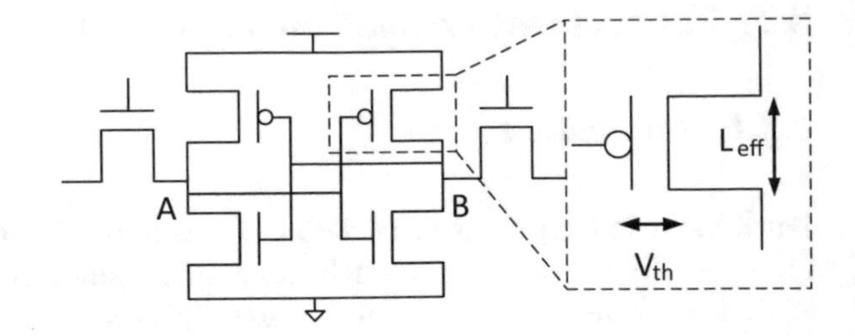

Fig. 8.1 Example of a weak-PUF: state of an SRAM cell can be used as a response bit [17]

Other examples of weak-PUFs include the following:

(i) Butterfly PUF [18]: It uses cross-coupled NAND gates to generate the response. Such PUFs are easy to implement on a field-programmable gate array (FPGA), thereby providing protection for FPGA-based designs.
(ii) Coating PUF [19]: It comprises a matrix material with random distributed dielectric components. Sensors are used to determine the capacitance of the matrix material, and their values are used as responses. This PUF can detect tampering attacks. When an attacker tries to peel off the coat, the capacitance changes. The sensors detect this change, and thus, detect the tampering attack.
(iii) Resistive PUFs [20, 21]: This PUF uses the resistance value of the power lines of an IC as the response of PUF. A specialized circuit is used to extract digital values from the resistance values.

Strong-PUFs

In this type of PUF, the number of CRPs generated is exponential in the number of components, and thus this type of PUF has a large number of CRPs.

Example—Arbiter-based PUF. One example of a strong-PUF is an arbiter-based PUF shown in Fig. 8.2 [22]. An arbiter-PUF consists of N stages, where N is the number of bits in the challenge. Each stage consists of a pair of multiplexers, whose inputs are connected to the outputs of the previous stage as shown in Fig. 8.2. The inputs of the first stage are tied together. The output of the last stage is fed to a D latch, which acts as an "arbiter."

When a challenge is applied to the arbiter-PUF, two paths are selected. A rising edge is then applied to the input of the arbiter-PUF. Due to process variations, the relative speed of the two paths will be different in different chips. Consequently, the latch may hold a 0 or 1. The output of the latch is the response bit. The number of path pairs, and hence the response bits is exponential in the number of stages (or challenges). Thus, this PUF is considered as a strong-PUF.

The disadvantage of the arbiter PUF is that an adversary can model the PUF by obtaining a polynomial number of challenge–response pairs using linear-delay models. This way he can predict the response of the PUF [23]. One can solve this problem by introducing nonlinearity in PUF structure, thereby preventing an adversary from modeling the PUF.

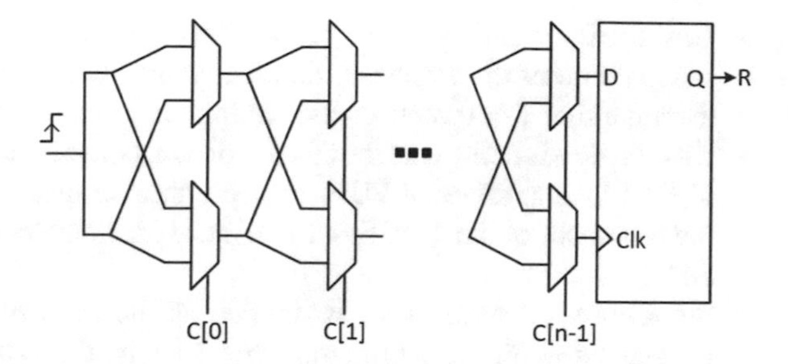

Fig. 8.2 Example of a strong-PUF: arbiter PUF [22]. Based on the challenge, two circuit paths are created. The response bit "R" can be 1 or 0 and depends up on which path is slower

Some of the modified arbiter-PUF architectures include the following:

(i) Feedforward arbiter PUFs [24]: In this type of arbiter-PUF, the number of stages exceeds the number of challenge bits. The output of some of the internal stages act as select lines for some of the other stages. This increases the non-linearity in the PUF circuit, preventing an adversary from modeling the PUF using linear-delay assumptions.
(ii) XOR-based arbiter PUF [23]: An XOR network is used to combine the responses of multiple feedforward arbiter PUFs. The output of this XOR network serves as the response of the PUF. The XOR network introduces nonlinearity into the PUF circuit, preventing an adversary from modeling the PUF.
(iii) Lightweight PUFs [25]: This PUF is a modified version of the feedforward arbiter PUF. To increase the nonlinearity, the challenge bits are fed to an XOR network, whose output drives the individual stages of the arbiter. Furthermore, an XOR network is used to combine the responses of multiple feedforward arbiter PUFs. The output of this XOR network serves as the response of the PUF.

Apart from arbiter-based strong-PUFs, one can also use randomly distributed glass particles on a transparent token as a strong-PUF. This PUF is called an optical PUF [26]. The challenge to this PUF is the angle of incidence of the light. The response of this PUF is the interference pattern. One needs to convert these optical signals to electrical signals, which increases the cost of this PUF, leading to practicality issues.

8.2.3 Security Metrics

One can use the following security metrics to quantify the security of PUFs:

- **Uniqueness** is defined as the Hamming distance between the responses from PUFs in two different circuits upon applying the same challenge. This metric helps one uniquely differentiate an IC from other ICs containing the same PUF structure. Its ideal value should be 50 % because one can then differentiate maximum number of ICs. Note that sometimes this metric is called inter-Hamming distance.

- **Uniformity** is defined as the proportion of 1's and 0's in a response. It ensures the randomness of the response. Its ideal value should be 50 % because any affinity toward either 1 or 0 reduces the randomness in the response.
- **Bit-aliasing** is defined as the affinity of a response bit toward either 0 or 1. Because of bit-aliasing, different PUFs may produce similar response bits. Consequently, the responses of these PUFs will be more predictable. Ideally, the value should be 50 %.
- **Steadiness**, or robustness, is defined as the ratio of response bits that remain unchanged at different time intervals. Ideally, the value for steadiness should be 100 %. Note that this metric is different from the uniformity metric. Steadiness ensures that the responses are stable across different time intervals; uniformity ensures that responses are random, making them unpredictable. Note that sometimes this metric is called intra-Hamming distance.

A comprehensive analysis of the metrics used to evaluate the security of PUFs can be found in [27].

8.2.4 Applications and Protocols

User Authentication Using a Strong-PUF

One can use the strong-PUF to authenticate a user. The protocol is as follows:

1. A strong-PUF is manufactured.
2. The authentication server obtains the strong-PUF. It then applies a set of randomly generated challenges and records the corresponding responses. This set of challenge–response pairs (CRPs) is used to create the CRP table.
3. The user obtains the strong-PUF.
4. Whenever the user wants to be authenticated, he sends a request to the server.
5. The server randomly picks a challenge from the CRP table and sends it to the user.
6. The user applies this challenge to his strong-PUF and obtains the response. This response is sent to the server.
7. The server checks if the received response matches with the one from the CRP table. If so, the user is authenticated; otherwise, the user is not authenticated. Though, in theory, a perfect match is required, in practice a "close" match suffices in order tolerate errors.
8. The server deletes the used CRP from the table.

The last step is needed because an attacker can record the response, while the user sends the correct response, and reuse it later to spoof the server if the server sends the same challenge. Such an attack is known as the replay attack. Since there is a finite number of CRPs in the CRP table, there is a non-negligible probability that the server may reuse the same challenge if the CRP is not deleted. Hence, the used CRPs are deleted to avoid such attacks.

Since the server stores only a finite number of CRPs and a CRP is deleted after being used, the above protocol becomes obsolete when the server runs out of CRPs. To avoid such problems, researchers have proposed to store a compact model of the PUF on the server [28]. The server can use this compact model to produce CRPs on the fly.

Key Generation Using a Weak-PUF

Since a weak-PUF generates unique responses per chip, such responses can be used to generate secret keys [23]. Such secret keys can be used by other security primitives such as an encryption engine or a hash engine. Weak-PUF generated keys are used for certified execution of software on a processor [29]. Weak-PUFs can also be used to prevent piracy of IC designs (See Sect. 8.3 for more details).

8.2.5 Challenges

Attacks on PUFs

Several attacks on PUFs have been proposed in the literature [30–32]. These attacks try to build a simulation model of the PUF, especially a strong-PUF, by monitoring several CRPs. Such attacks use machine learning techniques. Researchers have also developed side-channel attacks on PUFs. In this type of attack, a simulation model of PUF is developed based on its power or delay characteristics. This model is then used to mimic the PUF.

Reliability Issues

The response of the PUFs can vary due to environmental conditions, such as temperature and voltage, and due to aging effects. For example, in the case of arbiter PUF, a change in operating voltage changes the delays of the transistors, which in turn affects the response bits. Such changes in response bits at run time impact the usage of PUFs in security applications. For instance, when a weak-PUF is used to generate secret keys, a change in the response bit results in a different key. Thus, PUFs are required being error-prone.

To make the response of a PUF more reliable, encoding schemes and "helper data" are provided [33–35]. Such schemes tolerate errors and improve security by not leaking sufficient amount of secret information.

8.2.6 Beyond PUFs—Public PUFs (PPUFs)

PPUF is a variant of PUF. Its simulation models are made public [36–38], unlike a PUF whose simulation models are hidden from the attacker. Although an attacker can simulate the PPUF on a given challenge to obtain a response, the simulation time is too large (e.g., several years) compared to the time it takes to apply a challenge and obtain its response on the PUF primitive (e.g., a few nanoseconds).

A PPUF using XOR gates, as shown in Fig. 8.3, is constructed in [36]. Because of process variations, different gates will have different delays. The simulation time of the XOR PPUF is exponential in the number of rows of gates [36]. This PPUF uses three values: the previous input to the gates in the bottom row, the current input to those gates, and the output sampling time. When a server wants to authenticate a user, it will send these three values. A user can apply the previous and current inputs to the XOR gates, sample the output, and send it back to the server within a stipulated time. The server can verify the output through simulation using the publicly available simulation model of the user's PPUF. Here, the server simulates only a predetermined subset of the PPUF, but not the entire PPUF. This subset is known only the server. An attacker can only simulate the PPUF, as he does not have the PPUF

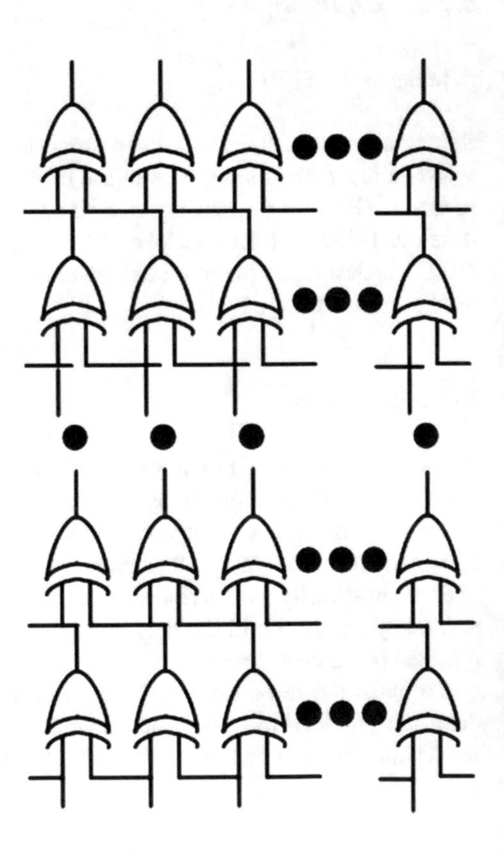

Fig. 8.3 Example of a PPUF constructed using a canonical XOR circuit [17]

circuit. However, since the simulation time is exponential in the number of devices, he cannot predict the correct response within the stipulated time. Thus, an attacker cannot break the security offered by PPUF.

Similar to XOR gates, one can also use emerging technology devices to implement PPUFs [37, 38]. These implementations have a smaller overhead when compared to the XOR-based implementation.

A PPUF can implement two-party security protocols such as authentication, key exchange, bit commitment, and time stamping. One cannot use PUF to implement many of these protocols as it requires both the parties to know the challenge–response pairs a priori.

8.2.7 Sensor PUFs

The sensor physical unclonable function was originally proposed by Rosenfeld et al. [39]. Rosenfeld described a technique to build a PUF which accepts light as another input to the challenge–response generation mechanism. The uniqueness of the sensor PUF is determined by nonhomogeneous coatings over the photodiode light sensing elements.

A sensor PUF builds upon the concept of the PUF by introducing a sensed quantity as another challenge input. The PUF is fully defined by a challenge–sensor–response triple rather than a challenge–response pair. The concept of a sensor PUF can be interpreted as the utilization of a standard PUF while taking advantage of the noise associated with variable operating conditions.

An ideal implementation of a sensor-PUF should exhibit the following properties:

1. **Stability.** Given a fixed challenge and a fixed sensor input, the response bits should be the same across all operating conditions for an IC.
2. **No leakage.** No challenge–sensor–response triple should leak information about any other triple.
3. **Manufacturer resistance.** The manufacturer should have no control over the response of the PUF due to the limits of the manufacturing process. Therefore it should be infeasible to generate two PUFs with identical responses.

Sensor PUFs may similarly be classified as weak or strong, although the number of sensor inputs cannot be considered as part of the challenge space in this respect. Rather, the number of distinct sensor inputs defines the sensor resolution and can be as simple as a binary value (i.e., whether a physical quantity exceeds some threshold) or as complicated as the whole input space of an image sensor [40].

8.2.8 Other Reads

A detailed survey of different types of PUFs is provided in [3, 16, 41].

8.3 IP Protection

8.3.1 Motivation

While the IC design flow spans many countries, not all countries have strict laws against intellectual property theft. Some of the few exceptions are countries such as the USA and Japan [42–45]. Thus, every IC/IP designer bears an additional responsibility to protect his/her design. If a designer can harden the functionality of an IC while it passes through the different, potentially untrustworthy phases of the design flow, these attacks can be thwarted [6, 7]. Or, a designer should at least be able to track down the source of piracy, enabling him to file a litigation against the attacker [12]. Techniques that enable a designer to achieve these objectives are collectively called IP protection techniques.

8.3.2 Classification

IP protection techniques can be classified as:

Active (or) Passive: Active metering techniques provide a way for the designer or the IP owner to control, modify, enable, or disable the target design or IC [6, 46–49]. For instance, combinational logic locking uses additional logic gates (XOR/XNOR) to lock the functionality of an IC (see Sect. 8.3.5 for more details).

Passive metering techniques, unlike active metering techniques, do not control, enable, or disable the design [13, 50, 51]. Instead, they enable a designer to identify the design or an IC.

Intrinsic (or) Extrinsic: Intrinsic techniques involve modifying the design [46]. For instance, watermarking techniques are intrinsic techniques as they modify the design to include the watermark, enabling a designer to verify and claim ownership of the design.
Extrinsic techniques do not modify the design. Instead, additional components are added [6, 48, 49]. An example of an extrinsic technique is burn-in fuses, where serial numbers for an IC are embedded. Such serial numbers enable a designer to track the manufactured ICs.

Reproducible (or) Unclonable IDs: As the name indicates, IDs that can be reproduced by an attacker are called reproducible IDs. Examples include indented serial numbers, digitally stored serial numbers, processor serial numbers [52], and burn-in fuses. Note that these IDs are easy to probe: An attacker can depackage the chip, delayer it, and insert probes to read these IDs.

IDs that cannot be copied by an attacker are unclonable IDs. Examples include different types of PUFs, as discussed in Sect. 8.2. These IDs, unlike reproducible IDs, cannot be easily reverse engineered. Unfortunately, the hardware cost of these IDs are greater than that of the reproducible IDs.

8.3.3 Watermarking

Watermarking is a passive, intrinsic, and reproducible IP protection technique. A designer's signature is embedded into the design artifact [53]. Later, if a designer suspects that an attacker has used his design illegaly, the designer can later reveal the watermark during litigation and claim ownership of an IC/IP. Watermarks can be embedded during different synthesis steps of the design flow: high-level [11], logic, and physical synthesis [54]. One can also embed watermarks for FPGA designs [55].

During high-level synthesis, one needs to map the variable to a register and an operation to an operator, while optimizing for area, power consumption, and performance. However, there are several possible choices of mapping, resulting in the same optimal solution. The choice of variable to register mapping and operation to operator mapping can serve as a watermark [11].

Similarly, during logic synthesis one needs to encode the different states into Boolean values, while optimizing for area, power consumption, and performance. When there are multiple optimal solutions, the choice of encoding can act as a watermark [54].

As most of these synthesis steps involve graph partitioning, one may encode the watermark as constraints during graph partitioning [56, 57]. For instance, one can constrain a set of nodes to be in the same partition. Alternately, a watermark can constrain the number of edges (edge-cuts) spanning the partitions.

Consider embedding a watermark in the graph shown in Fig. 8.4. This graph has 16 nodes and 31 edges. The number of possible watermarking solutions for different number of pairs and the quality of the corresponding solutions are depicted in

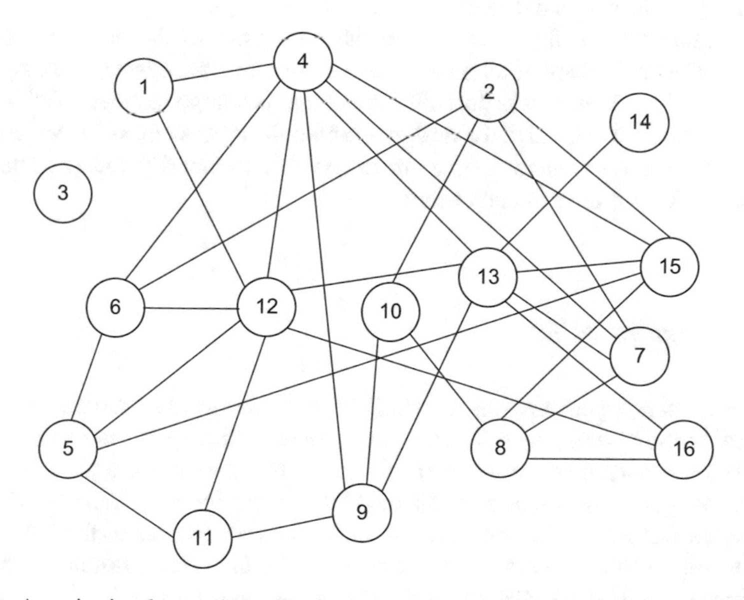

Fig. 8.4 A motivational example for IP watermarking based on graph partitioning. *Source* [56]

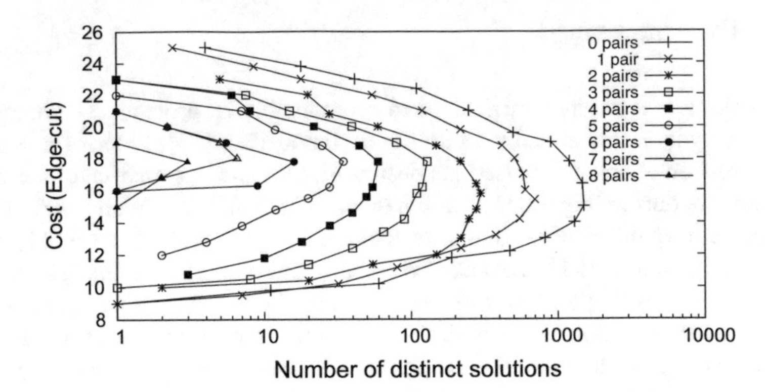

Fig. 8.5 Watermarking: number of possible watermarks vs quality of solutions for the graph when the following pairs of vertices are merged together: (16, 14), (6, 2), (16, 4), (9, 8), (5, 16), (9, 4), (11, 10), (9, 4). *Source* [56]

Fig. 8.5. While there is only one solution for an edge-cut value of 9 (and hence, this is not a good watermark constraint), there are 37 different solutions for an edge-cut value of 13. There is a trade-off between number of possible solutions and the output quality.

A watermark should have the following characteristics [11]:

1. The watermark should not alter the functionality of the design. For example, in case of embedding the watermark during high-level synthesis, the watermark embedded as a choice of variable-to-register mapping should yield the same functionality as the original design.
2. The watermark should clearly prove the ownership of the designer. In other words, the probability of an attacker embedding the same watermark signature should be very low. This is possible when there is a large number (10^{80}) of optimal solutions, from which the designer randomly selects one as his watermark.
3. Apart from the designer, no one should be able to identify and/or remove the watermark from the original design.

8.3.4 *Fingerprinting*

While watermarking enables one to identify that the design has been pirated, it does not reveal the source of piracy. To solve this problem, fingerprinting has been introduced. Here, along with the watermark of the designer, the signature of the buyer (for instance, his public key) will be embedded into the design [12]. When challenged, the designer can reveal the watermark to claim the ownership and the buyer's signature to reveal the source of piracy. For example, the power, timing, or thermal fingerprint of an IC is revealed on applying a set of input vectors.

Similar to watermarking, fingerprinting can also be applied during high-level, logic, and physical synthesis [12]. Conventionally, weak-PUFs are used as fingerprints of an IC. Signatures extracted from static random access memory cells in the IC can act as fingerprints [17].

Fingerprinting techniques are evaluated using the same set of metrics used to evaluate watermarking techniques, with one additional metric: the number of possible fingerprints (watermarks) should be large enough to differentiate signatures of different buyers.

8.3.5 Logic Locking

Logic locking[1] hides the functionality and the implementation of a design by inserting additional gates into the original design. In order for the design to exhibit its correct functionality (i.e., produces correct outputs), a valid key has to be supplied to the locked design. The gates inserted for locking are the *key gates*. Upon applying a wrong key, the locked design will exhibit a wrong functionality (i.e., produce wrong outputs).

EPIC [6] incorporates logic locking into the IC design flow, as shown in Fig. 8.6. In the untrusted design phases, the IC is locked, and its functionality is not revealed. Post fabrication, the IP vendor activates the locked design by applying the valid key. The keys are stored in a tamper-evident memory inside the design to prevent access to an attacker, rendering these key inputs inaccessible to an attacker.

Logic locking prevents attacks such as IP piracy and hardware Trojans. Since the design is locked by the designer, the foundry cannot use any copies or overproduced ICs without the secret keys. Furthermore, it prevents an attacker from analyzing the structural behavior of the design, thereby hindering Trojan insertion.

Logic locking techniques can prevent IP piracy, overbuilding and reverse engineering attacks. Some logic locking techniques offer protection against multiple attacks such as IP piracy attacks along with Hardware Trojan insertion [58, 62, 63].

Classification

Logic locking techniques can be broadly classified as:

Sequential logic locking. In sequential logic locking, additional logic (invalid) states are introduced in the state transition graph [14, 47, 62]. The state transition graph is modified in such a way that the design reaches a valid state only on the application of a correct sequence of key bits. If the key is withdrawn, the design, once again, ends up in a black state.

---

[1] Researchers have previously used the terms "logic obfuscation" [6, 14] and "logic encryption" [58–60] for this purpose. However, echoing the call for consistent terminology by Plaza and Markov [61], we use the term "logic locking" in this chapter.

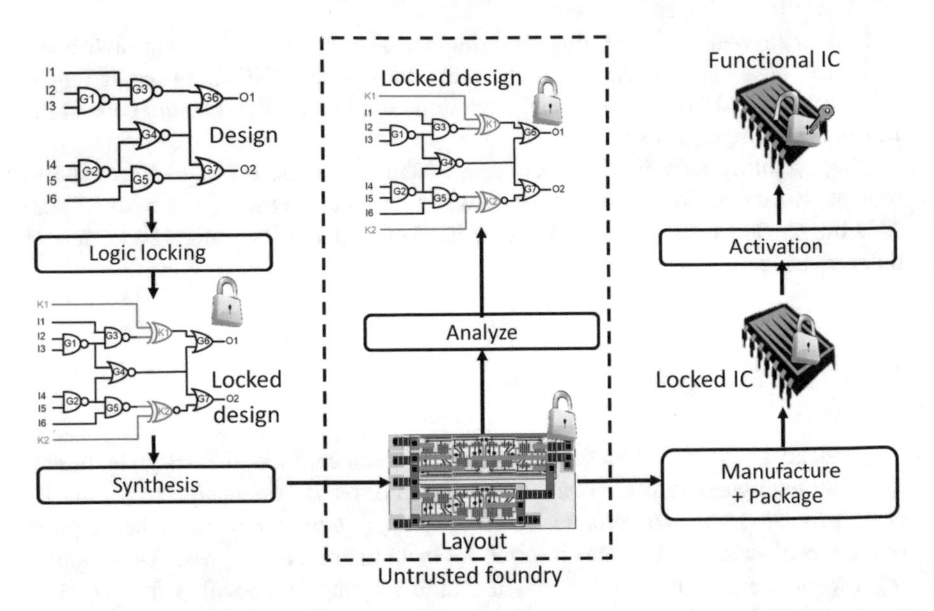

Fig. 8.6 IC design flow with logic locking [6]. The design is in the locked form in the untrusted design regime, shown as *dotted lines*. Upon fabrication, the IC is activated by applying the secret key. The attacker in the untrusted foundry can reverse engineer the design, but he can only obtain the locked design. This prevents an attacker from pirating the design and/or identifying safe places to insert Trojans

Another sequential locking approach is to withhold a part of the design and replace it with programmable logic. This way, the IP owner hides a part of the design from the rogue elements during manufacturing stages [64, 65]. The withheld design is later configured into programmable logic. The circuit will function correctly only when these elements are configured/programmed correctly. However, the introduction of programmable memory elements into the circuit will incur significant performance overhead.

Combinational logic locking. In combinational logic locking, also referred to as logic encryption or logic locking in literature, different logic gates are inserted in a circuit to conceal the functionality of a design. These elements include XOR/XNOR gates [6, 7, 59, 60], AND/OR gates [58], muxes [60, 61] or a combination of these basic elements [66]. One of the inputs to these gates serves as a key input, which is a newly added signal driven by a tamper-evident memory unit. Unless the correct key is loaded onto the on-chip memory, a design will not work correctly.

Example: Consider the example design shown in Fig. 8.7a. Three key gates (K0, K1, and K3) are added to the design to lock it. The correct key is 000.

The insertion of the gates is done after logic synthesis and before physical synthesis. The design can be then be resynthesized or resized. In case of logic locking using XOR/XNOR gates, if the key gates were left as such without any other modifications to the circuit, the key bits could be extracted by inspecting if a key gate is

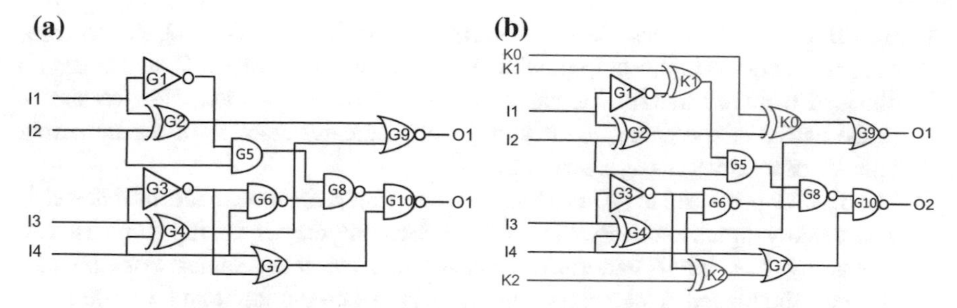

Fig. 8.7 Logic locking: **a** The original design. **b** Design locked by randomly inserting key gates

XOR or XNOR. To eradicate such a simple deduction analysis between the key gate types and the key values, following post-processing steps can be applied:

1. The netlist can be synthesized such that the XOR/XNOR key gates are replaced with other gates like AND/OR/NAND, etc., rendering it difficult to identify the key gates as XOR/XNOR.
2. The existing inverters in the design can be absorbed into the key gates, changing their polarity in a manner oblivious to the attacker; similarly, additional inverters can be added next to the key gates to change the polarity of the key gates.

Activation of the Locked ICs.

In logic locking, a circuit is activated/unlocked after fabrication by loading the key onto the on-chip tamper-evident memory. The activation can be conducted either prior to or post-manufacturing test. When a chip is to be activated prior to the manufacturing test (pre-test activation), secure communication infrastructure is needed so to load the key remotely onto the chip. EPIC [7], Secure split test [67, 68] are platforms, which make use of public key cryptography protocols to load a key onto the chip. However, it is also possible to conduct the manufacturing test first, using dummy key values, and later on load the key onto the chip (post-test activation) in a secure facility.

Metrics

A logic locking technique should:

1. Prevent an attacker from deducing the key, and
2. Produce wrong outputs on applying a wrong key.

Preventing an attacker from deducing the key. Multiple attacks have been presented against existing logic locking techniques. The objective of an attacker is find out the key used for locking the circuit [61, 66, 69–71]. Based on the capabilities of an attacker, these attacks can be broadly classified into two types:

1. **An attacker with access to input–output pairs** [61, 66, 69–71]. An attacker can gain access to input–output pairs by exercising an activated IC, which can be obtained from the market. He can buy this IC from the market. He also access to the netlist of the encrypted design, which he can obtain from the untrusted foundry or by reverse engineering the IC.
   In the attack proposed in [70], an attacker determines the input patterns that sensitizes the key inputs to observable points such as outputs or scan flip-flops. He can then apply those inputs patterns to determine the secret key values. For example, consider the locked design shown in Fig. 8.7b, where an input pattern 1000 will sensitize K0 to O1. Thus, an attacker can determine the value of K0. Similarly, inputs patterns 0111 and X010 sensitize K1 and K2 to O1 and O2, respectively.[2]
   In the attack proposed in [71], Boolean Satisfiability (commonly known as SAT) solvers are used to identify the key values such that render the output of the locked design identical to that of the original design for all inputs.
2. **An attacker with access only to test input data and test responses** [61, 69]. An attacker at the testing facility has access to the test data, which are provided to him by the designer. A hill-climbing search based attack makes use of test stimuli and responses to recover the secret key [61, 69]. If an attacker gets access to the scan chains, he can launch the attack even without the netlist.

Producing wrong outputs on applying a wrong key. When the key gates are inserted randomly in a circuit, there is no guarantee that an incorrect key produces incorrect output for all the input patterns. Consequently, the resultant logic locking technique is considered weak.

For example, consider the design shown in Fig. 8.7b. When the input pattern is X000, the correct output pattern is 00. However, even on applying any incorrect key, the design still produces the correct output.

To solve this problem, researchers developed an analogy between the problem of logic locking and IC testing principles such as fault activation, propagation, and masking [59, 60]. Based on this analogy, a fault impact metric was developed. The key gates are inserted at places with the highest fault impact. This technique maximizes the Hamming distance between the correct output and the incorrect outputs on applying a random incorrect key. XOR/XNOR key gates are combined with mux key gates to achieve higher Hamming distance at outputs [66].

Dupuis et al. [58] propose a logic locking scheme that inserts AND/OR key gates in a circuit to minimize the number of low-controllability locations, which makes it difficult to insert Hardware Trojans in the circuit.

8.3.6 Metering

It is a set of tools, methodologies, and protocols used to track a manufactured IC. In passive metering, a part of an IC's functionality is used for metering [13]. The

[2] *X refers to a don't care value. It can be freely set to either 1 or 0.*

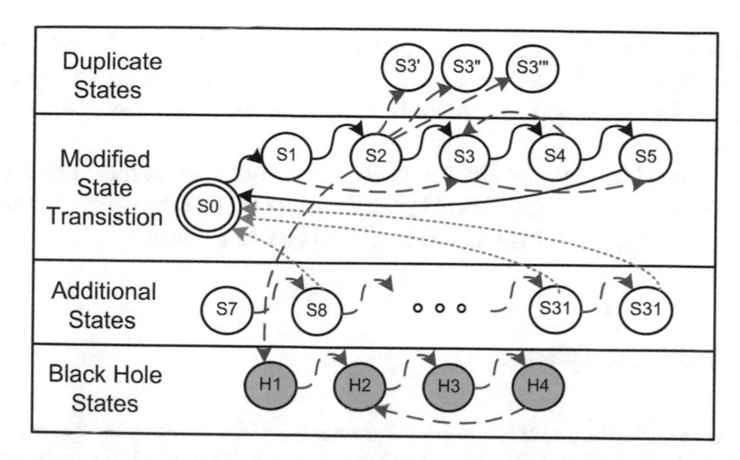

Fig. 8.8 Hardening a controller. Approach 1: Existing states are replicated [46]. Approach 2: State transitions are modified [14, 72–74]. Approach 3: Additional states are added [47, 50, 75]. Approach 4: Black-hole states are added [47, 50, 75]. S0 through S6 are the states in the original FSM. All the other states are added for obfuscation. Solid edges are the state transitions in the original FSM. Dashed edges are state transitions from an invalid state to a valid state, on applying the valid key. Dotted edges are the state transitions from a valid state to an invalid state, on applying an invalid key or when key is withdrawn

identified ICs are matched against their record in a database. This will reveal unregistered ICs or overbuilt ICs. In active metering, parts of the IC's functionality can be only accessed, locked, or unlocked by the designer and/or IP rights owners [47]. The difference between metering and locking is that while metering uses a unique unlock key per IC, locking just locks the IC.

Example: Figure 8.8 shows an example finite state machine (FSM) of a controller and how it has been hardened using several metering techniques. Each node represents a state. The edges represent the state transition. The solid edges represent the state transitions in the original design. The dotted edges represent the ones added for metering.

A controller can be hardened by adding extra states and/or transitions in the FSM. The controller can be hardened by:

1. **State replication**: Some states in the original FSM may be replicated [46]. For example, in Fig. 8.8, the state S2 has been replicated three times. Only the original state S2 has an outward transition; none of the other states has an outward transition. On applying an incorrect key, the design enters into one of the replicated states. Consequently, it enters into a lock-down state.
2. **Additional transitions**: Additional transitions between the states of the original design are added into the design [14, 72–74]. For example, in Fig. 8.8, the transition from S1 to S3 and S3 and S5 are added. On applying an incorrect key, the design will skip the S2 and S4 states, thus exhibiting a different, and thus, wrong functionality.

3. **Additional states**: Extra states are added into the design [47, 50, 75]. For example, in Fig. 8.8, states S7 through S31 are the additional states. On applying an incorrect key, the design enters into one of these states and eventually enters the reset state.
4. **Black-hole states**: An invalid key leads the design into invalid states via illegal transitions, and eventually into black-hole states, where the design is stuck [47, 50, 75]. In Fig. 8.8, H1 through H4 are the black-hole states.

8.3.7 Split Manufacturing

Leading fabless semiconductor companies such as AMD and research agencies such as Intelligence Advanced Research Projects Agency (IARPA) have proposed split manufacturing to protect against IP piracy and trojan insertion [15, 76]. In split manufacturing, the layout of the design is split into the Front End Of Line (FEOL) layers and Back End Of Line (BEOL) layers which are then fabricated separately in different foundries. The FEOL layers consist of transistors and other lower metal layers (≤M4) and the BEOL layers consist of the top metal layers (>M4). Post fabrication, the FEOL and BEOL wafers are aligned and integrated together using either electrical, mechanical, or optical alignment techniques. The final ICs are tested upon integration of the FEOL and BEOL wafers [15, 76]. The asymmetrical nature of the metal layers facilitates split manufacturing. The top BEOL metal layers are usually thicker and have a larger pitch than the bottom FEOL metal layers [77]. Hence, a designer can easily integrate the BEOL and FEOL wafers.

Figure 8.9 shows a possible split manufacturing aware IC design flow. A gate-level netlist is partitioned into blocks which are then floorplanned and placed. The transistors and wires inside a block form the FEOL layers. The top metal wires connecting the blocks and the IO ports form the BEOL layers. The BEOL and FEOL wires are assigned to different metal layers and routed such that the wiring delay and routing congestion are minimized. The layout of the entire design is split into two—one layout just contains the FEOL layers and the other layout just contains the BEOL layers. The two layouts are then fabricated in two different foundries. In one embodiment, the FEOL layout is first fabricated and then sent to a trusted second foundry where the BEOL layout is built on top of it [15]. In another embodiment, the fabricated FEOL and BEOL layouts are obtained by the system integrator, and are then integrated by using electrical, mechanical, or optical alignment techniques and tested for defects [76].

Split manufacturing aims to improve the security of an IC as the FEOL and BEOL layers are fabricated separately and combined post fabrication. This prevents a single foundry (especially the FEOL foundry) from gaining full control of the IC. For instance, without the BEOL layers, an attacker in the FEOL foundry can neither identify the safe places within a circuit to insert trojans nor pirate the designs without the

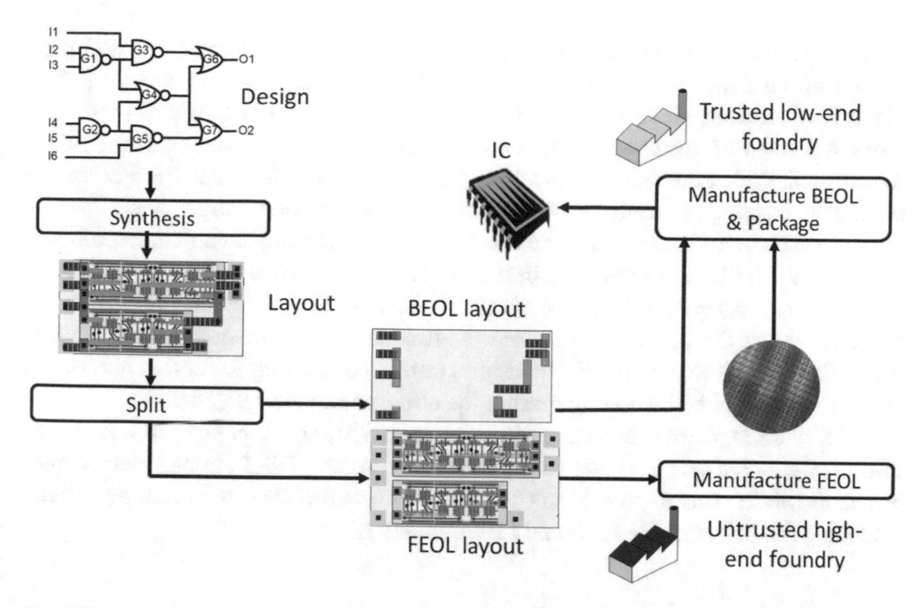

Fig. 8.9 Split manufacturing: the layout of a design is split into two Front End-of-Line (FEOL) and Back End-of-Line (BEOL) parts. These two parts are manufactured at two different places. This prevents the attacker at the FEOL foundry from accessing the BEOL part. With only an incomplete design, an attacker can neither pirate the design nor identify safe places insert hardware Trojans in it

BEOL layers. The economic benefit of split manufacturing comes from performing the low cost BEOL layer fabrication in-house and outsourcing the expensive FEOL layer fabrication [15, 76, 78].

Practicality of Split Manufacturing

Even though two foundries are involved in manufacturing a single IC, split manufacturing should not degrade the manufacturing quality of an IC. Recently, researchers have proved the practicality of split manufacturing by manufacturing an FPGA in two different foundries [79]. One can also leverage the 3D manufacturing technology where security sensitive components can be placed in one layer and manufactured in a trusted low-end foundry; and other components of the design can be placed in another layer and manufactured in an untrusted high-end foundry [80]. Feasibility of split manufacturing for analog designs has also been demonstrated [81].

Attacks and Defenses for Split Manufacturing

The attacker in the foundry should not be able to determine the missing BEOL connections from the FEOL connections. Naive split manufacturing is vulnerable to *proximity attack*. This attack exploits the heuristic that floorplanning and placement (F&P) tools use to reduce the wiring (delay) between the pins to be connected [82]—place the partitions close by and orient the partitions. This heuristic of most F&P tools is a security vulnerability that can be exploited by an attacker in the FEOL foundry who does not have access to the BEOL layers. Thus, the attacker simply makes the missing connections between the two closest compatible[3] pins.

To thwart such attacks, a fault-analysis-based pin swapping technique is proposed in [83]. The idea here is to find the best set of pins to swap such that when an attacker performs the proximity attack, the obtained netlist will be different from that of the original netlist; the difference between the two netlists can be quantified via Hamming distance of the outputs. Furthermore, instead of splitting the design at M4, one can also split at M1, thereby increasing the effort of an attacker [84].

In [85], an algorithm to select wires for the BEOL layers is provided. A formal notion of an attacker's inability to figure out the missing BEOL connections is provided. However, this approach has a significant performance overhead, potentially offsetting the benefits of a high-end FEOL foundry.

8.4 Conclusion

In this chapter, we elaborated on two important security techniques: PUFs and IP protection techniques. Weak-PUFs have already found applications in securing FPGA designs [86]. Several companies such as Verayo [87] and Intrinsic ID [88] are trying to commercialize strong-PUFs as well. While reliability challenges still exist for PUF circuits, PPUFs, PUFs using emerging technology devices, and sensor PUFs provide a wide variety of applications.

In case of IP protection techniques, a designer can use a technique depending upon the threat model he faces. For instance, if he does not have access to a BEOL foundry, he can pursue logic locking. Otherwise, he can pursue split manufacturing to protect his design, as logic locking requires one to store keys on the chip. However, note that split manufacturing requires the designer to trust the end user. Thus, one needs to carefully select an IP protection technique that suits their business model.

References

1. Kocher, P., Jaffe, J., Jun, B.: Differential power analysis. Advances in cryptology (CRYPTO 99). Lect. Notes Comput. Sci. **1666**, 388–397 (1999)
2. SEMI. Innovation is at risk as semiconductor equipment and materials industry loses up to $4 billion annually due to IP infringement (2008). www.semi.org/en/Press/P043775

[3] Two pins are compatible if one pin is the output of a gate or an input port, and the other pin is an input of a gate or an output port.

3. Herder, C., Yu, M.-D., Koushanfar, F., Devadas, S.: Physical unclonable functions and applications: a tutorial. Proc. IEEE **102**(8), 1126–1141 (2014)
4. Guin, Ujjwal, DiMase, Daniel, Tehranipoor, Mohammad: Counterfeit integrated circuits: detection, avoidance, and the challenges ahead. J. Electron. Test. **30**(1), 9–23 (2007)
5. Rostami, M., Koushanfar, F., Karri, R.: A primer on hardware security: models, methods, and metrics. P. IEEE **102**(8), 1283–1295 (2014)
6. Roy, J.A., Koushanfar, F., Markov, I.L.: EPIC: ending piracy of integrated circuits. IEEE/ACM Design, Automation and Test in Europe, pp. 1069–1074 (2008)
7. Roy, J.A., Koushanfar, F., Markov, I.L.: Ending piracy of integrated circuits. Computer **43**(10), 30–38 (2010)
8. Karri, R., Rajendran, J., Rosenfeld, K., Tehranipoor, M.: Trustworthy hardware: identifying and classifying hardware Trojans. IEEE Comput. **43**(10), 39–46
9. Top 5 Most Counterfeited Parts Represent a $ 169 Billion Potential Challenge for Global Semiconductor Market. http://press.ihs.com/press-release/design-supply-chain/top-5-most-counterfeited-parts-represent-169-billion-potential-cha
10. DARPA. Defense Science Board (DSB) study on High Performance Microchip Supply (2005). www.acq.osd.mil/dsb/reports/ADA435563.pdf
11. Koushanfar, Farinaz, Hong, Inki, Potkonjak, Miodrag: Behavioral synthesis techniques for intellectual property protection. ACM Trans. Des. Autom. Electron. Syst. **10**(3), 523–545 (2005)
12. Caldwell, A.E., Choi, H.-J., Kahng, A.B., Mantik, S., Potkonjak, M., Qu, G., Wong, J.L.: Effective iterative techniques for fingerprinting design IP. IEEE Trans. Comput.-Aided Des. Integr. Circuits Syst. **23**(2), 208–215 (2004)
13. Koushanfar, F., Qu, G., Potkonjak, M.: Intellectual Property Metering. Information Hiding, Workshop (2001)
14. Chakraborty, R.S., Bhunia, S.: HARPOON: an obfuscation-based soc design methodology for hardware protection. IEEE Trans. Comput.-Aided Des. Integr. Circuits Syst. **28**(10), 1493–1502 (2009)
15. Intelligence Advanced Research Projects Activity. Trusted Integrated Circuits Program. https://www.fbo.gov/utils/view?id=b8be3d2c5d5babbdffc6975c370247a6
16. Rhrmair, U., Devadas, S., Koushanfar, F.: Security Based on Physical Unclonability and Disorder. Introduction to Hardware Security and Trust, pp. 65–102 (2012)
17. Holcomb, D.E., Burleson, W.P., Fu, K.: Power-up SRAM state as an identifying fingerprint and source of true random numbers. IEEE Trans. Comput. **58**(9), 1198–1210 (2009)
18. Guajardo, Jorge, Kumar, Sandeep S., Schrijen, Geert-Jan, Tuyls, Pim: FPGA intrinsic PUFs and their use for IP protection. Cryptographic Hardware Embed. Syst. **4727**, 63–80 (2007)
19. Tuyls, P., Schrijen, G.-J., Kori, B., van Geloven, J., Verhaegh, N., Wolters, R.: Read-proof hardware from protective coatings. Cryptographic Hardware Embed. Syst. **4249**, 369–383 (2006)
20. Helinski, R., Acharyya, D., Plusquellic, J.: A physical unclonable function defined using power distribution system equivalent resistance variations. ACM/IEEE Design Automation Conference, pp. 676–681 (2009)
21. Helinski, R., Acharyya, D., Plusquellic, J.: Quality metric evaluation of a physical unclonable function derived from an IC's power distribution system. ACM/IEEE Design Automation Conference, pp. 240–243 (2010)
22. Gassend, B., Clarke, D., van Dijk, M., Devadas, S.: Silicon physical random functions. ACM Conference on Computer and Communications Security, pp. 148–160 (2002)
23. Suh, G.E., Devadas, S.: Physical unclonable functions for device authentication and secret key generation. IEEE/ACM Design Automation Conference, pp. 9–14 (2007)
24. Lee, J.W., Lim, D., Gassend, B., Suh, G.E., van Dijk, M., Devadas, S.: A technique to build a secret key in integrated circuits for identification and authentication applications. IEEE Internationall Symposium on VLSI Circuits, pp. 176–179 (2004)
25. Majzoobi, M., Koushanfar, F., Potkonjak, M.: Lightweight secure PUFs. IEEE/ACM International Conference on Computer-Aided Design, pp. 670–673 (2008)

26. Pappu, R., Recht ,B., Taylor, J., Gershenfeld, N.: Physical one-way functions. Science **297**(5589), 2026–2030 (2002)
27. Maiti, A., Gunreddy, V., Schaumont, P.: A Systematic Method to Evaluate and Compare the Performance of Physical Unclonable Functions (2011). https://eprint.iacr.org/2011/657.pdf
28. Devadas, S.: Non-networked RFID PUF authentication. U.S. Patent 8 683 210, U.S. Patent Appl. 12/623 045 (2008)
29. Suh, G.E., O'Donnell, C.W., Devadas, S.: Aegis: a single-chip secure processor. IEEE Des. Test Comput. **24**(6), 570–580 (2007)
30. Rührmair, U., Sehnke, F., Sölter, J., Dror, G., Devadas, S., Schmidhuber, J.: Modeling attacks on physical unclonable functions. ACM Conference on Computer and Communications Security, pp. 237–249 (2010)
31. Schuster, D.: Side-channel analysis of physical unclonable functions (PUFs). PhD Dissertation, Technische Universität München (2010)
32. Wei, S., Wendt, J.B., Nahapetiany, A., Potkonjak, M.: Reverse engineering and prevention techniques for physical unclonable functions using side channels. IEEE/ACM Design Automation Conference, pp. 1–6 (2014)
33. Devadas, S., Yu, MDM.: Secure and robust error correction for physical unclonable functions. IEEE Des. Test 99 (2013)
34. Paral, Z., Devadas, S.: Reliable and efficient PUF-based key generation using pattern matching. IEEE International Symposium on Hardware-Oriented Security and Trust, pp. 128–133 (2011)
35. Yin, C.-E., Qu, G.: Improving PUF security with regression-based distiller. IEEE/ACM Design Automation Conference, pp. 1–6 (2013)
36. Nathan Beckmann and Miodrag Potkonjak. Hardware-based public-key cryptography with public physically unclonable functions. Information Hiding, pp. 206–220 (2009)
37. Rajendran, J., Rose, G.S., Karri, R., Potkonjak, M.: Nano-PPUF: a memristor-based security primitive. IEEE Computer Society Annual Symposium on VLSI, pp. 84–87 (2012)
38. Ruhrmair, U., Chen, Q., Stutzmann, M., Lugli, P., Schlichtmann, U., Csaba, G.: Towards electrical, integrated implementations of SIMPL systems. Information Security Theory and Practices. Security and Privacy of Pervasive Systems and Smart Devices, vol. 6033, pp. 277–292 (2010)
39. Rosenfeld, K., Gavas, E., Karri, R.: Sensor physical unclonable functions. IEEE International Symposium on Hardware-Oriented Security and Trust, pp. 112–117
40. Cao, Y., Zalivaka, S.S., Zhang, L., Chang, C.-H., Chen, S.: CMOS image sensor based physical unclonable function for smart phone security applications. International Symposium on Integrated Circuits, pp. 392–395 (2014)
41. Maes, R., Verbauwhede, I.: Physically Unclonable Functions: A Study on the State of the Art and Future Research Directions, pp. 3–37. Towards Hardware-Intrinsic, Security (2010)
42. Council Decision 96/644/EC of 11 November 1996 on the extension of the legal protection of topographies of semiconductor products to persons from the Isle of Man (2015). http://eur-lex.europa.eu/legal-content/EN/TXT/?uri=celex:31996D0644
43. Law on the Circuit Layout of a Semiconductor Integrated Circuits (Act No. 43 of May 31, 1985, as last amended by Act No. 50 of June 2, 2006) (2015)
44. Malbon, J., Lawson, C., Davison, M.: A Commentary. Edward Elgar Publishing, The WTO Agreement on Trade-Related Aspects of Intellectual Property Rights (2014). ISBN 9781845424435
45. Government Printing Office. The Copyright Law of the United States and Related Laws Contained in Title 17 of the United States Code (2012). ISBN 9780160795084
46. Alkabani, Y., Koushanfar, F., Potkonjak, M.: Remote activation of ICs for piracy prevention and digital right management. In: Proceedings of IEEE/ACM International Conference on Computer-Aided Design, pp. 674–677 (2007)
47. Alkabani, Y., Koushanfar, F.: Active Hardware Metering for Intellectual Property Protection and Security, pp. 291–306. USENIX, Security (2007)
48. Huang, J., Lach, J.: IC activation and user authentication for security-sensitive systems. IEEE International Workshop on Hardware-Oriented Security and Trust, pp. 76–80 (2008)

49. Roy, J.A., Koushanfar, F., Markov, I.L.: Protecting bus-based hardware IP by secret sharing. ACM/IEEE Design Automation Conference, pp. 846–851 (2008)
50. Koushanfar, F., Qu, G.: Hardware metering. IEEE/ACM Design Automation Conference, pp. 490–493 (2001)
51. Lofstrom, K., Daasch, W.R., Taylor, D.: IC identification circuit using device mismatch. IEEE International Solid-State Circuits Conference, pp. 372–373 (2000)
52. Pentium III serial numbers. http://www.pcmech.com/article/pentium-iii-serialnumbers/
53. Kahng, A.B., Lach, J., Mangione-Smith, W.H., Mantik, S., Markov, I.L., Potkonjak, M., Tucker, P., Wang, H., Wolfe, G.: Watermarking techniques for intellectual property protection. IEEE/ACM Design Automation Conference, pp. 776–781 (1998)
54. Kahng, A.B., Mantik, S., Markov, I.L., Potkonjak, M., Tucker, P., Wang, H., Wolfe, G.: Robust IP watermarking methodologies for physical design. IEEE/ACM Design Automation Conference, pp. 782–787 (1998)
55. Lach, J., Mangione-Smith, W.H., Potkonjak, M.: FPGA fingerprinting techniques for protecting intellectual property. IEEE Custom Integrated Circuits Conference, pp. 299–302 (1998)
56. Wolfe, G., Wong, J.L., Potkonjak, M.: Watermarking graph partitioning solutions. IEEE/ACM Design Automation Conference, pp. 486–489 (2001)
57. Alpert, C.J., Kahng, A.: Recent Directions in Netlist Partitioning. Integration, The VLSI journal (1995)
58. Dupuis, S., Ba, P.-S., Di Natale, G., Flottes, M.L., Rouzeyre, B.: A novel hardware logic encryption technique for thwarting illegal overproduction and hardware Trojans. IEEE International On-Line Testing Symposium, pp. 49–54 (2014)
59. Rajendran, J., Pino, Y., Sinanoglu, O., Karri, R.: Logic encryption: a fault analysis perspective. In: Proceedings of the IEEE/ACM Design, Automation and Test in Europe, pp. 953–958 (2012)
60. Rajendran, J., Zhang, H., Zhang, C., Rose, G.S., Pino, Y., Sinanoglu, O., Karri, R.: Fault analysis-based logic encryption. IEEE Trans. Comput. **64**(2), 410–424 (2015)
61. Plaza, S.M., Markov, I.L.: Solving the third-shift problem in ic piracy with test-aware logic locking. IEEE Trans. Comput.-Aided Des. Integr. Circuits Syst. **34**(6), 961–971 (2015)
62. Chakraborty, R.S., Bhunia, S.: Security against hardware Trojan through a novel application of design obfuscation. IEEE/ACM International Conference on Computer-Aided Design, pp. 113–116 (2009)
63. Colombier, B., Bossuet, L.: Survey of hardware protection of design data for integrated circuits and intellectual properties. IET Comput. Digital Tech. **8**(6), 274–287 (2014)
64. Baumgarten, A., Tyagi, A., Zambreno, J.: Preventing IC piracy using reconfigurable logic barriers. IEEE Des. Test Comput. **27**(1), 66–75 (2010)
65. Khaleghi, S., Da Zhao, K., Rao, W.: IC piracy prevention via design withholding and entanglement. Asia-Pacific Design Automation Conference, pp. 821–826 (2015)
66. Lee, Y.-W., Touba, N.A.: Improving logic obfuscation via logic cone analysis. IEEE Latin-American Test Symposium, pp. 1–6 (2015)
67. Contreras, G.K., Rahman, M.T., Tehranipoor, M.: Secure split-test for preventing ic piracy by uuntrusted foundry and assembly. IEEE International Symposium on Defect and Fault Tolerance in VLSI and Nanotechnology Systems, pp. 196–203 (2013)
68. Roy, J.A., Koushanfar, F., Markov, I.L.: Protecting bus-based hardware ip by secret sharing. In: Proceedings of IEEE/ACM Design Automation Conference, pp. 846–851 (2008)
69. Plaza, S.M., Markov, I.L.: Protecting Integrated Circuits from Piracy with Test-aware Logic Locking (2014)
70. Rajendran, J., Pino, Y., Sinanoglu, O., Karri, R.: Security analysis of logic obfuscation. IEEE/ACM Design Automation Conference, pp. 83–89 (2012)
71. Subramanyan, P., Ray, S., Malik, S.: Evaluating the Security of Logic Encryption Algorithms. IEEE International Symposium on Hardware Oriented Security and Trust, pp. 137–143 (2015)
72. Chakraborty, R.S., Bhunia, S.: Hardware protection and authentication through netlist level obfuscation. IEEE/ACM International Conference on Computer-Aided Design, pp. 674–677 (2008)

73. Chakraborty, R.S., Bhunia, S.: Security against hardware trojan through a novel application of design obfuscation. IEEE/ACM International Conference on Computer-Aided Design, pp. 113–116 (2009)
74. Chakraborty, R.S., Bhunia, S.: RTL hardware ip protection using key-based control and data flow obfuscation. IEEE International Conference on VLSI Design, pp. 405–410 (2010)
75. Koushanfar, Farinaz: Provably secure active IC metering techniques for piracy avoidance and digital rights management. IEEE Trans. Inf. Forensics Secur. **7**(1), 51–63 (2012)
76. Jarvis, R.W., McIntyre, M.G.: Split manufacturing method for advanced semiconductor circuits. US Patent no. 7195931 (2004)
77. FreePDK45:Metal Layers. http://www.eda.ncsu.edu/wiki/FreePDK45:Metal_Layers
78. Jagasivamani, M., Gadfort, P., Sika, M., Bajura, M., Fritze, M.: Split fabrication obfuscation: metrics and techniques. IEEE Symposium on Hardware Oriented Security and Trust (2014)
79. Hill, B., Karmazin, R., Otero, C.T.O., Tse, J., Manohar, R.: A split-foundry asynchronous FPGA. IEEE Custom Integrated Circuits Conference, pp. 1–4 (2013)
80. Valamehr, J., Sherwood, T., Kastner, R., Marangoni-Simonsen, D., Huffmire, T., Irvine, C., Levin, T.: A 3-D split manufacturing approach to trustworthy system development. IEEE Trans. Comput.-Aided Des. Integr. Circuits Syst. **32**(4), 611–615 (2013)
81. Vaidyanathan, K., Liu, R., Sumbul, E., Zhu, Q., Franchetti, F., Pileggi, L.: Efficient and secure intellectual property (IP) design for split fabrication. IEEE Symposium on Hardware Oriented Security and Trust (2014)
82. Naveed, A.: Sherwani. Springer Publications, Algorithms for VLSI Physical Design Automation (2002)
83. Rajendran, O., Sinanoglu, J., Karri, R.: Is split manufacturing secure? IEEE Design, Automation and Test in Europe Conference, pp. 1259–1264 (2013)
84. Vaidyanathan, K., Das, B.P., Sumbul, E., Liu, R., Pileggi, L.: Building trusted ICs using split fabrication. IEEE Symposium on Hardware Oriented Security and Trust (2014)
85. Imeson, F., Emtenan, A., Garg, S., Tripunitara, M.: Securing Computer Hardware Using 3D Integrated Circuit (IC) Technology and Split Manufacturing for Obfuscation. USENIX Security (2013)
86. Altera. Altera Reveals Stratix 10 Innovations Enabling the Industrys Fastest and Highest Capacity FPGAs and SoCs. http://newsroom.altera.com/press-releases/nr-altera-stratix10.htm
87. Verayo, P.: Physical unclonable function. http://www.verayo.com/tech.php
88. Intrinsic ID. Physical unclonable function. https://www.intrinsic-id.com/technology/physically-unclonable-functions-puf/

Chapter 9
A Systematic Approach to Fault Attack Resistant Design

Nahid Farhady Galathy, Bilgiday Yuce and Patrick Schaumont

9.1 Introduction to Fault Attacks

Electronic systems are subject to temporary and permanent faults caused by imperfections in the manufacturing process as well as by anomalies of the environment. Fault effects in electronics have been intensively studied in the context of system reliability as well as error resiliency. However, faults can also be used as a hacking tool. In a fault attack, an adversary injects an intentional fault in a circuit and analyzes the response of that circuit to the fault. The objective of a fault attack is to extract cryptographic key material, to weaken cryptographic strength, or to disable the security. Unlike some other attacks, such as power-based or electromagnetic-based side-channel analysis, fault attacks do not require complex signal measurement equipment. The threat model of a fault attack assumes an adversary who can influence the physical environment of the electronic system—a condition that holds for a large class of embedded electronics such as smart cards, key fobs, access controls, embedded controllers, and so on. Fault attacks have been studied since the turn of the century, and today a great variety of methods are available to attack all forms of cryptography [3, 4, 16, 17].

A generic solution against faults is to use redundancy, such as by replicating the hardware implementation, by repeating computations, or by applying data error-coding techniques. The idea of redundancy is to tolerate sporadic faults by ensuring that at least part of the circuit obtains a correct result. The advantage of fault tolerant design is that it can handle (within some limits) any fault regardless of the

N.F. Galathy (✉) · B. Yuce · P. Schaumont
Virginia Tech, Blacksburg, USA
e-mail: farhady@vt.edu

B. Yuce
e-mail: bilgiday@vt.edu

P. Schaumont
e-mail: schaum@vt.edu

S. Bhunia et al. (eds.), *Fundamentals of IP and SoC Security*,
DOI 10.1007/978-3-319-50057-7_9

fault location and fault timing in the circuit. However, fault tolerant design using redundancy is expensive. Spatial redundancy multiplies the hardware cost, and time redundancy reduces the performance, each by a factor of several times. Full fault tolerance is therefore only available to systems that can afford over-design. Despite the costly overhead, most of these fault tolerant solutions are still not applicable to the fault attack problem because in these designs, the fault is assumed to be random and sporadic.

In fault attacks, faults are injected by an adversary rather than by nature. The adversary is intelligent and determined, rather than random and indifferent. The adversary also makes specific assumptions about the objectives of the fault attack, and about the algorithm being cryptanalyzed. Indeed, because of the widespread adoption of cryptographic standards, these assumptions are quite reasonable. This means that the objective of a fault attack is quite specific: the objective is to extract a secret key. In this chapter, we assume that rendering a circuit inoperable is not a valid objective for a fault attack. Instead, such an attack belongs to a class of attacks known as *denial-of-service*. We will concentrate instead on fault attacks that extract a cryptographic key.

Another common assumption of the adversary is that many design details of the cryptographic implementation are known. Indeed, by using basic reverse engineering techniques, the adversary may learn the execution schedule of a cryptographic algorithm as it operates clock cycle by clock cycle, or the meaning of memory locations and registers used by the digital circuit. Knowledge of such design details is often helpful for a fault attack, and therefore the worst-case assumption is to assume that the adversary is fully knowledgeable about the implementation details of a cryptographic design.

The question we wish to address in this chapter is the following: *How to systematically build a fault attack resistant design?* To answer this question, we first provide an analysis of a successful fault attack requirements. The fault attack analysis lead to two main contributions of this chapter. First, it differentiates the intentional fault injection from random sporadic faults. Second, it provides an insight for designing a fault attack resistant design. Although a designer cannot prevent a fault attack from happening, a designer can control the effects of injected faults. By suitable design techniques, it is therefore possible to create circuits that are harder to attack using common fault attack techniques.

The chapter is organized as follows. In Sect. 9.2, we review common fault injection techniques, and their effects on digital circuits. In Sect. 9.3, we will describe the four essential steps that an adversary has to take, in order to complete a fault attack. In Sect. 9.4, we review several common fault analysis methods, and their requirements with respect to fault injection. In Sect. 9.5, we combine the insights from fault injection (Sect. 9.3) with those from fault analysis (Sect. 9.4) to define fault attack resistant design techniques. Finally, Sect. 9.6 concludes the chapter.

9.2 Fundamental Fault Attack Requirements

In a digital circuit, a fault is manifested through a temporary or permanent change in the correctness of computations. Faults in digital systems originate from a variety of causes, related to manufacturing issues as well as to environmental issues. Of primary interest to fault attacks are *intentional* faults caused by an adversary, as opposed to faults that have a random, uncontrolled cause.

A successful fault attack is composed of a fault measurement and the fault analysis process. The fault analysis process is based on the information leaked while building the fault model. The aim of the attacker is to be able to inject an intentional fault, using a series of techniques to manipulate the environmental conditions of a circuit, that results in the desired fault model.

In this section, the emphasis is on the requirements for fault measurement. Two preliminary concepts to understand fault measurement, are *fault model* and *fault injection*. We will first explain these two, and then explain how they relate to fault measurement.

9.2.1 Fault Model

The fault characteristics resulting from a fault injection are commonly captured in a *fault model*, which is also the starting point of various cryptanalytic methods. A fault model expresses the important fault characteristics: the location of the fault within the circuit, the number of bits affected by the fault, and the fault effect on the bits (stuck-at, bit-flip, random, set/reset).

Table 9.1 lists four common fault effects: chosen bit fault, single bit fault, byte fault, and random fault. For example, in chosen bit fault model, the attacker must precisely select the location of the faulty bit and change its value to either 0/1.

Table 9.1 Fault models assumed by an adversary

	Fault location	Number of bits	Fault effect
Chosen bit fault	Precise	1	Set/reset
Single bit fault	Loose	1	Stuck-at
Byte fault	Loose	8	Stuck-at
Random fault	Loose	Any	Random

9.2.2 Fault Injection Techniques

As mentioned, the objective of the attacker is to build the fault model for a successful post-processing of the information. There are several fault injection tools and techniques for building the fault model. The following are six possible mechanisms of fault injection.

- **Clock Glitches** are used to shorten the clock period of a digital circuit during selected clock cycles [2, 36]. If the instantaneous clock period decreases below the critical path of the circuit, then a faulty value will be captured in the memory or state of the circuit. An adversary can inject a clock glitch by controlling the clock line of the digital circuit, triggering a fault in the critical path of the circuit. If the adversary knows the circuit structure, he or she will be able to predict location of the circuit faults. Glitch injection is one of the least complicated methods of fault injection, and therefore it can be considered as a broad threat to secure circuits.
- **Voltage Starving** can be used to artificially lengthen the critical path of a circuit, to a point where it extends beyond the clock period [5]. This method is similar to injection of clock glitches, but it does not offer the same precise control of fault timing.
- **Voltage Spikes** cause an immediate change in the logic threshold levels of the circuit [3]. This changes the logic value held on a bus. Voltage spikes can be used, for example, to mask an instruction read from memory while it is moving over the bus. Similar to clock glitches, voltage spikes have a global affect and affect the entire circuit.
- **Electromagnetic Pulses** cause Eddy currents in a chip, leading to erroneous switching and isolated bit faults [30]. Using special probes, EM pulses can be targeted at specific locations of the chip.
- **Laser and Light Pulses** cause transistors on a chip to switch with photoelectric effects [39]. Through focusing of the light, a very small area of the circuit can be targeted, enabling precise control over the location of the fault injection.
- **Hardware Trojans** can be a source of faults as well. This method requires that the adversary has access to the circuit design flow, and that the design is directly modified with suitable trigger/fault circuitry. For example, recent research reports on an FPGA with a backdoor circuit which disables the readback protection of the design [33].

The fault injection mechanism determines the timing of the fault, the duration of the fault (transient or permanent), and the fault intensity (weak/strong).Together, these characteristics enable the adversary to select a specific fault model, which is needed as the starting point of cryptanalysis by fault injection. The method of fault injection also influences the difficulty of performing it. Depending on the level of tampering required with the actual circuit, one distinguishes noninvasive, semi-invasive [34], and invasive attacks. Table 9.2 illustrates the relation between the aforementioned six possible fault injection mechanisms, along with the fault models resulting from their use.

Table 9.2 Fault injection mechanisms and associate fault effects

Injection	Fault characteristic			Fault Model[a]				Invasiveness
	intensity	Timing	Duration	Chosen bit	Single bit	Byte	Random	
Glitches	Variable	Precise	Transient			•	•	Noninvasive
Starving	Variable	Loose	Transient			•	•	Noninvasive
Spikes	Fixed	Precise	Transient				•	Noninvasive
EM Pulse	Variable	Precise	Transient		•	•	•	Noninvasive
Laser Pulse	Fixed	Precise	Transient	•	•	•	•	Semi-invasive
Trojans	Fixed	Precise	Permanent	•	•	•	•	Invasive

[a]• means the Fault Model can be generated with this fault injection method

9.3 Fault Measurement

A fault model and a fault injection mechanism to trigger the fault model are two essential ingredients of a fault attack. But to apply them in a successful fault attack, we need to consider a larger scope. Figure 9.1 shows that a successful fault attack consists of two steps, fault measurement and fault analysis. The fault injection is part of the fault measurement phase, while the fault model is a building block in fault analysis.

Indeed, Fig. 9.1 shows both the requirements of a successful fault attack and the principles of the fault-attack resistant design. From the adversary's point of view, each step of the pyramid should be followed in order. An adversary first needs to choose a fault model and fault analysis technique based on the target cryptosystem, Then, he needs to obtain exploitable faults by following the steps of fault measurement, from fault injection access to fault observation.

From the designer's side, the steps of this pyramid should be considered while designing a fault attack resistant device. For each step, the designer should evaluate the costs and benefits of securing the design against this step. Using the evaluation results, the designer is able to make design decisions to prevent the adversary from building the required fault model. Next, we explain the steps of fault measurement and demonstrate them using a case study.

9.3.1 *Fault Measurement Steps*

The reality of fault measurement is more complicated than injecting a fault. First of all, the adversary needs to be able to physically inject a fault. The fault injection also needs to have the desired effect, and result in an exploitable fault, that results in the required fault model. Finally, the exploitable fault needs to be observable. Following is a more comprehensive definition of these four levels.

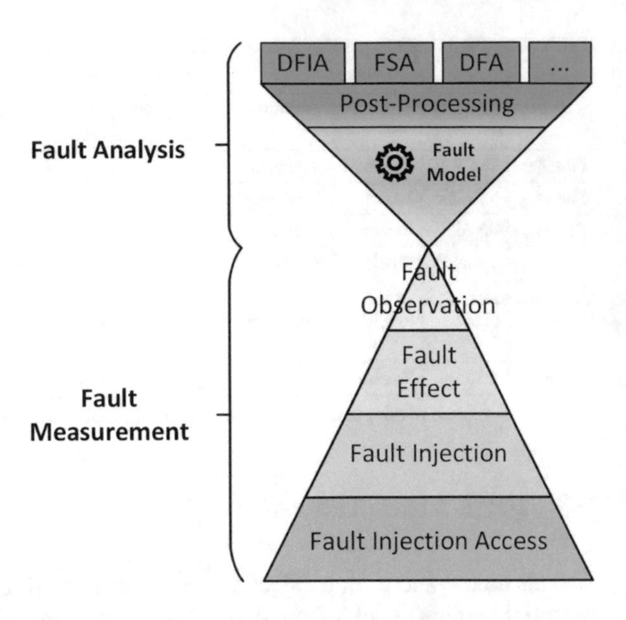

Fig. 9.1 A fault attack requires fault measurement and fault analysis

- *Fault Injection Access*: The first and foremost step of the fault measurement is getting physical access to the device under test (DUT). For example, an adversary needs to control external clock and supply voltage ports of DUT for clock and voltage glitching attacks, respectively [2]. Similarly, the adversary must have physical access to chip surface for laser and electromagnetic pulse-based fault attacks [35]. In addition, an adversary may also need to control data inputs and outputs of DUT. The amount of physical access needed for each attack is different.
- *Actual Fault Injection*: The second step of the fault measurement is disturbing the operation of DUT by applying a physical stress on it. The applied physical stress pushes DUT out of its normal operating conditions and causes faulty operation. Based on the chosen fault injection method, the adversary can control the timing, location, and intensity of the applied physical stress. Each value of these three parameters affects DUT differently and causes different faults in DUT operation. Therefore, the adversary needs to carefully set these parameters to create an exploitable fault in DUT operation. In clock glitching, for example, the adversary causes setup time violation by temporarily applying shorter clock cycles. The adversary can control the timing and length (i.e., intensity) of the applied shorter clock cycle. However, there is no control on the location of the applied physical stress (i.e., a shorter clock cycle) in this case because clock is a global signal for DUT.
- *Fault Effect*: The third step is creating a fault effect on DUT operation as a consequence of the applied physical stress. The fault effect can be defined as the logical (or digital) effect of the applied physical stress on DUT operation. For example, an applied clock glitch might create 2-bit faults at the fault injection point. Similarly, a laser pulse might affect 1-bit of DUT. On the other hand, it is also possible to not

create any fault effect even though a physical stress is applied on DUT. The adversary has a limited and indirect control on the fault effect through controlling the physical stress. The fault effect depends on various factors such as circuit implementation, used fault injection method, applied physical stress, etc. The adversary may need to apply several physical stresses with different parameters to create the desired fault effect on DUT operation [13, 22].

- *Fault Observation*: The final step toward the fault measurement is to observe the effects of the fault injection in the output of the block cipher or the algorithm under the attack. The authors of [41], show that there are methods that can compute the probability of success of a fault injection attempt in this phase using observability analysis. This method basically computes the probability of observability from the point of fault injection to the output of the cipher. In this work, we use a probability-based observability analysis method for the probability of propagating an exploitable fault to the output. Observability analysis, which is widely used in VLSI test area, reflects the difficulty of propagating the value of a signal to primary outputs [38].

9.3.2 Fault Measurement Case Study

In this section, we apply the ideas of fault measurement to a specific example. Figure 9.2 shows a simple combinational logic of two rounds of a hypothetical block cipher. Each round is composed of an SBOX module and an XOR gate. SBOX is a substitution module that obscures the relationship between the secret key and the output. To explain the fault model and fault injection steps, we assume that the intentional fault model required for an attack is to inject a random fault into the output of the combinational logic in round 9 shown in Fig. 9.2. The injected fault must then be propagated through round 10 and be observable by the adversary in the output of round 10.

Each combinational block requires a certain *propagation delay* (T_{pd}) to compute its output value. For the correct operation of the circuit, combinational block outputs must settle to their final values and remain stable at least some *setup time* (t_{su}) before the sampling clock edge. Therefore, the *clock period* (T_{clk}) must satisfy the following equation for all paths from input registers to output registers:

$$T_{clk} \geq T_{pd} + T_{su} \tag{9.1}$$

This equation specifies the *setup time constraints* of a circuit. The setup time constraint of the longest (i.e., critical) path determines the minimum clock period for the circuit. Applying a shorter clock period than this value will fail the setup time constraints.

In this case, we inject faults into the operation of a circuit by violating its setup time constraints. *Setup time violation* is a widely used low-cost fault injection mech-

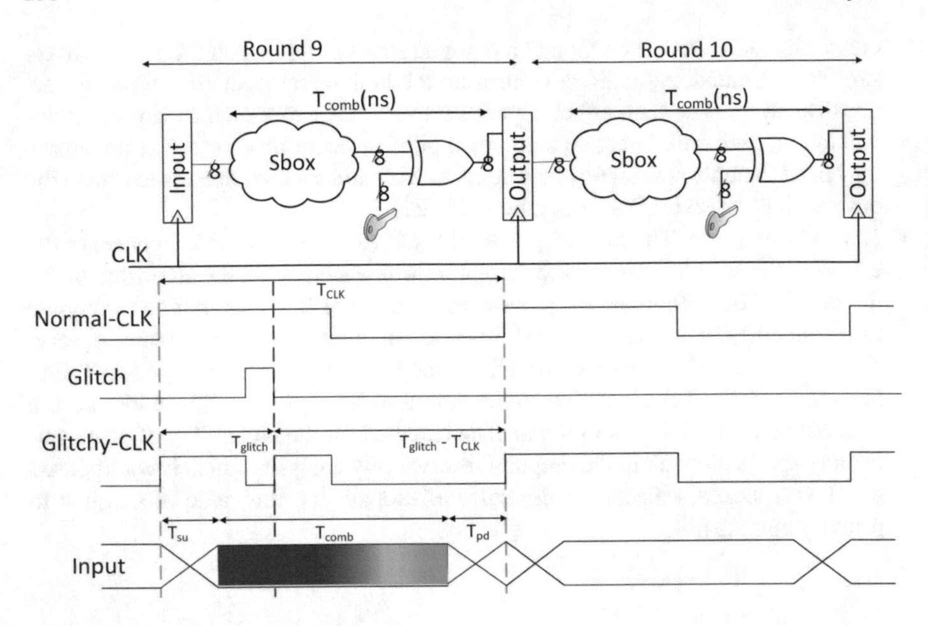

Fig. 9.2 Sample combinational logic

anism [1]. In the following paragraphs, we explain fault measurement process for fault injection using setup time violation.

- *Fault Injection Access*: In synchronous circuits the data is processed by combinational blocks, which are surrounded by input/output registers. The data is captured when the sampling edge of the clock signal arrives at the registers. The attacker must have access to the clock signal that is driving this circuit.
- *Actual Fault Injection*: An adversary can cause setup time violation via clock glitches. Figure 9.2 shows the effect of a glitch on the clock signal. As shown, the glitch signal is XORed with the normal clock. During round 9, a clock glitch will temporarily shorten the clock cycle period from T_{clk} to T_{glitch}, thereby causing timing violation of the digital logic.
- *Fault Effect*: When the *glitch period* (T_{glitch}) violates timing constraint of a path, the output value of this path is captured before its computation is completed. As shown in Fig. 9.2, the computation of the data in the combinational logic is not yet finished and it is captured as output due to the glitchy clock. Therefore, the captured value can be faulty.
- *Fault Observation*: To observe the effects of fault injection, the output of round 10 must be different from the correct value. Therefore, the effect of injected fault into the output of round 9 must be propagated through round 10 without being masked. If the value of output 9 is different from its correct value and is not masked by the round 10 operations, the fault measurement process is successful.

9.4 Fault Analysis

In this section, we will explain the pyramid in Fig. 9.1 from the adversary's point of view. Any fault attack consists of two phases: a *fault measurement* phase and a *fault analysis* phase.

Based on the adversary's access to the target device and the information required to attack a specific block cipher, the adversary aims for a fault model. Therefore, choosing the fault model is a part of the fault analysis process. Then, in the measurement process, the adversary goes through the four steps mentioned in Sect. 9.3 to build the fault model using actual measurements. In this section, we explain the fault attack process with three different fault models. All of the example attacks are on the advanced encryption standard (AES) algorithm. The details of this algorithm are explained in the following section.

9.4.1 Advanced Encryption Standard

The AES algorithm consists of 10 rounds. The first 9 rounds have 4 main operations, SBOX, ShiftRows(SR), MixColumns(MC), and AddRoundKey(ADK). Round 10 omits the MixColumn operation. Figure 9.3 shows the structure of the AES algorithm. In this figure, P is the applied plaintext to the AES algorithm, S_{10} is the intermediate state variable for round 10. K_{10} is the key for round 10 and C represents the ciphertext. The faulty value of the variable x is shown by x'.

9.4.2 Differential Fault Analysis

DFA is one of the most studied types of attack on cryptographic systems such as RSA [9], DES [6], and AES. DFA assumes that the attacker is in possession of the device and is able to obtain two faulty and fault free ciphertexts for the same plaintext.

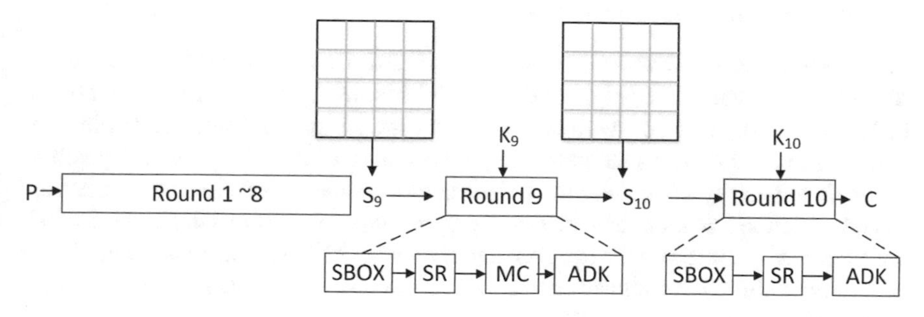

Fig. 9.3 AES structure

DFA also assumes that the attacker is aware of some characteristics of the injected fault. There are many proposed types of DFA attack on the AES algorithm [7, 27, 29]. These attacks are based on different fault models and choose various methods of injection techniques based on the fault model. In this section, we will explain the steps of a simple electromagnetic pulse-based DFA attack on AES, which is proposed by Dehbaoui et al. [11].

Fault Model: This attack adopts Piret's fault model [28]. This fault model requires an adversary to induce one byte fault in AES state between the start of round 9 and the MixColumns operation.

Fault Measurement: The DUT for the attack is a RISC microcontroller running AES algorithm. In this attack, the adversary injects faults by means of transient electromagnetic (EM) pulses. A small magnetic foil is used apply EM pulses without any physical contact to DUT. The adversary can control the timing, energy, and position of the applied EM pulses. Using different combinations of these three parameters, the adversary can affect only one byte of the computation and can select the affected byte. As a result, an adversary can induce the exploitable faults using this setup.

Fault Analysis: Due to the MixColumns in round 9, one byte fault in the beginning of round 9 will cause 4 faulty bytes in the ciphertext. Therefore, we can find 4 bytes of the key of round 9. Assuming that the attacker only injects fault into one byte, the C and C' differ in four bytes. There are 255×4 possible values for these four bytes which is saved in a list *D*. For each key guess, the adversary should compute the value of these four bytes using inverse equations of AES operations. The key is potentially a correct candidate if the computed value is in the list *D*. The adversary should continue injecting fault in the same location until only 1 key remains as the candidate.

9.4.3 Fault Sensitivity Analysis

Fault sensitivity analysis (FSA) attack is proposed by Li et al., in CHES 2010 [22]. This attack is based on the fact that fault behavior is biased and can have a data dependency on a secret value in a circuit.

Fault Model: If an adversary gradually increases the fault intensity, a circuit can reach a point at which the output of the circuit becomes faulty. This threshold point is called *fault sensitivity*. For setup time violation, the critical path delay determines the fault sensitivity point. As the path delay distribution of a circuit is data-dependent, the fault sensitivity of the circuit is also data-dependent. Therefore, the FSA fault model is defined as a dependency between the input values of the SBOX to their fault sensitivity. Based on the experiments shown in [22], the input values with larger Hamming weight, have longer critical path delays as well. In this attack, the target of the fault injection is round 10 of AES.

Fault Measurement: Based on the fault model, the authors of [22] choose the setup time violation or clock glitch for their measurements. The assumption is that the attacker is in physical possession of the device, therefore, he has access to the external clock signal. The fault injection should be in the output of round 10 in the AES algorithm for each byte of the ciphertext.

To inject the actual fault, the attacker increases the intensity of the fault injection in the last round, gradually by increasing the frequency of the applied external clock. As explained in Sect. 9.3, increasing the clock frequency, the timing paths of the circuit will eventually be violated. Therefore, there is a moment that the output of round 10 of the block cipher is not correct anymore. To build the fault model, the adversary should apply several input values to the block cipher and record the clock frequency corresponding to fault sensitivity for each input data.

Fault Analysis: For FSA, the effect of fault bias is the data dependency of fault sensitivity. The adversary first inverts the ciphertext to round 10 input (S_{10}) using a key guess. Then, he estimates the effect of fault bias as the Hamming Weight of round 10 input $HW(S_{10})$. In this step, the attacker uses the Pearson correlation coefficient to find the key guess for which the fault sensitivity is strongly correlated to $HammingWeight(S_{10})$ for all inputs.

9.4.4 *Differential Fault Intensity Analysis*

Differential Fault Intensity Analysis (DFIA) is proposed by Ghalaty et al. [13]. This attack is also based on the concept of biased fault behavior, but has a different perception of the biased fault from the FSA attack.

Fault Model: Fault intensity is the strength by which a circuit is pushed outside of its nominal operating conditions with the intent of inducing a fault. For example, when faults are introduced using clock glitches, then the fault intensity corresponds to the length of clock cycle that is obtained as a result of the glitches. The target of fault injection for DFIA is the output of round 9 of AES. The fundamental assumption of DFIA on fault model relies on the fact that a small change in fault intensity will result in a small change in fault behavior.

Fault Measurement: The target of fault injection for this attack is the output of round 9 in AES algorithm. The adversary can determine the timing of fault injection by power analysis methods such as the methods mentioned in [19] to find the start and end of each round of AES. The fault injection process is similar to FSA attack. In this attack, first, the attacker applies an input to the AES algorithm. Then, he gradually increase the fault intensity by increasing the clock frequency at the output of round 9. Due to the biased effect of the fault injection and non-uniform distribution of timing paths in the circuit, the number of faulty bits increases by increasing the fault intensity.

Fault Analysis: To estimate the small change, the adversary computes the input of round 10 (S'_{10}, S''_{10},...), by inverting the faulty ciphertexts and key guess for several fault intensity levels. Then, he computes the distance between the hypothesized intermediate variables by using the Hamming Distance function.

The fault bias assumption for DFIA enables the use of a distinguisher that looks for the smallest change. Unlike the previous techniques, DFIA can combine fault behaviors collected at multiple fault intensities. Hence, the complete fault bias characteristic of a circuit can be exploited. Based on the assumption of fault attack, the error values are close to each other for the correct key guess. For wrong key guesses, the distance between injected error values will be random due to the non-uniform behavior of the SBOX module. Therefore, the distinguisher function simply chooses the key that shows the minimal distance between intermediate variables.

9.5 A Systematic Organization of Fault Attack Countermeasures

In this section, we systematically organize the existing *fault attack countermeasures* with respect to the steps of *fault measurement*. We list a few examples of countermeasures for each step of the fault measurement and explain their main principles. Figure 9.4 shows the listed countermeasures and the steps of fault measurement together.

As discussed in Sect. 9.2, the requirements of a successful attack are fault measurement and fault analysis. To build the required fault model for further fault analysis, the attacker must go through the steps of the fault measurement pyramid from bottom to top. He first has to obtain access to the injection device and finally observe

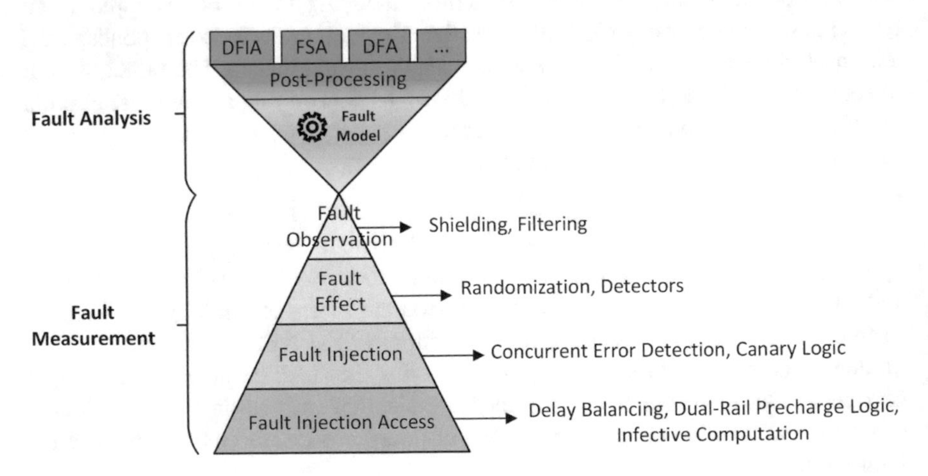

Fig. 9.4 List of countermeasures and the steps of fault measurement

the effects of the injected fault at the output of the system. An interruption in any of these steps would lead to an unsuccessful fault measurement and therefore a failure in the fault attack process. From the designer's point of view, to build a fault attack resistant design, he should be able to thwart any of the steps of the pyramid and prevent the attacker from progressing to the next step.

Accordingly, the designers of fault attack resistant hardware systems should aim at thwarting the steps of *fault measurement*. Therefore, for each step of the pyramid given in Fig. 9.4, the options for securing the design against this step should be evaluated at the design-time. Using the formulation in Fig. 9.4, the designer can address each step independently. Depending on the security requirements and design constraints, it is also possible to combine different countermeasures that are designed for different steps of the *fault measurement*. Each countermeasure brings an overhead on the design while increasing the level of security. Thus, there is always a tradeoff between the cost and security of a design. The designers can use the pyramid given in Fig. 9.4 for a better and more systematic security evaluation of the countermeasures while comparing cost-security tradeoff of different countermeasure options.

Next, we provide a survey of existing countermeasures against the *fault attacks*.

9.5.1 Countermeasures for Fault Injection Access

The countermeasures thwarting this step aim at preventing an adversary to gain physical access to the device so that the adversary cannot apply any physical stress on DUT.

Shielding

The main principle of shielding is covering sensitive parts of DUT by a protective layer to make these parts physically inaccessible. For example, a passive metal layer can be used to make laser or electromagnetic pulse attacks harder. In this case, an adversary needs to remove the shield to gain physical access to the sensitive parts of DUT. Similarly, active shields, metal meshes that covers the sensitive circuit parts, can also be employed for this purpose. In this case, random sequences of data are continuously fed to the mesh and the values of the sequences are checked after they pass through the mesh. If a disconnection or modification is detected, the DUT does not operate correctly anymore [3]. An ultraviolet (UV)-resistant dye can also be used for protection against fault attacks that use UV light [4]. These countermeasures makes fault injection access step of the fault measurement harder.

The main problem of shielding is its cost [4]. In addition, shielding has some physical limitations. For example, a laser pulse can pass through the gaps in a shield [39]. Alternatively, an adversary can make the shield ineffective by applying a laser pulse from back-side of DUT instead of applying it from the protected front-side.

Filtering

The main principle of filtering is to reduce or filter out the effects of the external physical stress by placing on-chip components between some external pins (e.g., power supply pin) and the internal circuitry of DUT. For example, some devices have built-in voltage regulators, which first conditions the external supply voltage and then applies the conditioned supply voltage to the internal circuitry [40]. The voltage regulators filter out some noise and glitches at the external supply voltage. However, the filtering capabilities of a voltage regulator depends on its design and the load capacitance. Therefore, some glitches are able to pass the regulator and affect the internal circuitry. The problems of this countermeasure are its cost and physical capabilities. The voltage regulator can only filter glitches with specific parameters. Therefore, an adversary might create exploitable faults by applying a physical stress that is outside of filtering capabilities of the regulator.

9.5.2 Countermeasures for Fault Injection

The purpose of the countermeasures at this step is detecting the dangerous physical changes in DUT's environment. After detecting a dangerous change, they produce an *alarm* signal to indicate possible fault in the DUT operation.

Randomization

In the second step of the fault measurement, an adversary attempts to inject exploitable faults in DUT operation by applying a physical stress on the device with certain timing, location, and intensity parameters. Randomization aims at creating randomness on one of these parameters to make fault injection harder. For example, in most fault attacks, an adversary needs to inject faults in a certain point of the executed cryptographic algorithm. Therefore, the adversary must be able to predict when this certain point of the algorithm is executed. Randomization can make this prediction harder by using random clock cycles or a random execution strategy. Timing randomness at the clock cycle level can be introduced by using a random internal clock signal which is generated based on a random bit sequence and the external clock signal [20]. However, this solution degrades the performance as only some of the external clock cycles are used for actual computation. A random execution strategy, in which the operations of a cryptographic algorithm are executed in a random order [8, 25], can also be used as a means of randomization.

Although randomization increases the required time for a successful fault attack, it does not provide a perfect protection. For instance, Van Woudenberg et al. presented an optical fault injection attack on a microprocessor protected by an unstable internal clock [39]. In their attack, they use a pattern-based trigger generator, which can detect when the target algorithm executed by observing the power consumption

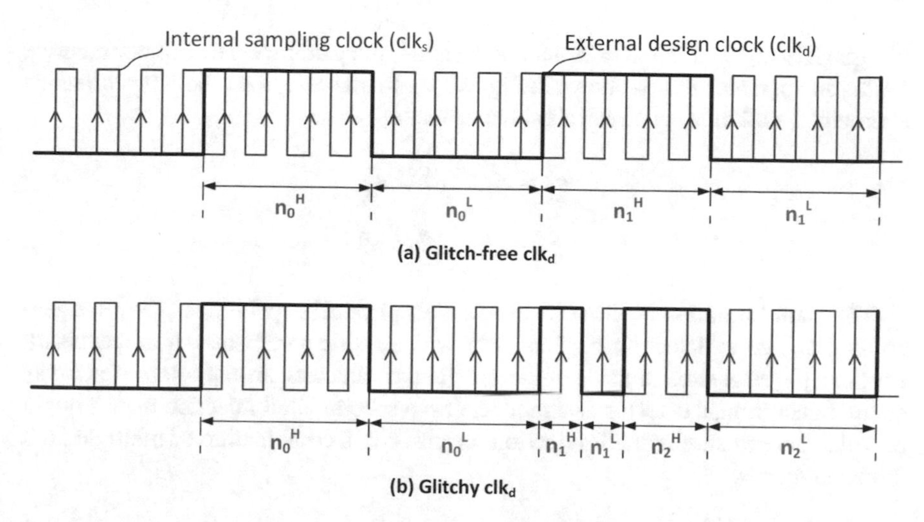

Fig. 9.5 **a** Operation of the Clock Monitor in the case of a Glitch-Free External Clock. **b** Operation of the Clock Monitor in the case of a Glitchy External Clock

of the microprocessor. In addition to this weakness, randomization also brings a timing overhead because DUT does useful computation during only some of the clock cycles.

Detectors

The fault injection can also be thwarted using detectors that detect anomalies in the physical environment of DUT. These detectors can sense the changes in the voltage, light, temperature, and clock frequency. After the detection of an anomaly, an *alarm* signal is raised and the required security action is taken by the circuit.

For example, Luo et al. proposed a clock monitor that detects if there is an anomaly in the clock signal of a circuit and raises an alarm [24]. The proposed clock monitor relies on the fact that a clock glitch creates irregularity in the clock signal. To detect such irregularity, they sample the external design clock (clk_d) with a faster internal sampling clock (clk_s) as illustrated in Fig. 9.5. For each cycle i of the external clock clk_d, they measure the length of high phase (n_i^H) and low phase (n_i^L) using counters. Then, they compare the measured parameters of two consecutive clock cycles i and $i+1$. If the parameters do not match, an *alarm* signal is raised. If there is no glitch in the external clock, the following equations are satisfied as it is shown in Fig. 9.5(a):

$$n_0^L = n_1^L$$

$$n_0^H = n_1^H$$

If a glitch is injected in the external clock signal, the parameters of two consecutive cycles do not match as it is shown in Fig. 9.5(b). In this case, the following equations is obtained, and thus, an *alarm* signal is generated:

$$n_0^L \neq n_1^L, \quad n_1^L \neq n_2^L$$

$$n_0^H \neq n_1^H, \quad n_1^H \neq n_2^H$$

The main limitation of the detectors is their physical capabilities. They are generally designed to detect physical stresses with specific parameters. If an adversary applies a physical stress outside of the specified parameters, an exploitable fault may occur. In addition, a detector designed against a specific fault injection means might be vulnerable to another fault injection means or to a combination of multiple fault injection means.

9.5.3 Countermeasures for Fault Effect

In this step, the countermeasures are designed to catch the fault effects caused by the physical stress applied in the fault injection step. The countermeasures monitor the values of the signals and generate an *alarm* signal in case of a faulty signal value.

Concurrent Error Detection (CED)

The main principle of the concurrent error detection (CED) is detecting the faults in parallel with the normal operation of DUT. Most of the proposed CED techniques follows the general scheme shown in Fig. 9.6 [26]. In this scheme, a design consists of three blocks: *operation*, *prediction*, and *checker*. The *operation* block takes inputs and produces outputs based on the DUT specification. The *prediction* block takes the same inputs as the *operation* block and predict some special characteristics of system outputs based on these inputs. The *prediction* block can be designed by

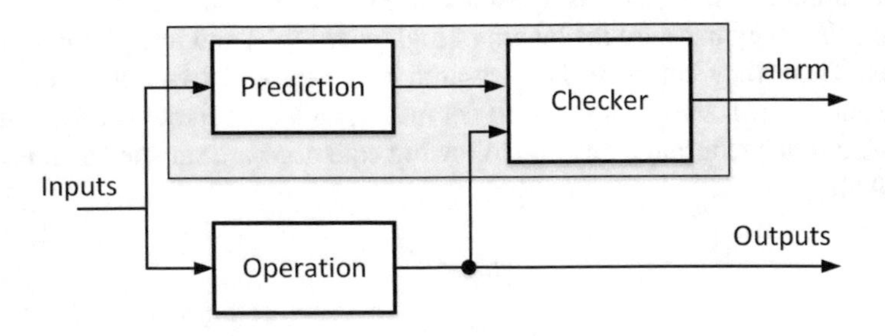

Fig. 9.6 A conceptual diagram for concurrent error detection (CED) countermeasure

utilizing hardware redundancy, time redundancy, information redundancy, or hybrid redundancy [15]. The *checker* block checks if the outputs of the *operation* block shows some special characteristics by comparing them with the outputs of *prediction* block. If the outputs of *operation* and *prediction* blocks do not match, an *error* signal is generated.

In the hardware redundancy, the *operation* block is duplicated as the *prediction* block. The *checker* block compares if the outputs of the *prediction* and *operation* blocks are the same. The area overhead of this technique is at least 100 % while its timing overhead is almost zero.

In time redundancy, the operation is computed twice with the same input and the results are compared to each other. In other words, the hardware of the *operational* block is also used as the *prediction* block. Although its area overhead is almost zero, the timing overhead of this technique is 100 %.

The common weakness of the hardware and time redundancy techniques is that the attacker can bypass them if the same fault can be injected into the *operation* and *prediction* blocks.

The information redundancy utilizes error detecting codes for the *prediction* block. In these techniques, some check bits are added to inputs and they are propagated with the inputs. After the computation of the outputs, the results are verified with the check bits from the *prediction* block. For example, using parity bits is a common example of these techniques. Similarly, robust codes and cyclic redundancy check (CRC) can also be used as a CED countermeasure. The information redundancy allows various area, time, and security tradeoffs based on the used error detecting code method [15].

The hybrid redundancy-based techniques combine the properties of the previous CED techniques. For instance, an operation can be followed by the inverse of this operation and the results these two operations can be compared to detect faults [18].

One of the common weaknesses of the above CED techniques is that they assume a uniform fault distribution. However, Guo et al. showed that an adversary with the capability of injecting biased faults might bypass most of these CED techniques [15]. Another common weakness is that an adversary might bypass a CED countermeasure by tampering with the final *checker* block of it.

Canary Logic

The fault effect caused by setup time violation [32] can be detected using Canary logic [31], which predicts timing errors through circuit-level timing speculation. In Canary logic, each flip-flop (FF) in the design is converted into a timing error-predicting Canary FF by adding a delay element, a shadow FF, and an XOR gate (Fig. 9.7). The input data of the shadow FF is the delayed version of the main FF. The timing errors are predicted by comparing the outputs main and shadow FFs via the XOR gate. The output of the XOR gate is used as an *alarm* signal indicating a timing error is about to occur. Because of the delay element, the shadow FF encoun-

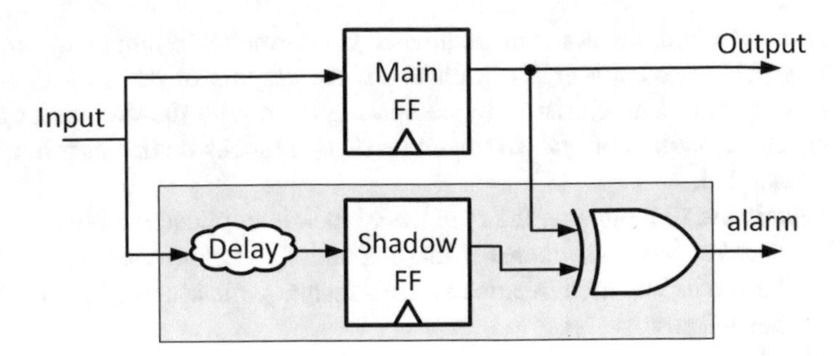

Fig. 9.7 Block diagram of the Canary FF

ters a timing error before the main FF. Therefore, the shadow FF protects the main FF against timing errors.

The main problem is this countermeasure is its area overhead. The area of a canary FF is at least two times larger than of a regular FF because of the additional logic. Therefore, replacing each regular FF with a canary FF brings a high area overhead. A weakness of the canary logic is that it cannot raise the *alarm* signal if a timing error occurs in both the main and shadow FFs [21]. Considering fault attacks are based on intentional fault injection than random faults, this case is likely to happen.

9.5.4 Countermeasures for Fault Observation

The main purpose of the countermeasures at this step is making a faulty output independent of the processed secret data. These countermeasures let an adversary to inject faults and let the injected fault to propagate the output of DUT. However, they guarantee that the adversary cannot exploit the faulty output in the fault analysis.

Delay Balancing

Delay balancing can be used to thwart FSA attacks that use setup time violation as the fault injection means. In these attacks, characterizing the data dependency of DUT's fault sensitivity (i.e., dynamic critical path) is the main issue for the adversary. Therefore, eliminating the factors that affect the data dependency of fault sensitivity is an effective countermeasure for such attacks. Ghalaty et al. [12] proposed a systematic delay balancing countermeasure to remove the dependency of the critical timing delay to the processed data values in the circuits. They propose a transformation that operates at two levels of abstraction, at netlist level and at gate level.

- Netlist level: The delay of the netlist must be independent of the input data.
- Gate level: The switching time of gates must be random during circuit evaluation, meaning that the switching distribution is uniform over the computation time of the circuit.

The proposed countermeasure is based on inserting delay elements in different paths of the circuit based on the statistical timing analysis of the circuit. The goal is to equalize the effective delay of each path in a circuit. The effective delay of a path is defined as the number of effective gates (i.e., AND and OR gates) in this path multiplied by their propagation delays. In the countermeasure, first, the effective delay from each output to each input within fan-in cone of it is computed. Then, the maximum effective delay is determined. Finally, delay elements are inserted into the input side of each path such that the sum of effective path delay and inserted buffer delay becomes equal to the maximum effective delay. The number of delay elements is inversely proportional to the length of the path. As a result, this countermeasure eliminates the data dependency of the critical path delay. Therefore, the adversary cannot exploit the faulty outputs. The main problem of this countermeasure is its area cost.

Dual-Rail with Precharge Logic (DPL)

Dual-rail with precharge logic (DPL) is a countermeasure that was originally proposed against side-channel attacks [10]. However, it is also inherently resistant against some fault attacks. The main principle of DPL is to make the power consumption independent of the processed data by consuming a constant amount of power at each cycle.

In DPL, every signal α is represented by two complementary wires (α_f, α_t). Every computation has two phases, namely, precharge and evaluation. In the precharge phase, all wires are initialized to the same value. Depending on the implementation, this initialization value is $(0, 0)$ or $(1, 1)$, called *NULL*0 and *NULL*1. These two values are *NULL* tokens that do not contain any meaningful information. In the evaluation phase, the actual computation takes place and *NULL* tokens alternate to *VALID* tokens: $(1, 0)$ or $(0, 1)$, called *VALID*0 and *VALID*1. These two values are *VALID* tokens that contain the value of the signal α. During evaluation, exactly one of the complementary wires is toggled.

The inherent fault attack resistance of DPL is based on the fact that a fault turns a *VALID* token into a *NULL* token [14]. The output value of a gate will be *NULL* if any input of it is a *NULL* token. Considering high diffusion capabilities of cryptographic algorithms, a *NULL* token will diffuses the very quickly while it is propagated to the outputs. As a result, the faulty output does not carry any information about secret data. As it is seen, faults are not detected in DPL. Instead, faulty values are allowed to propagate to the outputs, knowing that they are not exploitable by an adversary. Converting a single-rail logic into a dual-rail logic will bring both area and time overhead.

Infective Computation

The main principle of the infective computation is to make the faulty output look random (i.e., non-exploitable) [23]. This is achieved by propagating the fault effects

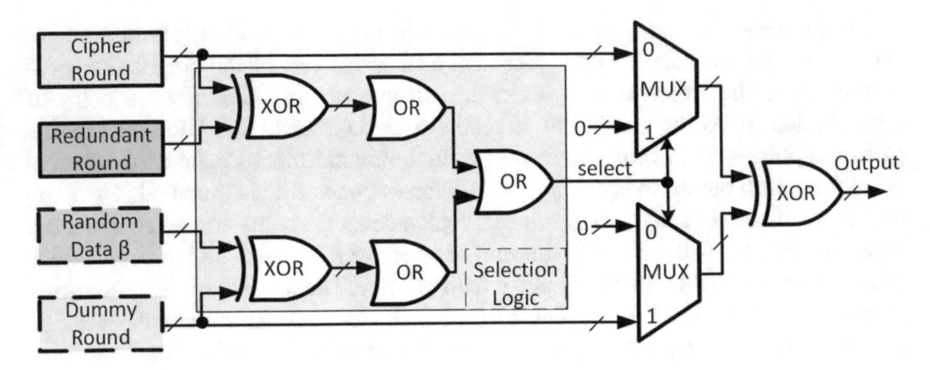

Fig. 9.8 A conceptual diagram for the infective computation countermeasure

to the whole computation with a diffusion scheme. The diffusion scheme has no effect if the fault injection is not successful. Infective computation techniques do not require checking procedures, and thus they do not alter the computation flow.

In CHES 2014, Tupsamudre et al. proposed an infective countermeasure for AES that utilize redundant and dummy rounds [37]. The conceptual diagram of their countermeasure is shown in Fig. 9.8. In the *cipher* and *redundant* rounds, the round function of AES (f_{AES}) is applied on the plaintext. Thus, each AES round is executed twice in this countermeasure. There are also *dummy* rounds that are randomly executed throughout the execution of the algorithm. In a dummy round, the AES round function is applied on a random data β and a dummy secret key k^d. The output of the dummy round is the random data β. After each computation of an AES round, a *selection logic* decides the output value. The *selection logic* computes if there is an error in any of the *dummy*, *redundant*, and *cipher* rounds by applying simple XOR and OR operations. Then, the output signal of the *selection logic* (i.e., *select* signal) selects the output value. In other words, the *select* signal activates/deactivates a diffusion mechanism. If the *select* signal is 0, the result of the cipher round is assigned to the output (Fig. 9.8). Otherwise, the result of dummy round, which is random and independent of the secret key, is assigned to output (Fig. 9.8). Therefore, the faulty output cannot be exploited by the adversary.

9.6 Conclusion

In this chapter, we presented a systematic approach for fault-attack resistant design. To this end, we proposed a hierarchy of steps that an attacker or a designer should go through to attack a device or build a secure device. The main objective of this classification is to clarify the difference between the fault attack resistant designs and fault tolerant design. The key difference between these two is that in presence of fault, the fault tolerant design must continue to work correctly, while the fault attack resistant deign must prevent secure data leakage. The proposed hierarchy, classifies the requirements of fault measurement and fault analysis for attacking a device into four steps. The interruption of any of these steps prevents the attacker from launching a successful fault attack. Therefore, the pyramid of fault measurement (Fig. 9.4) can be used by the designer to build a fault attack resistant design considering the costs and security coverage of countermeasures in each step. The designer can find the optimized combination of countermeasures to prevent fault measurement for an attacker. While the pyramid provides a road map for designers to apply fault attack countermeasures, it does not provide information on cost and security efficiency of each countermeasure. This is considered as an open research problem.

Acknowledgements This research was supported through the National Science Foundation Grant 1441710, and through the Semiconductor Research Corporation.

References

1. Agoyan, M., Dutertre, J.M., Naccache, D., Robisson, B., Tria, A.: When Clocks Fail: On Critical Paths and Clock Faults. In: Smart Card Research and Advanced Application, pp. 182–193. Springer (2010)
2. Balasch, J., Gierlichs, B., Verbauwhede, I.: An in-depth and black-box characterization of the effects of Clock Glitches on 8-bit MCUs. In: 2011 Workshop on Fault Diagnosis and Tolerance in Cryptography (FDTC), pp. 105–114 (2011)
3. Bar-El, H., Choukri, H., Naccache, D., Tunstall, M., Whelan, C.: The Sorcerer's Apprentice guide to fault attacks. Proc. IEEE **94**(2), 370–382 (2006). Feb
4. Barenghi, A., Breveglieri, L., Koren, I., Naccache, D.: Fault injection attacks on cryptographic devices: theory, practice, and countermeasures. Proc. IEEE **100**(11), 3056–3076 (2012). Nov
5. Barenghi, A., Bertoni, G.M., Breveglieri, L., Pellicciolі, M., Pelosi, G.: Injection technologies for fault attacks on microprocessors. In: Joye, M., Tunstall, M. (eds.) Fault Analysis in Cryptography. Information Security and Cryptography, pp. 275–293. Springer, Berlin (2012)
6. Biham, E., Shamir, A.: Differential fault analysis of secret key cryptosystems. In: Advances in CryptologyCRYPTO'97, pp. 513–525. Springer (1997)
7. Blömer, J., Seifert, J.P.: Fault based cryptanalysis of the advanced encryption standard (AES). In: Financial Cryptography, pp. 162–181. Springer (2003)
8. Bo, Y., Xiangyu, L., Cong, C., Yihe, S., Liji, W., Xiangmin, Z.: An AES chip with DPA resistance using hardware-based random order execution. J. Semicond. **33**(6), 065009 (2012)
9. Boneh, D., DeMillo, R.A., Lipton, R.J.: On the importance of eliminating errors in cryptographic computations. J. Cryptol. **14**(2), 101–119 (2001)
10. Danger, J.L., Guilley, S., Bhasin, S., Nassar, M.: Overview of dual rail with Precharge logic styles to thwart implementation-level attacks on hardware cryptoprocessors. In: 2009 3rd International Conference on Signals, Circuits and Systems (SCS), pp. 1–8. IEEE (2009)
11. Dehbaoui, A., Dutertre, J.M., Robisson, B., Orsatelli, P., Maurine, P., Tria, A.: Injection of transient faults using electromagnetic pulses-practical results on a cryptographic system. IACR Cryptol. ePrint Arch. **2012**, 123 (2012)
12. Ghalaty, N.F., Aysu, A., Schaumont, P.: Analyzing and eliminating the causes of fault sensitivity analysis. In: Proceedings of the Conference on Design, Automation & Test in Europe. p. 204. European Design and Automation Association (2014)
13. Ghalaty, N.F., Yuce, B., Taha, M., Schaumont, P.: Differential Fault Intensity Analysis. In: 2014 Workshop on Fault Diagnosis and Tolerance in Cryptography (FDTC), pp. 49–58. IEEE (2014)

14. Guilley, S., Sauvage, L., Danger, J.L., Selmane, N.: Fault injection resilience. In: 2010 Workshop on Fault Diagnosis and Tolerance in Cryptography (FDTC), pp. 51–65. IEEE (2010)
15. Guo, X., Mukhopadhyay, D., Karri, R.: Provably secure concurrent error detection against differential fault analysis. IACR Cryptol. ePrint Arch. **2012**, 552 (2012)
16. Joye, M., Tunstall, M. (eds.): Fault Analysis in Cryptography. Information Security and Cryptography. Springer, Berlin (2012)
17. Karaklajic, D., Fan, J., Verbauwhede, I.: A systematic M safe-error Detection in hardware implementations of cryptographic algorithms. In: 2012 IEEE International Symposium on Hardware-Oriented Security and Trust (HOST), pp. 96–101 (2012)
18. Karri, R., Wu, K., Mishra, P., Kim, Y.: Concurrent error detection schemes for fault-based side-channel cryptanalysis of symmetric block ciphers. IEEE Trans. Comput.-Aided Des. Integr. Circuits Syst **21**(12), 1509–1517 (2002)
19. Kocher, P., Jaffe, J., Jun, B., Rohatgi, P.: J. Cryptogr. Eng. **1**(1), 5–27 (2011)
20. Kömmerling, O., Kuhn, M.G.: Design principles for tamper-resistant Smartcard processors. In: USENIX Workshop on Smartcard Technology, vol. 12, pp. 9–20 (1999)
21. Kunitake, Y., Sato, T., Yasuura, H., Hayashida, T.: Possibilities to miss predicting timing errors in canary flip-flops. In: 2011 IEEE 54th International Midwest Symposium on Circuits and Systems (MWSCAS), pp. 1–4. IEEE (2011)
22. Li, Y., Sakiyama, K., Gomisawa, S., Fukunaga, T., Takahashi, J., Ohta, K.: Fault sensitivity analysis. In: Cryptographic Hardware and Embedded Systems, CHES 2010, pp. 320–334. Springer (2010)
23. Lomné, V., Roche, T., Thillard, A.: On the need of randomness in fault attack countermeasures-application to AES. In: 2012 Workshop on Fault Diagnosis and Tolerance in Cryptography (FDTC), pp. 85–94. IEEE (2012)
24. Luo, P., Fei, Y.: Faulty clock detection for crypto circuits against differential fault analysis attack. Cryptol. ePrint Arch. Report 2014/883. http://eprint.iacr.org/ (2014)
25. Markantonakis, K., Mayes, K.: Secure Smart Embedded Devices. Platforms and Applications. Springer, Berlin (2013)
26. Mitra, S., McCluskey, E.J.: Which concurrent error detection scheme to choose? In: Test Conference, 2000. Proceedings. International, pp. 985–994. IEEE (2000)
27. Moradi, A., Shalmani, M.T.M., Salmasizadeh, M.: A generalized method of differential fault attack against AES cryptosystem. In: Cryptographic Hardware and Embedded Systems-CHES 2006, pp. 91–100. Springer (2006)
28. Piret, G., Quisquater, J.J.: A differential fault attack technique against SPN structures, with application to the AES and KHAZAD. In: Cryptographic Hardware and Embedded Systems-CHES 2003, pp. 77–88. Springer (2003)
29. Quisquater, J.J., Samyde, D.: Electromagnetic analysis (EMA): measures and counter-measures for Smart Cards. In: Smart Card Programming and Security, pp. 200–210. Springer (2001)
30. Quisquater, J., Samyde, D.: Eddy current for magnetic analysis with active sensor. In: Esmart (2002)
31. Sato, T., Kunitake, Y.: A simple flip-flop circuit for typical-case designs for DFM. In: 8th International Symposium on Quality Electronic Design, 2007. ISQED'07, pp. 539–544. IEEE (2007)
32. Selmane, N., Guilley, S., Danger, J.L.: Practical setup time violation attacks on AES. In: Seventh European Dependable Computing Conference, 2008. EDCC 2008, pp. 91–96. IEEE (2008)
33. Skorobogatov, S., Woods, C.: Breakthrough silicon scanning discovers backdoor in military chip. In: CHES, pp. 23–40 (2012)
34. Skorobogatov, S.P.: Semi-invasive attacks—A new approach to hardware security analysis. Technical report. UCAM-CL-TR-630, University of Cambridge, Computer Laboratory (2005)
35. Skorobogatov, S.P., Anderson, R.J.: Optical fault induction attacks. In: Cryptographic Hardware and Embedded Systems-CHES 2002, pp. 2–12. Springer (2003)

36. Takahashi, J., Fukunaga, T., Gomisawa, S., Li, Y., Sakiyama, K., Ohta, K.: Fault injection and key retrieval experiments on an evaluation board. In: Joye, M., Tunstall, M. (eds.) Fault Analysis in Cryptography, pp. 313–331. Information Security and Cryptography, Springer, Berlin (2012)
37. Tupsamudre, H., Bisht, S., Mukhopadhyay, D.: Destroying fault invariant with randomization. In: Cryptographic Hardware and Embedded Systems–CHES 2014, pp. 93–111. Springer (2014)
38. Wang, L.T., Wu, C.W., Wen, X.: VLSI Test Principles and Architectures: Design for Testability. Academic Press (2006)
39. van Woudenberg, J., Witteman, M., Menarini, F.: Practical optical fault injection on secure microcontrollers. In: 2011 Workshop on Fault Diagnosis and Tolerance in Cryptography (FDTC), pp. 91–99 (2011)
40. Yanci, A.G., Pickles, S., Arslan, T.: Characterization of a voltage Glitch attack detector for secure devices. In: Symposium on Bio-inspired Learning and Intelligent Systems for Security, 2009. BLISS'09, pp. 91–96. IEEE (2009)
41. Yuce, B., Ghalaty, N.F., Schaumont, P.: TVVF: Estimating the vulnerability of hardware cryptosystems against timing violation attacks. In: 2015 IEEE International Symposium on Hardware Oriented Security and Trust (HOST), pp. 72–77. IEEE (2015)

Chapter 10
Hardware Trojan Attacks and Countermeasures

Hassan Salmani

10.1 Introduction

Reported by Ernst&Young LLP [1], modern devices like smart mobility and cloud computing are demanding more from silicon chips (e.g., lower power consumption for mobile devices and data centers, increasing integration of functions). System-on-chip (SoC) solutions that integrate increasing number of functions on a single chip serve as a primary way to address costumer demands. Semiconductor companies are integrating processor and memory cores with power management, graphic processors, a potentially large number of different wireless communications technologies (e.g., CDMA, GSM, WiFi, Bluetooth) and many other functions. With increasing the complexity of modern devices, proliferating specialized requirements, and decreasing time-to-market window, some companies are turning to third-part IP instead of designing in-house as a cost-effective approach. According to a new market research report of "Semiconductor (Silicon) IP Market by Form Factor (Integrated Circuit IP, SOC IP), Design Architecture (Hard IP, Soft IP), Processor Type (Microprocessor, DSP), Application, Geography and Verification IP - Forecast & Analysis to 2013–2020" published by MarketsandMarkets is expected to grow at a CAGR of 12.6 % from 2014 to 2020 and reach $5.63 billion in 2020 [2].

There are three main categories of IPs [3]: soft, firm, and hard—Fig. 10.1 depicting their relationships and tradeoffs. *Soft IP* blocks are specified using RTL or higher level descriptions. As a hardware description language (HDL) is process-independent, they are more suitable for digital cores. They are highly flexible, portable, and reusable, but not necessarily optimized in terms of timing and power. Presented at the layout level, *hard IP* blocks are highly optimized for a

H. Salmani (✉)
Howard University, Washington, D.C., USA
e-mail: hassan.salmani@howard.edu

S. Bhunia et al. (eds.), *Fundamentals of IP and SoC Security*,
DOI 10.1007/978-3-319-50057-7_10

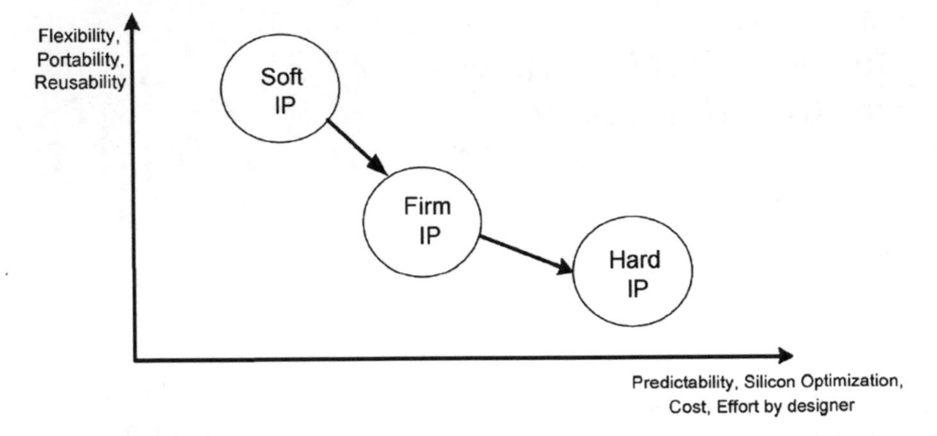

Fig. 10.1 Different types of IP blocks [3]

given application in a specific process. Their characteristics are already determined; however, this comes with high cost and lack of flexibility. *Firm IP* blocks are parameterized circuit descriptions so they can be optimized according to specific design needs. Firm IPs are between soft and hard IPs, being more flexible and portable that hard IPs, yet more predictable than soft IPs.

SoCs are widely used in modern embedded systems that are dedicated for an application(s) or specific part of an application or product or part of a larger system. Modern systems employ embedded systems to enhance their capabilities, dependability, or performance. For example, smart phones in the telecommunication industry have been an influential gadget providing variety of services including banking, gaming, and shopping in the palm of a hand. Or, modern airplanes contain advance avionics such as inertial guidance systems to meet safety requirements, and modern vehicles increasingly use embedded systems to maximize efficiency and reduce pollution. Embedded systems accounted for almost 1.6 trillion dollars of value and 7.1 billion unit shipments in 2010. International Data Corporation (IDC) predicted such systems are growing at a compound annual growth rate of 10 % and should reach $2.6 trillion in revenues and nearly 11.6 billion units by 2015 [4]. Furthermore, modern embedded systems constitute a considerable portion of sophisticated systems as shown in Fig. 10.2.

The typical design and manufacturing flow of SoC is divided into two phases: front end and back end [5]. The former phase starts with ideas or demands that should incorporate knowledge about the application area. They are translated into a design specification in terms of high-level requirements, such as function, throughput, and power consumption. Then, the design specification is transferred from the marketing person's mind, back of envelope or word processor document into machine-readable form. In the following, architectural exploration will try different combinations of processors, memories, and bus structures to find an implementation with good power and load balancing. A loosely timed high-level

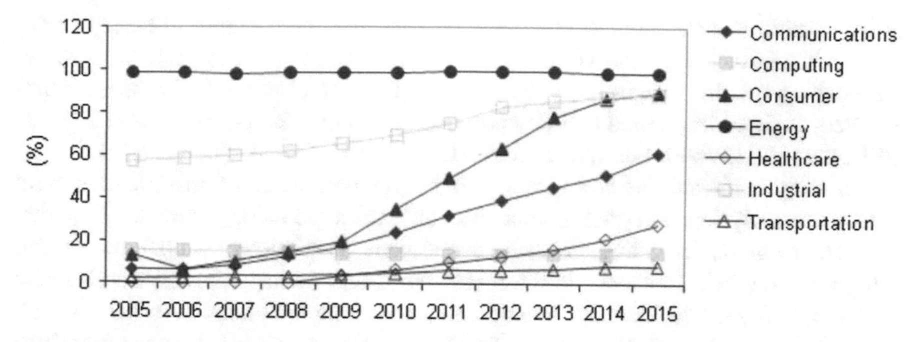

Fig. 10.2 Worldwide—Intelligent systems as percentage of total systems in each major industry (%) [4]

model is sufficient to compute the performance of an architecture. Detailed design will select IP providers for all of the functional blocks, or else they will exist from previous in-house designs and can be used without license fees, or else freshly written. Next, logic synthesis will convert from behavioral RTL to structural RTL. In the next phase, the back end phase, after RTL synthesis using a target technology library, a structural netlist with no gate delays is obtained. Place and route gives 2-D coordinates to each component, adds external I/O pads and puts wiring between the components. In the following, RTL annotated with actual implementation gate delays gives a precise power and performance model. If performance is not up to par, design changes are needed. The last step is design manufacturing. Fabrication of masks is commonly the most expensive single step (e.g., one million pounds), so must be correct first time. Fabrication is performed in-house by certain large companies (e.g., Intel, Samsung) but most companies use foundries (UMC, TSMC). At all stages (front and back end), a library of standard tests will be run every night and any changes that cause a previously passing test to fail (regressions) will be automatically reported to the project manager.

Today's SoC-based designs present a design team with five facets of complexity that range from functional complexity, architectural, and verification challenges to design team communication and deep submicron (DSM) implementation [6]. Facet one is functional complexity and the sheer amount of functional blocks contained in a modern system. The fact leads to the wide acceptance that designing these systems from scratch is far beyond the design productivity or capabilities of even the largest design teams. Therefore, some form of intellectual property (IP) reuse has become an inevitable part of SoC design. The second facet is the architectural challenge. When functional complexity of this scale is implemented in a short timeframe, there must be a detailed architecture set from the start of the project and rigorously adhered to throughout implementation. Due to increased design complexity and from the necessity to get the architecture right the first time, it is increasingly unacceptable, in terms of both time and money, to change the architecture midway, and equally unacceptable to "over-engineer" from a unit cost perspective. The verification challenge is the third facet and stems from the first—it is impossible to create broad functional content and increasingly difficult to integrate and verify it all. The verification challenge is believed to scale with somewhere between the square and the cube of the complexity of the block being verified. It should be noted that IP reuse does not solve the problem since IP must still be verified within the system context.

To bring hardware and software together at the earliest stage possible, one trend is to perform system integration on a "virtual" prototype using a simulation model. The challenge is to define just where software development stops and system integration begins. This leads to the fourth facet, and perhaps the biggest obstacle: the complexity of design team interactions and communications necessary to successfully undertake a SoC-based design. To achieve improvement in the first three facets, there needs to be interaction between system architects, algorithm designers, software developers, hardware designers, system integrators, and verification specialists. However, the reality is different—these organizational functions are rarely integrated together. More often, there are significant organizational, and sometimes physical, separation between these functions. The fifth facet is the set of issues that come with implementing a complex chip design in a DSM process technology. These include problems of timing closure, placement and routing, including avoidance of increasingly problematic physical effects such as crosstalk.

Notwithstanding the presence of design challenges, dependability is still one of key requirements of any computing device. The dependability of a system is based on the compliance of delivered services by the system with its functional specifications. The function of the system is described by functional specifications in terms of functionality and performance. The service delivered by the system, on the other hand, is its behavior as it is perceived by its user(s). A broad concept, dependability encompasses availability, reliability, safety, integrity, and maintainability attributes as described in Table 10.1 [7].

Security is more specific, focusing on availability, integrity, and confidentiality. System security demands availability for only authorized actions, integrity with improper meaning unauthorized, and confidentiality. Trust is the dependency of a system (system A) to another system (system B), through which the dependability of system A is affected by the dependability of system B. Trustworthiness in a system is the assurance that the system will perform as expected [7].

Along with all challenges and complexity of SoC design bringing forth as a constituent of a larger system, embedded systems also contain some inherent

Table 10.1 Dependability attributes [7]

Attribute	Definition
Availability	Readiness for correct service
Reliability	Continuity of correct service
Safety	Absence of catastrophic consequences on the users and the environment
Integrity	Absence of improper system alteration
Maintainability	Ability to undergo modification and repairs
Confidentiality	The absence of unauthorized disclosure of information

characteristics adversely impacting security. Embedded systems suffer from limited processing power so that advance defense mechanisms would not be applicable. Embedded systems typically operate on batteries, so limited power source is available. As increasing computation reduces system lifetime, limited power source is available for security measures. Physical exposure is typical of embedded systems, rendering them vulnerable to attacks that exploit physical proximity of attacker. Network connectivity of modern embedded systems provides opportunities to malicious entities to obtain or manipulate sensitive data.

The characteristics of embedded systems make them vulnerable to different types of attacks at different levels including the circuit level. At the circuit level, there are a wide range of physical and side-channel attacks that exploit design implementation and/or properties to lunch a security attack. These attacks are generally classified into noninvasive and invasive attacks. Noninvasive attacks involve observing and analyzing a device's side-channel signals to extract critical information. Timing attacks, fault injection techniques, power and electromagnetic analysis based attacks are some types of noninvasive attacks. Invasive attacks, such as reverse engineering or micro-probing, require immediate access to the device to observe, manipulate, and interfere with the system internals. This type of attacks is expensive and time consuming; therefore, they are usually implanted by organized groups may even supported by governmental entities.

10.2 IC Supply Chain Globalization and Threats

In early day of the semiconductor industry, a single company would often be able to design, manufacture, and test a new chip. However, the costs of building manufacturing facilities—more commonly referred "fab"—have gone extremely high. A fab could cost over $200 million dollars back in the 1980s; however, with employing advanced semiconductor manufacturing equipment to produce chips with ever-smaller features, a modern fab costs much more [8]. For example, in late 2012 Samsung made a new fab in Xian, China that cost $7 billion. It has been estimated that "[i]ncreasing costs of manufacturing equipment will drive the average cost of semiconductor fabs between $15 billion and $20 billion by 2020." [9].

Due to confluence of increasingly complex supply chains and cost pressures, the horizontal supply chain has become prevalent [10]. Figure 10.3 shows the percentage of design activities outsourced while chip level design constitutes about 40 % by 2005. Integrated Circuits (ICs) or chips are at the core of any modern computing system, and their security grounds the security of entire system. Notwithstanding the central impact of ICs security, malicious modification of IC circuit by untrusted parties has raised serious concerns for critical applications such as fail-safe military applications.

To address the issue, the Department of Defense and the National Security Agency of United States jointly funded a "Trusted Foundry" at an IBM semiconductor manufacturing facility in Vermont, US in 2004. The Trusted Foundry

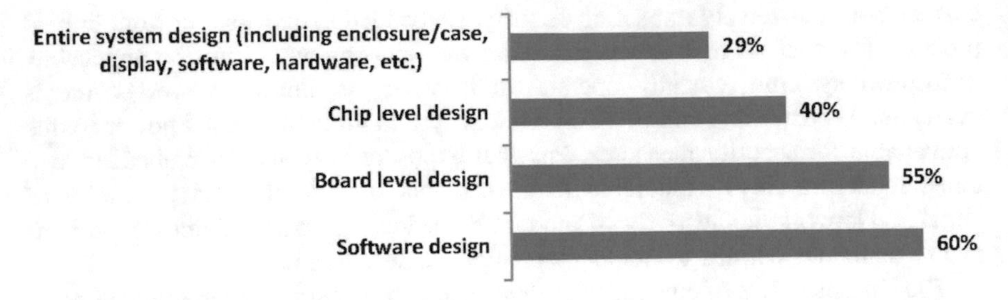

Fig. 10.3 Design activities outsourced [48]

program is "to ensure that mission-critical national defense systems have access to leading-edge integrated circuits from secure, domestic sources." Although the Trusted Foundry program is used to produce the most sensitive chips, these chips constitute only a small fraction of chips used for military applications. Department of Defense heavily relies on commercial supply chain to provide routers, navigation equipment, and most other electronics hardware—and therefore exposed to any associated vulnerabilities.

10.3 IC Design Flow and Vulnerabilities

A computer system development, as shown in Fig. 10.4, consists of several steps which are not necessarily performed in the same design house. The first step is to determine system specifications based on the customer's needs. A complex system

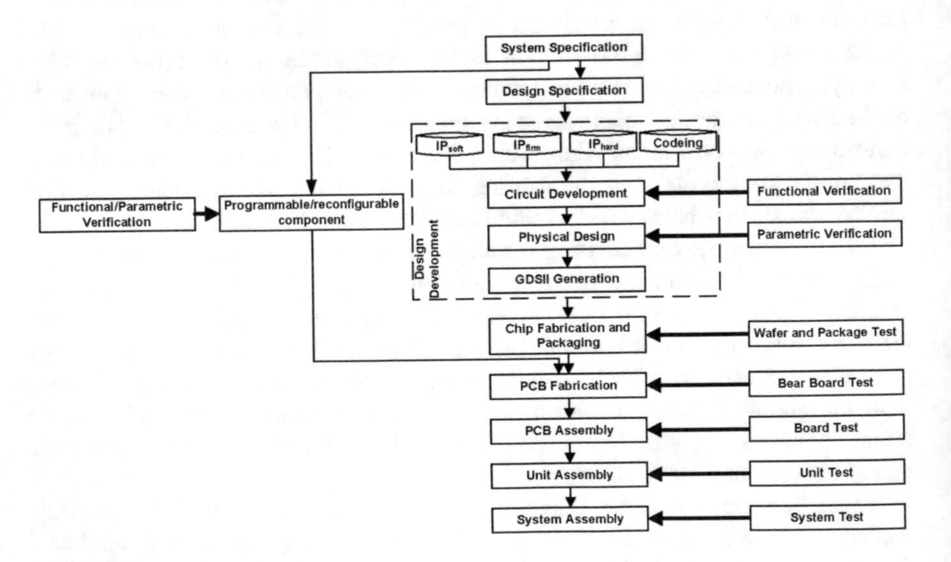

Fig. 10.4 System integration and test process

may require a variety of components like memories and chips with different applications and functionalities.

After providing the system specifications and choosing the structure of system and its required components, design development requires different tools. Each component demands specific attention to meet all the system specifications. To expedite system development and to reduce the final cost, outsourced alternatives have gradually replaced in-house processes. Third-party IP cores have displaced the in-house libraries of logic cells for synthesis. Commercial software has supplanted homegrown Computer Aided Design (CAD) tool software. In the next step, designed chips are signed-off for fabrication. Nowadays, most companies are fabless, outsourcing mask production and fabrication. Besides custom designs, companies can reduce total cost and accelerate system development using commercial-off-the-shelves (COTSs), reprogrammable modules, like micro-controllers, reconfigurable components, or field programmable gate arrays (FPGAs). Afterwards, they manufacture printed circuit boards (PCBs) and assemble system components on them. Finally, the PCBs are put together to develop units; the entire system is the integration of these units.

In each step, different verifications or tests are performed to ensure its correctness, as shown in Fig. 10.4. Functional and parametric verifications ascertain the correctness of design implementation in terms of service and associated requirements, like power and performance. Wafer and package tests after the fabrication of custom designs separate defective parts and guarantee delivered chips. The PCB fabrication is a photolithographic process and susceptible to defects; therefore, a PCB should be tested before placing devices on it. After the PCB assembly, the PCB is again tested to verify that the components are properly mounted and have not been damaged during the PCB assembly process. The tested PCBs create units and finally the system, which is also tested before shipping for field operation [11].

Each step of system development is susceptible to security breaches. An adversary may change system specifications to make a system vulnerable to malicious activities or susceptible to functional failures. As external resources, like third-party IPs and COTSs, are widely used in design process and system integration, adversaries may hide extra circuit(s) in them to undermine the system at a specific time or to gain control over it. The untrusted foundry issue is rooted in the outsourcing of design fabrication. Establishing a chip fabrication factory is extremely expensive and most semiconductor companies have become fabless in recent years. They ask foundries to fabricate their designs to reduce the overall cost. The third party, however, may change the designs by adding extra circuits, like back doors to receive confidential information from the chip, or altering circuit parameters, like wire thickness to cause a reliability problem in the field. The PCB assembly is even susceptible, as it is possible to mount extra components on interfaces between genuine components. In short, the cooperative system development process creates opportunities for malicious parties to take control of the system and to run vicious activities. Therefore, as a part of the system development process, security features should be installed to facilitate validation, and to unveil any deviation from genuine specifications.

10.4 Hardware Trojans Horses

The practice of outsourcing design and fabrication in the interest of economy, has raised serious national security concerns, since an adversary can subvert a design by adding extra circuits, called hardware Trojans [12]. In general, a hardware Trojan is defined as any intentional alteration to a design in order to alter its characteristics. A hardware Trojan has a stealthy nature and can alter design functionality under rare conditions. It can serve as a time bomb and disable a system at a specific time, or it can leak secret information through side-channel signals.

A Trojan may affect circuit AC parameters such as delay and power; it also can cause malfunction under rare conditions. As shown in Fig. 10.5, a hardware Trojan consists of Trojan payload and Trojan trigger. A functional Trojan takes inputs from some internal nets of the main circuit to the Trojan payload and re-stitches some other nets of the main circuit through Trojan payload to modify design functionality. The Trojan trigger determines the activation condition(s) under which the Trojan payload can propagate erroneous values into the main circuit.

The first detailed taxonomy for hardware Trojans was presented in [13, 14]. This comprehensive taxonomy lets researchers examine their methods against different Trojan types. Currently, the industry lacks metrics to evaluate the effectiveness of methods in detecting Trojans. Such metrics could foster a comprehensive taxonomy to help analyze Trojan detection techniques. Because malicious alterations to a chip's structure and function can take many forms, the Trojan taxonomy is decomposed into three main categories (see Fig. 10.6) according to their physical, activation, and action characteristics. Although Trojans could be hybrids of this classification (for instance, they could have more than one activation characteristic), this taxonomy captures the elemental characteristics of Trojans and is useful for defining and evaluating the capabilities of various detection strategies.

The physical characteristics category describes the various hardware manifestations of Trojans. The type category partitions Trojans into functional and parametric classes. The functional class includes Trojans that are physically realized through the addition or deletion of transistors or gates, whereas the parametric class refers to Trojans that are realized through modifications of existing wires and logic.

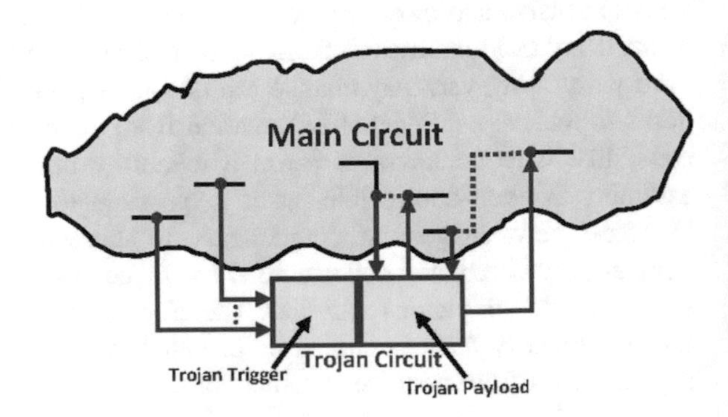

Fig. 10.5 Functional hardware Trojan implementation

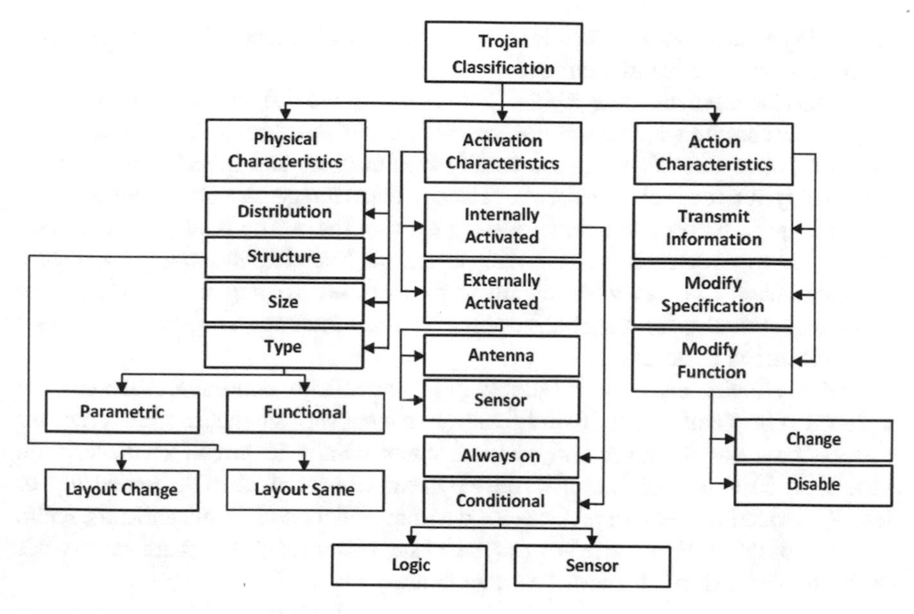

Fig. 10.6 Hardware Trojan taxonomy [13]

The size category accounts for the number of components in the chip that have been added, deleted, or compromised. The distribution category describes the location of the Trojan in the chip's physical layout. The structure category refers to the case when an adversary is forced to regenerate the layout to insert a Trojan, which could then cause the chip's physical form factor to change. Such changes could result in different placement for some or all design components. Any malicious changes in physical layout that could change the chip's delay and power characteristics would facilitate Trojan detection. Wang and colleagues identified current adversaries' capabilities for minimizing the probability of detection.

Activation characteristics refer to the criteria that cause a Trojan to become active and carry out its disruptive function. Trojan activation characteristics fall into two categories: externally activated (e.g., by an antenna or a sensor that can interact with the outside world) and internally activated (which are further classified as always on and condition based), as Fig. 10.6 shows. "Always on'" means the Trojan is always active and can disrupt the chip's function at any time. This subclass covers Trojans that are implemented by modifying the chip's geometries such that certain nodes or paths have a higher susceptibility to failure. The adversary can insert the Trojans at nodes or paths that are rarely exercised. The condition-based subclass includes Trojans that are inactive until a specific condition is met. The activation condition could be based on the output of a sensor that monitors temperature, voltage, or any type of external environmental condition (such as electromagnetic interference, humidity, altitude, or temperature). Alternatively, this condition could be based on an internal logic state, a particular input pattern, or an internal counter value. The Trojan in these cases is implemented by adding logic gates and/or flip-flops to the chip, and hence is represented as a combinational or sequential circuit.

Action characteristics identify the types of disruptive behavior introduced by the Trojan. The classification scheme, shown in Fig. 10.6, partitions Trojan actions into three categories: modify function, modify specification, and transmit information. The modify-function class refers to Trojans that change the chip's function by adding logic or by removing or bypassing existing logic. The modify-specification class refers to Trojans that focus their attack on changing the chip's parametric properties, such as delay when an adversary modifies existing wire and transistor geometries. Finally, the transmit-information class includes Trojans that transmit key information to an adversary.

Trojan circuits are sly, triggering only under rare conditions. Trojans are designed to be silent most of their lifetime, to have a very small size relative to their host designs, and to make only limited contributions to circuit characteristics. Analyzing the vulnerabilities of IC development process requires the knowledge of design, fabrication, and test processes. To ensure a client's IC is authentic, the entire design and fabrication process must be made trustworthy or manufactured ICs should be verified by clients for trustworthiness.

10.5 Hardware Trojans Detection and Prevention

Hardware Trojans have negligible effect on a circuit and rarely become fully activated. While considerable number work has been presented on hardware Trojan detection, they can be broadly categorized into two groups: side-channel signal analysis and logic-value analysis. Majority of work on Trojan detection based on side-channel analysis has focused on power and delay side-channel signals. To enhance Trojan detection resolution, some techniques have proposed embedding monitoring systems into main circuits to capture any abnormality in circuit performance or power consumption. And some other work also recommended design for hardware trust to magnify Trojan impact during authentication. In addition to side-channel based techniques, detection techniques based on logic-value analysis mainly focus on generating effective test patterns to fully activate Trojans and propagate design malfunction to primary outputs.

10.5.1 Trojan Detection Based on Side-Channel Analysis

Delay-Based Trojan Detection Techniques

In [15], it has shown that extra capacitance incurred by a hardware Trojan attribute to wire and gate capacitances would change the delay of path connected to Trojan payload or Trojan trigger. Simulation results for a Trojan (a minimum sized NAND

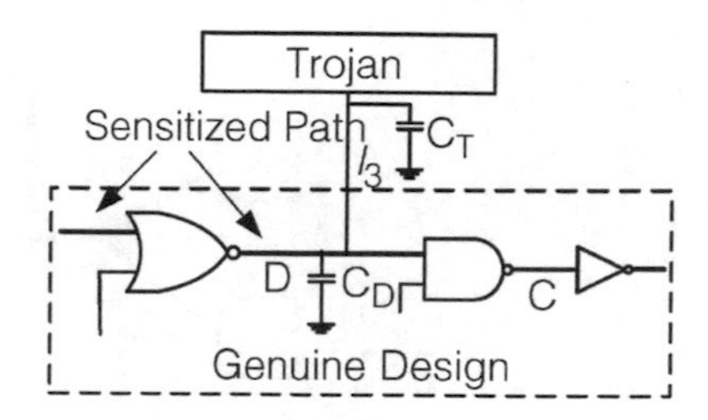

Fig. 10.7 A NAND Trojan connected to an internal net by a wire with the length of l_3 [15]

Table 10.2 The impact of different l_3 on the delay of a path in the original circuit [15]

	Without Trojan (ps)	Location 1 (ps)	Location 2 (ps)	Location 3 (ps)	Location 4 (ps)
Path delay	764.5	794.5	837.7	890.0	953.8
Increased delay	0	30	73.2	125.5	189.3

in 90 nm technology node) connected to the node *D* on a sensitized path in Fig. 10.7 show that the path's delay changes and variations directly depend on the length of distance l_3, Table 10.2 indicating the impact of the same Trojan placed in four different locations. The Trojan is placed in a further distance from Location 1 to Location 2, to Location 3, and to Location 4, and the results signify the further the location, the larger path delay. Based on these facts, a number of techniques have been proposed, and some of them are studied in the following.

Two major challenges of delay-based hardware Trojan detection are (1) to cover as many as circuit paths possible and (2) incur as low as possible cost. In [15], a "clock sweeping" technique is proposed to obtain path delay information without any additional hardware. With one example, Fig. 10.8 illustrates the technique. Clock sweeping involves applying a pattern over a range of clock frequencies from low to high, a common practice in industry used for speed binning of parts. The difference between two successive frequencies defines the sweep step size, i.e., $\Delta t = 1/f_i - 1/f_{i+1}$. With increasing the clock speed, sensitized paths whose delay is larger than the current clock period start to fail. The obtained start-to-fail clock frequency can indicate the delay of the paths sensitized by a pattern. For example, in the Fig. 10.8 the operation frequency of the circuit is f_0 correspond to the delay of A-D path. By increasing the clock frequency to f_1, the A-D path fails. If the frequency was increased to f_2, the A-E path would correctly operate in the Trojan-free circuit. However, the existence of a Trojan on the A-E path would increase its delay and an incorrect value will be captured at f_2 clock frequency. Collected logic values with corresponding frequencies and patterns are used to generate a signature for a circuit under authentication. Then the multidimensional scaling statistical analysis is used to authenticate the circuit under test.

In another work [16], an on-chip path delay measurement circuit using a shadow register is proposed to measure delay of selected register-to-register path delays, shown in Fig. 10.9. This method also uses a sweeping-clock-delay measurement technique to measure path delay. A Trojan can be detected when one or a group of

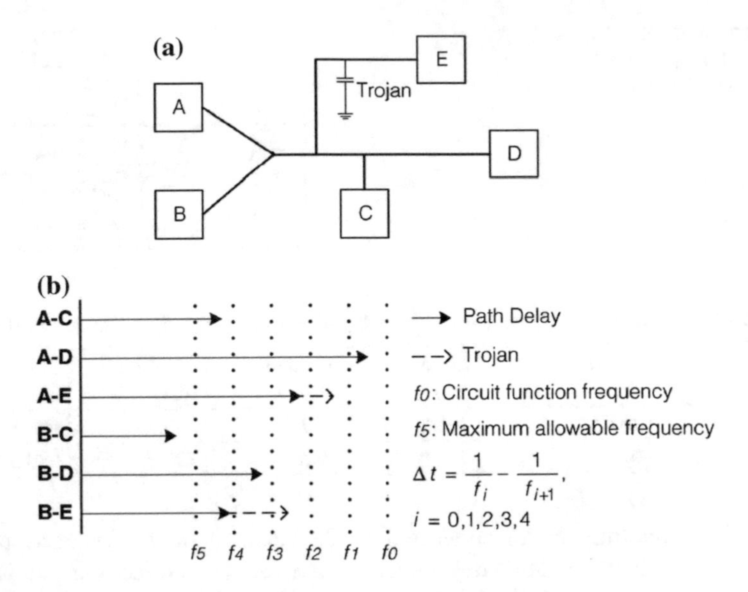

Fig. 10.8 **a** An example circuit. **b** Clock sweeping [15]

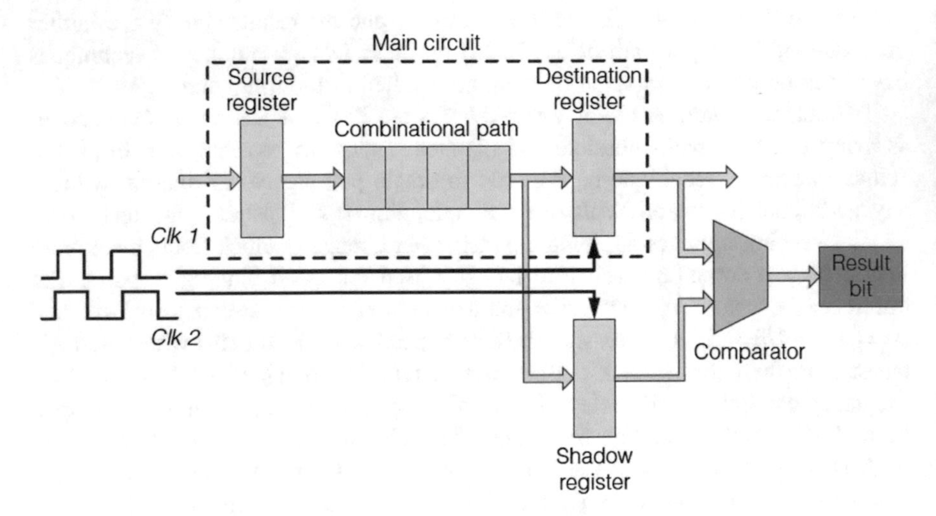

Fig. 10.9 The basic architecture for shadow register Trojan prevention scheme [16]

path delays are extended beyond the threshold determined by the process variations. The measurement circuit characterizes a selected path by measuring its exact delay. *CLK*1 is the main clock that drives all flop-flops in the main circuit. *CLK*2 is a clock with the same frequency as *CLK*1 but shifted and drives a show register whose input is the input of register at the end of path being characterized. By shifting *CLK*2, the exact delay of selected path is obtained with a precision of the skew step size whenever the comparison result is unequal.

Power-Based Trojan Detection Techniques

One of pioneer work in hardware Trojan detection is [17]. In this work, a set of patterns is applied to a batch of chips. With each pattern application, the power trace on each chip is collected. The chips are reverse engineered and inspected to ensure they are Trojan free. The collection of power traces from Trojan-free chips serves as a reference. After obtaining the reference, the same set of patterns is applied to a design under authentication and its power traces are collected. The power traces are compared with the reference and any measurable difference beyond a specific threshold flags Trojan existence.

One of major challenges with power-based techniques is process variations. Manufactured chips of one circuit, although have the same functionality, present different characteristics in terms of transistor parameters such voltage threshold and channel length due to limited accuracy of manufacturing equipment. Variations are broadly categorized into inter-chip and intra-chip variations. Inter-chip variations show a slight shift in parameters from a chip to another chip. On the other hand, intra-chip variations imply random process variations inside a chip where voltage threshold of a transistor is reduced while that of nearby one increased. As a result, two manufactured chips by the same company may present noticeable difference in their power consumption. The difference may be too high such that Trojan contribution into power consumption might be masked due to process variations. A number of techniques have been proposed to reduce the impact of process variations and enhance power-based Trojan detection techniques.

To mitigate the impact of process variations, a multi-supply transient-current integration methodology is proposed in [18]. While the Fig. 10.10 presents the concept, a set of random patterns are applied to both a chip without Trojan and a chip under authentication. While the vertical axis presents charge (Q), the integration of current over time (t), any measureable different above a predefined threshold (D(t)) indicates Trojan existence. The technique benefits the fact that the impact of intra-chip process variations will be canceled over the time by activating different portions of a circuit by applying random test patterns. Furthermore, the technique does not incur any area overhead.

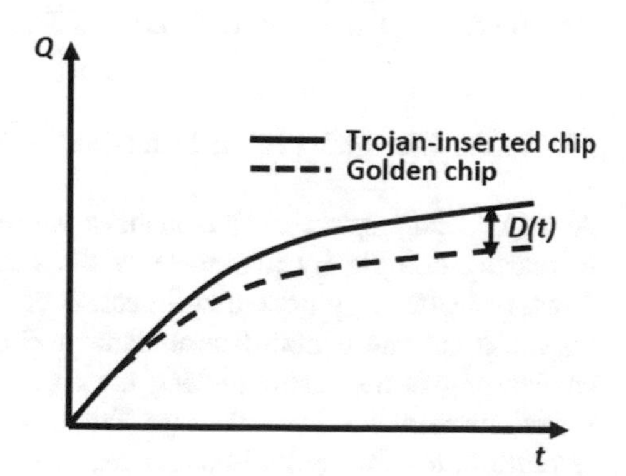

Fig. 10.10 Current integration method [18]

Multiple Excitation of Rare Switching (MERS) is a side-channel-aware test generation approach to increase Trojan detection sensitivity [19]. To make side-channel analysis successful in detecting Trojans, MERS is to maximize the switching activity in the Trojan circuit and to minimize the switching activity in other parts of the circuit so that the relative switching effect is maximized. The basic idea of MERS is that if a rare switching repeated N times where N is sufficiently large, the chances of switching in a Trojan associated with that rare node would significantly increases. To obtain a side-channel-aware test pattern set for a circuit, random patterns are applied to its Trojan-free netlist and switching activity of signals with low switching activity monitored. Patterns are then ranked based on exercising rare-triggering signals, and test patterns activating the largest number of rare nodes are selected. To minimize switching activity across the circuit, Hamming distance between any two patterns is calculated. Based on this calculation, test patterns are reordered such that any two consecutive test patterns generate minimum total switching activity.

In some other approaches including [20–23], on-chip sensors are embedded to capture abnormality in circuit power consumption. In [20] a novel ring oscillator network technique in ASICs is proposed. The ring oscillator network serves as a power supply monitor by detecting fluctuations in characteristic frequencies due to malicious modifications (i.e., hardware Trojans) in the circuit under authentication. The frequency of a ring oscillator is dependent on the power supply, thus the malicious addition or omission of gates may be detected by measuring changes in the frequency. In another work [21], circuit paths in a design are reconfigured into ring oscillators 1 (ROs) by adding a small amount of logic. To minimize the overhead, an algorithm is proposed to configure the circuit paths into ROs such that the number of secured gates is maximized. Rad et al. [22] proposes an on-chip array of current sensors to monitor supply current inside an IC to improve the sensitivity of power-based Trojan detection techniques. In a similar work in [23], authors employ a different kind of current sensor. It should be noted that the current measurement circuitry in [22] is off-chip while in [23] the whole measurement structure is on-chip. Karimian et al. [24] uses ring oscillators (ROs) to gather measurements of ICs and investigates several classification approaches with incorporating a genetic algorithm and the principal component analysis to distinguish between Trojan-inserted ICs and Trojan-free ICs with minimum error rate.

Multi-side Channel Trojan Detection Techniques

A major challenge for side-channel based Trojan detection techniques is the increasing complexity and scale of the state-of-the-art technology. With scaling down of technology node and limitation of manufacturing equipment, circuit mask imprecisions cause non-determinism in chip characteristics. This brings forth a challenging issue: distinguishing the characteristic deviations because of process variations and alterations due to Trojan insertion. To enhance Trojan detection resolution, several multiple-parameter side-channel analyses have been proposed.

In [25], the intrinsic relationship between dynamic current (I_{DDT}) and maximum operating frequency ($F_{\max}$) of a circuit is used to isolate the effect of a Trojan circuit from process noise. Figure 10.11a, b show average I_{DDT} and $F_{\max}$ values for an 8-bit ALU circuit (c880 from ISCAS-85 benchmark suite) obtained from simulation in HSPICE for 100 chips which lie at different process corners. The process corners are obtained by only considering inter-die variations on transistors' voltage threshold. A combinational Trojan (8-bit comparator circuit) is inserted on a non-critical path in c880; therefore, Trojan impact can be only observed on I_{DDT} and it does not affect $F_{\max}$. As shown in Fig. 10.11a, the spread in I_{DDT} due to variations easily masks the effect of the Trojan. The problem becomes more severe with decreasing Trojan size or increasing variations in device parameters in scaled technologies. Figure 10.11b indicates $F_{\max}$ for each process corner. While $F_{\max}$ is used for calibrating the process corner of the chips, the delay of any path in the circuit can be used for this purpose.

To distinguish Trojan contribution from process variations impact, the intrinsic relationship between I_{DDT} and $F_{\max}$ can be utilized to differentiate between the original and tampered versions. The plot for I_{DDT} versus $F_{\max}$ for the ISCAS-85 c880 circuit is shown in Fig. 10.11c. It can be observed that two chips (e.g., Chip$_i$ and Chip$_j$) can have the same I_{DDT} value, one due to the presence of Trojan and the other due to process variation. By considering only one side-channel parameter, it is not possible to distinguish between these chips. However, the correlation between I_{DDT} and $F_{\max}$ can be used to distinguish malicious changes in a circuit under process noise. The presence of a Trojan will cause the chip to deviate from the trend line. As seen in Fig. 10.11c, the presence of a Trojan in Chip$_i$ causes a variation in I_{DDT} when compared to a golden chip (Chip$_k$), while it does not have similar effect on $F_{\max}$ as induced by process variation, i.e., the expected correlation between I_{DDT} and $F_{\max}$ is violated by the Trojan.

In another work [26], some methods based upon post-silicon multimodal thermal and power characterization techniques are presented to detect and locate IC Trojans. The approach first estimates the detailed post-silicon spatial power consumption using thermal maps of the IC, and it then applies the two-dimensional principal component analysis to extract features of the spatial power consumption. Finally, it uses statistical tests against the features of authentic ICs to detect the Trojan.

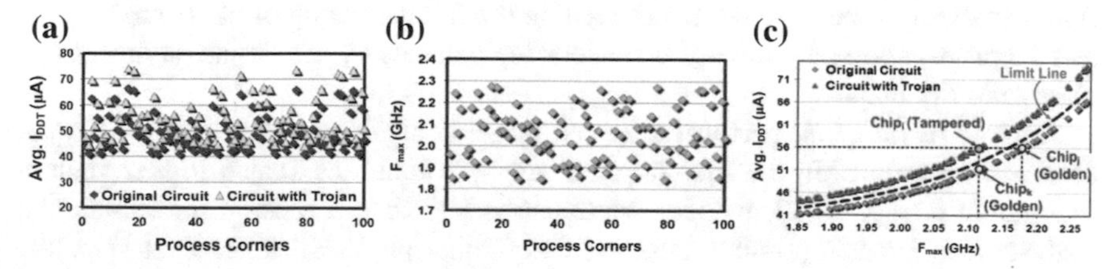

Fig. 10.11 **a** Average I_{DDT} values at 100 random process corners (with maximum variation of ±20 % in inter-die $\mathrm{V_{th}}$) for c880 circuit. The impact of Trojan (8-bit comparator) in I_{DDT} is masked by process noise. **b** Corresponding $F_{\max}$ values. The $F_{\max}$ versus I_{DDT} plot can help identify Trojan-containing ICs under process variations [25]

To accurately characterize real-world ICs, experiments are performed in presence of 20–40 % CMOS process variations to gate lengths, widths and oxide thickness which can hide Trojans. The results reveal detection of Trojans with 3–4 orders of magnitude smaller power consumptions than the total power usage of the chip, while it scales very well because of the spatial view to the ICs internals by the thermal mapping.

A unified formal framework for integrated circuits (ICs) Trojan detection that can simultaneously employ multiple noninvasive side-channel measurement types (modalities) is presented in [27]. First, the IC Trojan detection for each side-channel measurement is being formally defined and their complexity being analyzed. Then, a new sub-modular formulation of the problem objective function is devised. Based on the objective function properties, an efficient Trojan detection method with strong approximation and optimality guarantees is introduced.

In a different work, power and delay traces of a circuit obtained by non-destructive measurements create a system of equations to be solved using linear programming and singular-value decomposition with imposed measurements errors [28]. By the solution, gates in the circuit are characterized in terms of leakage current, switching power, or delay, i.e., a scaling factor of nominal value considering manufacturing variability for each gate is determined. Then Trojan detection is performed using constraint (equation) manipulation. For Trojan detection, three heuristic techniques are investigated: (i) statistical analysis; (ii) constraint manipulations; and (iii) comparison with technological and physical laws. In the first technique, the variable residuals and errors in individual equations are analyzed. In the second technique, additional constraints are imposed on linear programming formulation and the objective function is manipulated in a nonlinear program such that any added circuity will be isolated. The last technique compares the GLC results to the relative characteristics of the gates with respect to the well-established physical design and technological laws. A similar work is also presented in [29].

10.5.2 Trojan Detection Based on Logic Value

While majority of work has been focused on Trojan detection based on side-channel signal analysis, some little work has been on the full activation of hardware Trojans and the propagation of generated erroneous logic values by the Trojan payload to an observation point.

Authors in [30] first perform a Trojan target analysis and then apply a Trojan detection procedure. In the first step, the analysis identifies Trojan trigger vectors (q), shown in Fig. 10.12, whose occurrence is less than a specific threshold. The analysis also isolates possible nets used as Trojan payload. In the next step, the Trojan detection procedure generates a specific set of test vectors to produce rare-triggering vectors and propagate erroneous logic values to an observation point. Trojan test vectors are combined with traditional test patterns, such as stuck-at fault test patterns, and applied during design testing.

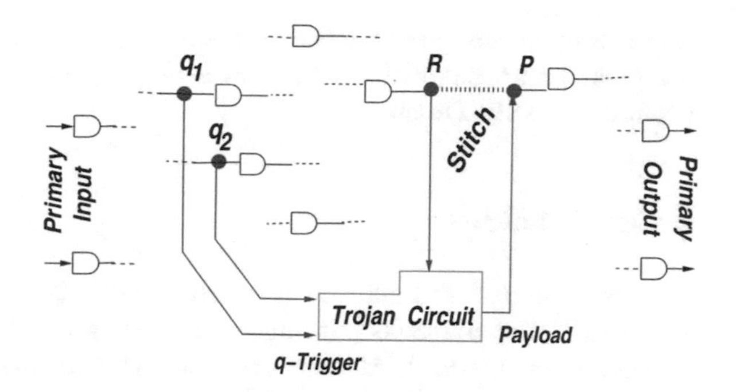

Fig. 10.12 Trojan circuit model [30]

The use of combinatorial testing is explored for hardware Trojan detection in [31] and its efficiency is studied against a various implementation of hardware Trojan in the AES cryptographic algorithm and in comparison with random-based testing. Assuming an upper limit of triggering complexity in terms of the number of gates used, the degree of changes in the physical characteristics of the circuit, and the number of input signals that can be combined for realizing the triggering logic, an attacker opts for a rather rare combination of a small number of input signal values (pattern) for hardware Trojan activation. Combinatorial testing-based approaches utilize theoretical results from combinatorics that can compact significantly the test suite size (i.e., the number of different test inputs applied) under specific assumptions.

Assuming that that the attacker is self-limited to use only k of the available n input signals, where $n \ll k$, as to define the activation sequence. In this case, a test suite comprising all the $2^k \times \binom{n}{k}$ possible input signal combinations will reveal the presence of the Trojan. On the other hand, Table 10.1 shows, how small size of a combinatorial test suit can cover a large number of test patterns where $n = 128$ and t indicates the length of test vector.

10.5.3 Trojan Detection Without a Golden Model

A fundamental limitation of majority of existing hardware Trojan detection techniques is the assumption of a golden IC (GIC) existence as a reference during the IC authentication. However, the existence and identification of a GIC can never be guaranteed: (1) if the hardware Trojan is inserted in the GDSII file, or if the foundry alters the mask to insert a hardware Trojan, then all the ICs will be infected, and (2) if an IC passes a testing procedure which is thorough and more aggressive than traditional testing, it still cannot be guaranteed to be a GIC because the hardware Trojan may be triggered by a rare uncovered event during test. Two groups of approaches have been investigated to tackle this limitation. A group has practiced some design techniques to detect hardware Trojans after design manufacturing, and the other group has mainly analyzed switching activities in gate-level netlist to capture hardware Trojans.

Design Techniques

A sensor-assisted self-authentication framework is proposed in [32], shown in Fig. 10.13. The framework incorporates on-chip detection sensors prior to fabrication and integrates them with the layout. The sensors are design-dependent and are found by an optimization procedure which decomposes the netlist into a set of "similar" sequences of logic gates which are frequently instantiated. The optimization procedure decomposes the timing graph representing the design's netlist into a set of frequently instantiated "delay features." Each delay feature is essentially a variation-aware expression with known sensitivities to unknown parameters such as process variations at the die-to-die and within-die levels. Each instantiation of a delay feature corresponds to an identified sequence of gates and interconnects in the netlist which have similar sensitivities to parameter variations. Therefore, the sensor generation procedure essentially decomposes the netlist into a set of "similar" sequences in which two sequences are similar if the changes in their delays are very close (i.e., less than a specified and small error tolerance). A graphical example of similar sequences is illustrated on the left-hand-side of Fig. 10.13.

At the post-silicon stage, a sensor-assisted self-authentication process is applied for each chip. Two "delay fingerprints" are generated. One corresponds to the on-chip delays of the integrated sensors. The other corresponds to the on-chip delays of a set of arbitrary-selected design paths. The delay fingerprints of the sensors are used to predict a delay range for each considered path. An actual on-chip delay range is also obtained for each path, using an on-chip delay measurement mechanism. For each considered path, a correlation analysis is then conducted between its predicted delay range and its actual delay range. If a hardware Trojan is inserted (in either the

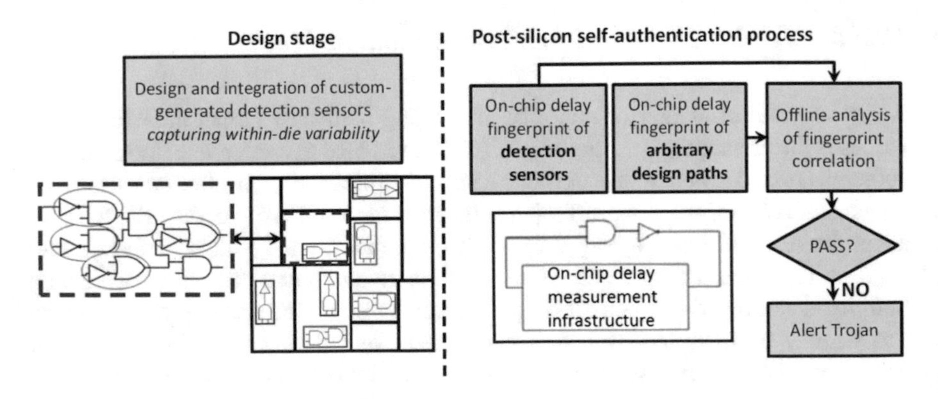

Fig. 10.13 Overview of self-authentication Trojan detection framework [32]

sensors or the design path), then its presence can be detected by observing a poor correlation between these two delay ranges. The post-silicon self-authentication process is shown on the right-hand-side of Fig. 10.13.

In another work [33], a temporal self-referencing approach is proposed for detecting sequential Trojan. The approach compares the current signature of a chip at two different time windows to completely eliminate the effect of process noise, thus providing high detection sensitivity for Trojans of varying size. The effectiveness of the technique is that the transient current "signature" of a Trojan-free circuit should remain constant over different time windows when the circuit undergoes the same set of state transitions multiple times. However, in a Trojan-infected circuit, the current signature varies over multiple time windows for the same set of state transitions of the original circuit, due to uncorrelated state transitions in the Trojan.

Front-End Circuit Switching Activity Analyses

The unused circuitry identification (UCI) technique is one of the first such techniques which distinguishes minimally used logic from the other parts of the circuit [34]. First, UCI creates a data-flow graph for a circuit. Nodes of graph are signals (wires) and state elements and its edges indicate data flow between the nodes. Based on this data-flow graph, UCI generates a list of all direct and indirect signal pairs where data flows from a source signal to a sink signal. In the following, UCI simulates the HDL code using design verification tests to find the set of data-flow pairs where intermediate logic does not affect the data that flows between the source and sink signals. UCI centers on the fact that the HT circuitry mostly remains inactive within a design, and hence such minimally used logic can be distinguished from the other parts of the circuit.

VeriTrust [35] flags suspicious circuitries by identifying potential trigger inputs used in HTs, based on the observation that these inputs keep dormant under non-trigger condition and hence are redundant to the normal logic function of the circuit. In order to detect the redundant inputs, it first performs functional testing and records the activation history of the inputs in the form of sums-of-products (SOP) and product-of-sums (POS). Then it further analyzes these unactivated SOPs and POSs to find the redundant inputs. However, because of the functional verification constraints, VeriTrust can see several unactivated SOPs and POSs and thus regard the circuit to be potentially infected resulting in false positives.

FANCI [36] applies Boolean function analysis to flag suspicious wires in a design which have weak input-to-output dependency. For each input in the combinational logic cone of an output wire, a control value (CV), which represents the percentage impact of changing an input on the output, is computed. If the mean of all the CVs is lower than a threshold, then the resulting output wire is considered malicious. This is a probabilistic method where the threshold is computed with some heuristic to achieve a balance between security and the false positive rate. A very low threshold may result in a high false positive rate by considering most of the wires (even non-malicious ones) as malicious, whereas a high threshold may actually result in false negatives by considering a HT related (malicious) wire to be not malicious.

An information-theoretic approach for Trojan detection has been proposed in [37]. It basically estimates the statistical correlation between signals in a circuit for Trojan detection with the use of OPTICS clustering algorithm. To study the correlation between the signals, inputs patterns are applied and a weighted graph of design created. While the technique presents full coverage for selected benchmarks, the accuracy of technique highly depends on observing enough activity on each signal for studying signals correlation and presented results indicated nonzero false positive rate. Furthermore, the application of the technique for large circuits may require considerable processing time and memory usage. In another effort, a score-based classification method is presented for identifying hardware Trojans [38]. The proposed technique extracts Trojan characteristics introduced at Trust-HUB [39] and defines an incremental metric to isolate some of the Trojan nets from the rest of circuit.

10.6 Trojan Prevention

After circuit synthesis and during physical design, placement tools spread cells such that circuit routability is guaranteed and circuit constrains in terms of power, performance, and size are met. This often leaves small gaps between cells, it is impossible to fill 100 % of the area with regular standard cells in VLSI designs. After completing placement and routing, designers usually fill the empty spaces with filler cells or decoupling capacitor (DECAP) cells to reduce design rule check (DRC) violations attributed to the base layers and ensure power rail connection. However, filler cells do not have functionality. If designers want to make some changes, well known as Engineering Change Order (ECO), the filler cells could be deleted and the empty spaces can be utilized for new gates. On the other hand, intelligent attackers can identify and remove some filler cells for Trojan insertion, because removing these non-functional filler cells does not change the original functionality of circuit.

In [40], the built-in self-authentication (BISA) technique is introduced to fill unused spaces in a circuit layout by functional filler cells, called BISA cells, instead of non-functional filler cells. These BISA cells are connected together to form a combinational circuit, the BISA circuit, that is independent from the original circuits. The BISA circuit is designed so that stuck-at patterns can test all its gates, thus any change on BISA cells will be detected. Furthermore, BISA cells are the same as standard cells that the circuit uses, thus identifying these cells will be extremely difficult. Thus, BISA can be used to prevent Trojan insertion or make Trojan insertion extremely difficult.

Figure 10.14 shows the structure of BISA consisting of a test pattern generator module (TPG), BISA circuit under test, and output response analyzer (ORA). In

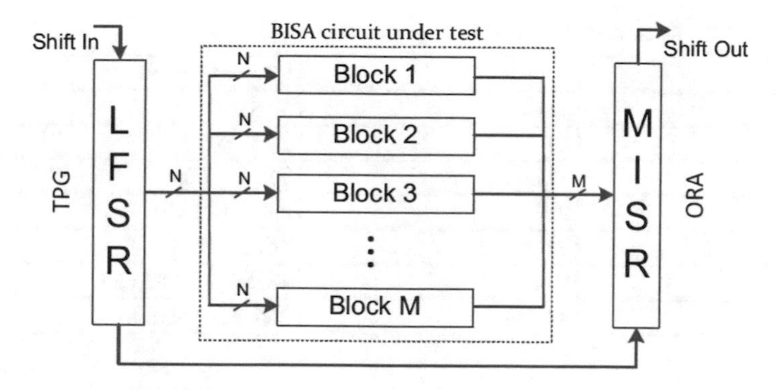

Fig. 10.14 BISA structure [40]

this paper, the linear feedback shift register (LFSR) is used as TPG and the multiple input signature response (MISR) as ORA. The output of ORA is used as signature to detect hardware Trojans. The BISA circuit under test is composed of all BISA cells which are inserted into unused spaces. The smaller combinational circuit with fewer gates is, the higher test coverage is. Therefore, the BISA circuit is divided into a number of smaller combinational logic blocks, called BISA blocks. Each BISA block can be considered as an independent combinational logic block. Figure 10.15 shows application of BISA to System05 in 90 nm technology node.

Table 10.3 shows BISA effectiveness under ten attacks. In the system05 circuit, 418 BISA cells are inserted to fill unused spaces. LFSR and MISR with size of 32 are used to form the BISA structure. 616 ATPG patterns can reach 99.65 % testable coverage. When 500 patterns from LFSR are applied, the stuck-at fault test coverage is 81 %. In Table 10.3, case 0 shows the result for the genuine BISA result.

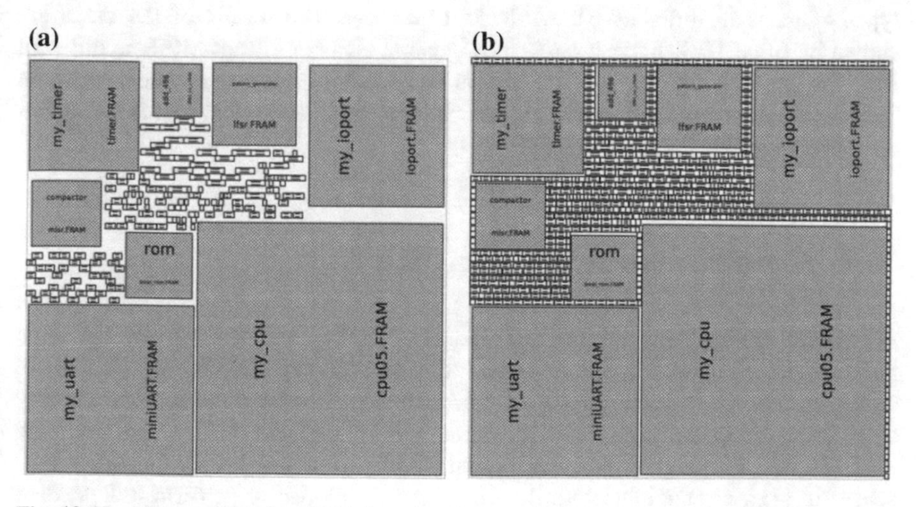

Fig. 10.15 **a** System05 before BISA insertion. **b** System05 after BISA insertion [40]

Table 10.3 Test suite strength and search space reduction [2]

Strength	Suit size	Covered pattern
t = 2	11	32,512
t = 3	37	2,731,008
t = 4	112	170,688,000
t = 5	252	8,466,124,800
t = 6	720	347,111,116,800
t = 7	2,462	12,099,301,785,600
t = 8	17,544	366,003,879,014,400

Table 10.4 Attack summary [40]

Case	Type	Attack description	Signature
0	Genuine	None	0712022D
1	Removal	Remove a leaf cell	0DA8936E
2	Removal	Remove an internal cell	157F4929
3	Removal	Remove a leaf cell	0ED740FC
4	Removal	Remove an internal cell	D5E2706E
5	Removal	Remove an internal cell	43D51D83
6	Change	Change a leaf cell OR3X1 to AND3X1	F230864
7	Change	Change a leaf cell AOI222X1 to OAI2223X1	157F4929
8	Change	Change a leaf cell AOI222X1 to OAI222X1	F39C3B1E
9	Change	Change a leaf cell AND3X1 to NAND2X1	157F4929
10	Change	Change a leaf cell NAND4X1 to NAND3X1	0B17041F

Five kinds of gates are selected to be removed from different BISA blocks separately. In addition, another five types of gates are selected to be changed to other types of gates in different BISA blocks separately. The results of ten cases are shown in Table 10.3. In each case, the signature generated from MISR is different from the genuine signature, which shows that BISA has detected these attacks. In Table 10.3, an internal cell means it has children cells, and a leaf cell is a cell that does not have children cells. (Table 10.4)

10.7 Circuit Vulnerability Analysis

Although there has been a significant amount of work on hardware Trojan detection and prevention, no systematic approach to assess the susceptibility of a circuit to Trojan insertion has been developed. Sections in a circuit with low controllability and observability are considered potential areas for implementing Trojans. This necessitates a thorough circuit analysis to identify potential Trojan locations. Presented in [41], a comprehensive flow has been developed to perform independent

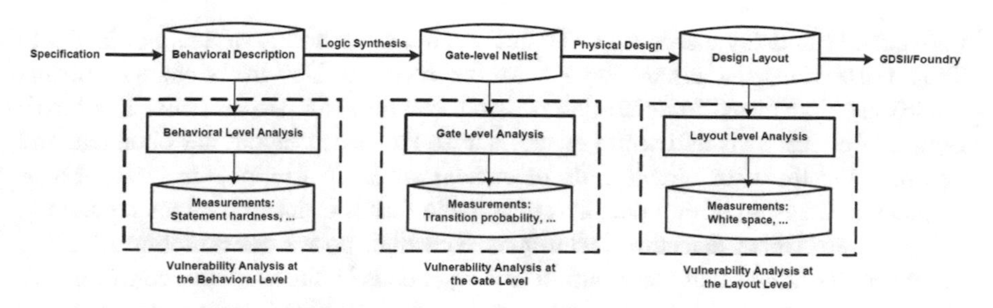

Fig. 10.16 The circuit vulnerability analysis flow [41]

design vulnerability analysis at behavioral, gate, and layout levels, shown in Fig. 10.16. The vulnerability analysis at the behavioral level begins with a circuit described in VHDL language and determines the hardness of executing each statement of code and the observability of circuit signals. At the gate level, to measure a Trojan's resiliency to power and delay side-channel analyses, the transition probability of every net and the delay of the longest path to which the net belongs are determined. At the layout level, the vulnerability analysis screens a circuit layout to find possible locations for Trojan cell placement and their distributions.

At the behavioral level, a circuit is stated in the form of concurrent and sequential statements. HDL constructs, such as loop and condition blocks, direct the execution order of these statements. The circuit's data and control flows determine the hardness of executing a statement (statement hardness) and the observability of internal signals at circuit outputs. A behavioral level circuit is vulnerable to Trojan insertion when statement hardness is high or observability is low. A Trojan at the behavioral level can change a statement that is rarely executed or carry out an attack through a signal with very low observability.

At the gate level, circuits are susceptible to hardware Trojans realized by the addition or deletion of gates. Gate-level Trojans can cause functional modification or parametric deviation, under rare conditions. To withdraw Trojan effects from established testing techniques, an adversary can exploit hard-to-detect areas (e.g., nets) in a circuit to implement a Trojan. Hard-to-detect areas are defined as areas not testable by established fault-testing techniques (stuck-at, transition delay, path delay, and bridging faults) or not having a noticeable impact on circuit side-channel signals (transient power and delay). To recognize hard-to-detect areas, power analysis, delay analysis, and structural analysis are performed to identify nets with low-transition probability, part of non-critical paths, and untestable by regular structural testing.

The distribution of Trojan cells across a circuit layout is a deterministic factor in Trojan impact on circuit side-channel signals. Trojan cells placed tightly in a particular area could have more impact on circuit power consumption as there would be greater localized switching activity in the area. On the other hand, the loose distribution of Trojan cells requires long wire connections for Trojan inputs and outputs and between Trojan cells. Hence, loose Trojans could affect the circuit performance or delay distribution. To analyze the vulnerability of a circuit layout to Trojan insertion, the circuit layout in the form of Design Exchange Format (DEF) file is screened to determine possible locations for Trojan cells. The distribution of circuit cells and white spaces across the circuit layout are obtained, and potential locations for circuit cells placement are then determined. White spaces adjacent to areas with high density are suitable places to insert Trojans resilient to power-based Trojan detection techniques. The high power consumption of dense areas can mask the small contribution of Trojan cells to circuit power consumption. Furthermore, the availability of white spaces across a circuit layout also makes it easier to insert Trojans resilient to delay-based Trojan detection techniques. Placing Trojan cells close to their driving cells reduces induced capacitances due to Trojan wire connections.

10.8 Design for Hardware Trust

Trojan detection resolution in power-based techniques depends on two main factors: Trojan activation and original circuit activation. Increasing Trojan activity and reducing original circuit activity enhance Trojan detection resolution. It has been proven that there is high correlation between the number of switching activity in a circuit and the number of switching activity at the output of scan flip-flops. A scan flop-flop (or scan cell) is a modified flip-flop whose data input is multiplex with another input signal (scan-in) that is accessible through a primary input port. The scan flip-flop is used to increase circuit testability by providing access to its internal parts. Scan flip-flops are grouped and chained to create shift registers, referred as the scan-chain architecture. In a test mode, the shift registers initialize the internal parts of circuit, the circuit switches to a functional mode, and then it switches back to the test mode after one or more clock cycles. High correlation between the switching activity of circuit and the switching activity of scan flip-flops drives two complementary techniques proposed in [42] and [43].

Salmani et al. [42] introduces a dummy scan flip-flop insertion technique to increase the probability of Trojan activation. As Trojan triggers are connected to circuit nets with low-transition probability to realize a latent circuit that is activated under a rare condition, the dummy scan flip-flop insertion technique identify such nets and re-stitch them through dummy scan flip-flops, shown in Fig. 10.17. The dummy scan flip-flops do not change circuit functionality in the functional mode. They are only used during the testing mode to provide immediate access to low-transition nets to increase their transition probability.

In a complementary work in [43], a scan-cell reordering technique is proposed to control switch activity in each portion of circuit. The scan-cell reordering technique forms the scan chains such that adjacent scan cells in a region of circuit's physical layout are connected to each other, shown in Fig. 10.18. Without incurring any area overhead, the technique provides control on each portion of circuit. By integrating the dummy scan-cell insertion and scan-cell reordering techniques for Trojan

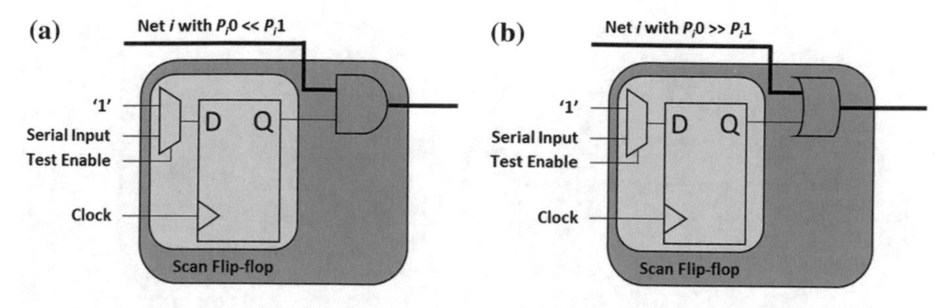

Fig. 10.17 The dummy flip-flop structures when (a) $P_i0 << P_i1$, and (b) $P_i0 >> P_i1$ [42]

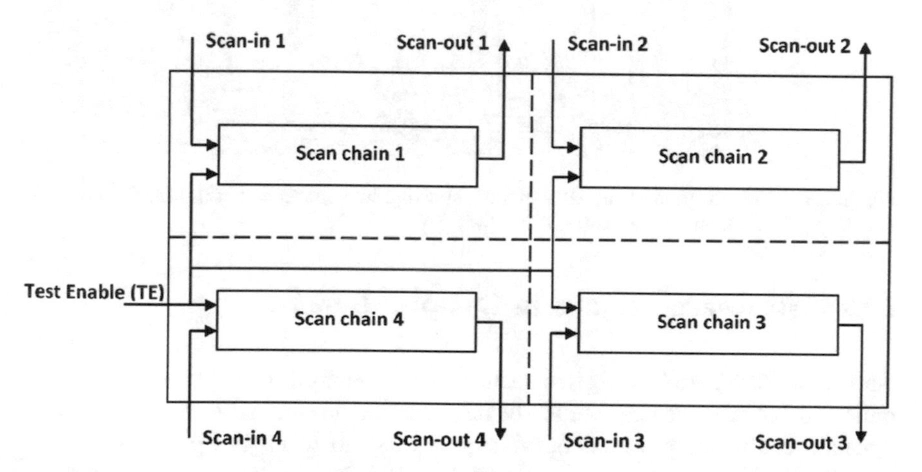

Fig. 10.18 Layout-aware scan-cell reordering concept [43]

detection purpose, it is possible to shut down a part of circuit and reduce circuit switching activity using the scan-chain reordering technique meanwhile increase Trojan activity using the dummy scan flip-flop technique.

Infrastructure IP for SoC security (IIPS) [44] is another approach to incorporate security into a design for its protestation against (1) scan-based attack for information leakage through low-overhead authentication; (2) counterfeiting attacks through integration of a Physical Unclonable Function (PUF); and (3) hardware Trojan attacks through a test infrastructure for trust validation. Figure 10.19 presents IIPS's block diagram consisting of a Master Finite State Machine (M-FSM) that controls the working mode of IIPS, a Scan Chain Enabling FSM (SE-FSM) to provide individual control over activation of scan chains in the SoC, and a clock control module to generate necessary clock and control signals for performing ScanPUF authentication and path delay-based hardware Trojan detection. Regarding hardware Trojan detection, IIPS enables the clock sweeping technique for Trojan detection through monitoring of delay shift by observing the latched value under clock sweep.

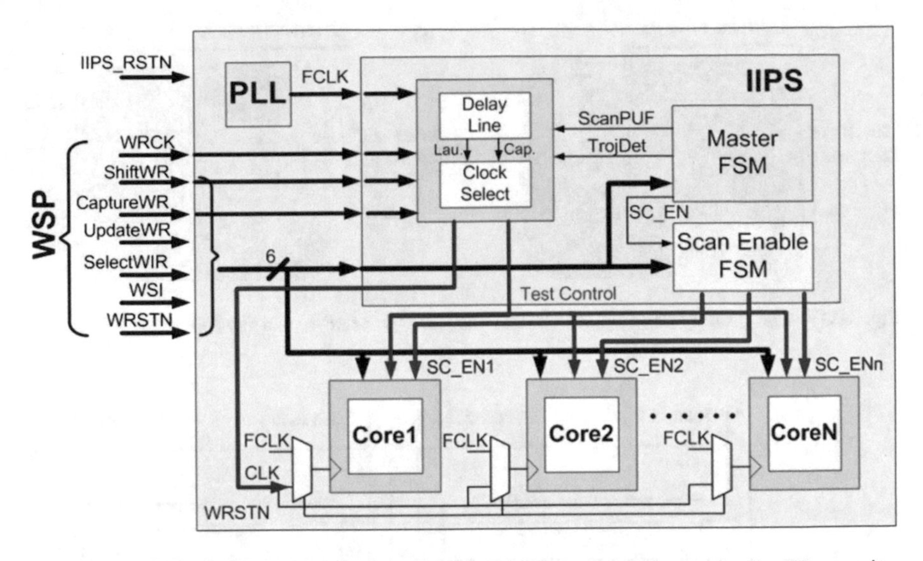

Fig. 10.19 Block diagram of the IIPS module showing interconnection with other IP cores in a SoC using SoC boundary scan architecture [44]

10.9 Hardware Trojans in Complex Designs

Authors in [45] have investigated security threats in third-party power management units. Power management units (PMUs) offer a host of services ranging from dynamic control of power rails, voltage scaling, to managing power states, and modern intelligent power management IPs (PMIPs) are to increase energy efficiency and provide flexible power management in high-end Multiprocessor System-on-Chips (MPSoCs). On the other hand, a Trojan-inserted PMIPs may cause serious issues such as data corruption, denial of service, or degrade the system performance and energy efficiency. For example, a malicious increase in the supply voltage can cause a surge in the peak power and chip temperature, leading to thermal throttling or functional errors due to chip failure. Jayashankara Shridevi et al. [45] studied two specific attack models, namely PMU-Voltage driven Immunity Reduction and Unhealth Syndrome (P-VIRUS) and DROWSY.

P-VIRUS inconspicuously manipulates the supply voltage request made by a processor for a set frequency, leading to improper voltage-frequency assignments, it adversely impacts the energy efficiency and performance of the MPSoC. For example, the voltage supplied by the PMU is less than the requested voltage. As a result, functional errors are raised due to timing violation rooted in the higher delay caused by the lower supply voltage. DROWSY tampers with the sleep and wake up requests of the on-chip components, affecting the availability of on-chip resources. DROWSY can be realized as delaying the sleep signal that causes extra power consumption, delaying the wake up signal that causes resource unavailability, or abrupt transition of blocks to sleep states that results in loss of data and performance degradation.

Hardware Trojans in network-on-chip (NoC) is another serious challenge in complex designs. In [46], it is shown that a hardware Trojan can mask itself as transient errors which can only be activated under very specific conditions to avoid detection. Induced transient errors in data can exploit vulnerabilities created by the fault-tolerant techniques. For example, intentional data corruption requires data retransmissions that may lead to a Denial-of-Service (DoS) attack by creating false congestion between the routers by consuming network resources. While the error may appear as benign for the system, this is intentionally created by HTs to create DoS attack and disrupt the system. In another work [46] it has studied DoS attacks in the NoC routers where a hardware Trojan can maliciously change the flit source/destination address or flit type information of a packet that has left the transmitter network interface (NI). If a Trojan payload modifies the destination address of a packet, that packet could be directed to an unauthorized IP core. A drop of the header flit or tail flit will result in the incomplete packet being retained in the router until some operation arrives to reset the router. To prevent limit hardware Trojan in NoCs, a collaborative dynamic permutation and flit integrity check method is proposed by [46] that is capable of examining the invariables of NoC to immediately terminate the detected HTs.

A novel technique with low-overhead security framework for a custom many-core router using Machine Learning techniques has been presented by [47]. While it is assumed that processing cores and memories are safe, and anomaly is included only through router. The attack corrupts the router packet by changing the destination address that results in traffic diversion, route looping, or core spoofing attack. To detect hardware Trojans in routers a "Golden Data Set" based on hardware feature analysis and anomaly insertion effects has been developed. The Golden Data Set considers Source Core, Destination Core, Packet Transfer Path, Distance, Dynamic Power Range, Execution Time Range, Clock Frequency, Supply Voltage and studies their correlation to reduce the complexity of proposed machine-leaning-based detection technique.

10.10 Challenges with Hardware Trojans

There have been significant efforts to address hardware Trojans, a very challenging issue in electronic chips. Although variety of techniques and methodologies has been proposed, majority of them are carrying certain assumptions or not well scalable to modern designs such that make their applicability limited. One of main assumptions is the existence of a golden model as a reference in side-channel based Trojan detection techniques. A golden model can be obtained by reverse engineering that is a costly and distractive process. Meanwhile, it may not provide a perfect reference because of variations in process parameters between chips. Furthermore, scalability of proposed techniques is not well studied as most experiments are performed on small circuits compared with industrial ones. In addition, it would be difficult to model hardware Trojans contrary to defects caused by manufacturing as hardware Trojans are intentional and malicious modification by nature. Another challenge in hardware Trojan detection is the lack of standard metrics to quantitatively determine the security of a design. The metrics make it possible to measure the effectiveness of different techniques and compare their strengths and weaknesses. Finally, although there have been some efforts on developing Trojans design, there is a need to more comprehensive trust benchmarks that different researchers can use to evaluate and compare their solutions.

References

1. Ernst&Young: Global semiconductor industry study. http://www.indabook.org/preview/37zGj6ryslpY95ZaVcT_D8zdtr5D-X-sWb6GMXsJzhQ,/Global-semiconductor-industry-study-report-Ernst-amp-Young.html?query=Cloud-Computing-Landscape-and-Research-Challenges
2. http://www.reportlinker.com/p 02070028-summary/Semiconductor-Silicon-IP-Market-by-Form-Factor-Integrated-Circuit-IP-SOC-IP-Design-Architecture-Hard-IP-Soft-IP-Processor-Type-Microprocessor-DSP-Application-Geography-and-Verification-IP-Forecast-Analysis-to.html
3. Saleh, R., Mirabbasi, S., Lemieux, G., Pande, P.P., Grecu, C., Ivanov, A.: System-on-Chip: Reuse and Integration. Proc. IEEE **94**(6), 1050–1069 (2006)
4. Morales, M., Rau, S., Palma, M.J., Venkatesan, M., Pulskamp, F., Dugar, A.: industry developments and models. Intelligent Systems: The Next Big Opportunity. International Data Corporation (2011)
5. Greaves, D.J.: System on Chip Design and Modelling. University of Cambridge, Computer Laboratory. Lecture Notes (2011)
6. Hardee, P.: The five facets of SoC design complexity. http://www.eetimes.com/document.asp?doc_id=1277891
7. Avizienis, A., Laprie, J., Randell, B., Landwehr, C.: Basic concepts and taxonomy of dependable and secure computing. IEEE Trans. Dependable Secure Comput. **1**(1), 11–33 (2004)
8. Villasenor, J.: Compromised By Design? Securing the Defense Electronics Supply Chain. The Center for Technology Innovation at Brookings (2013)
9. Johnson, B., Freeman, D., Christensen, D., Wang, S.T.: Market Trends: Rising Costs of Production Limit Availability of Leading-Edge Fabs. GARTNER, INC. http://www.gartner.com/DisplayDocument?doc_cd=238123. Accessed 1 Sept 2012
10. Villasenor, J., Tehranipoor, M.: The Hidden Dangers of Chop-Shop Electronics. In: IEEE Spectrum. http://spectrum.ieee.org/semiconductors/processors/the-hidden-dangers-of-chopshop-electronics (2013)
11. Wang, L., Wu, C., Touba, N.: VLSI Test Principles and Architectures: Design for Testability. Morgan Kaufmann Publishers (2006)
12. Adee, S.: The Hunt for the Kill Switch. In: IEEE Spectrum. http://www.spectrum.ieee.org/print/6171 (2008)
13. Tehranipoor, M., Koushanfar, F.: A survey of hardware Trojan taxonomy and detection. IEEE Des. Test Comput. **27**(1), 10–25 (2010)
14. Wang, X., Tehranipoor, M., Plusquellic, J.: Detecting malicious inclusions in secure hardware: challenges and solutions. In: Proceedings of the IEEE International Workshop on Hardware-Oriented Security and Trust, pp. 15–19 (2008)

15. Xiao, K., Zhang, X., Tehranipoor, M.: A clock sweeping technique for detecting hardware Trojans impacting circuits delay. IEEE Des. Test **30**(2), 26–34 (2013)
16. Li, J., Lach, J.: At-speed delay characterization for IC authentication and Trojan horse detection. In: Proceedings of IEEE International Symposium Hardware-Oriented Security and Trust, pp. 8–14 (2008)
17. Agrawal, D., Baktir, S., Karakoyunlu, D., Rohatgi, P., Sunar, B.: Trojan detection using IC fingerprinting. In: Proceedings of the Symposium on Security and Privacy, pp. 296–310 (2007)
18. Wang, X., Salmani, H., Tehranipoor, M., Plusquellic, J.: Hardware Trojan detection and isolation using current integration and localized current analysis. In: Proceedings of the International Symposium on Fault and Defect Tolerance in VLSI Systems, pp. 87–95 (2008)
19. Huang, H., Bhunia, S., Mishra, P.: MERS: statistical test generation for side-channel analysis based Trojan detection. In: ACM Conference on Computer and Communications Security (CCS), Vienna, Austria, 24–28 Oct 2016
20. Ferraiuolo, A., Zhang, X., Tehranipoor, M.: Experimental analysis of a ring oscillator network for hardware Trojan detection in a 90 nm ASIC. In: Proceedings of IEEE/ACM International Conference on Computer-Aided Design, pp. 37–42 (2012)
21. Rajendran, J., Jyothi, V., Sinanoglu, O., Karri, R.: Design and analysis of ring oscillator based design-for-Trust technique. In: Proceedings of IEEE VLSI Test Symposium, pp. 105–110 (2011)
22. Rad, R., Plusquellic, J., Tehranipoor, M.: A sensitivity analysis of power signal methods for detecting hardware trojans under real process and environmental conditions. IEEE Trans. Very Large Scale Integr. Syst. **18**(12), 1735–1744 (2010)
23. Narasimhan, S., Yueh, W., Wang, X., Mukhopadhyay, S., Bhunia, S.: Improving IC security against Trojan attacks through integration of security monitors. IEEE Des. Test Comput. **29** (5), 37–46 (2012)
24. Karimian, N., Tehranipoor, F., Rahman, M.T., Kelly, S., Forte, D.: Genetic algorithm for hardware Trojan detection with ring oscillator network (RON). 2015 IEEE International Symposium on Technologies for Homeland Security (HST), Waltham, MA, pp. 1–6 (2015)
25. Narasimhan, S., Du, D., Chakraborty, R.S., Paul, S., Wolff, F., Papachristou, C., Roy, K., Bhunia, S.: Hardware Trojan detection by multiple-parameter side-channel analysis. IEEE Trans. Comput. **62**(11), 2183–2195 (2013)
26. Hu, K., Nowrozy, A.N., Reday, S., Koushanfar, F.: High-sensitivity hardware Trojan detection using multimodal characterization. In: Proceedings of the Conference on Design, Automation and Test in Europe, pp. 1271–1276 (2013)
27. Koushanfa, F., Mirhoseini, A.: A Unified framework for multimodal submodular integrated circuits Trojan detection. IEEE Trans. Inf. Forensics Secur. **6**(1), 162–174 (2011)
28. Potkonjak, M., Nahapetian, A., Nelson, M., Massey, T.: Hardware Trojan horse detection using gate-level characterization. In: Proceedings of Design Automation Conference, pp. 688–693 (2009)
29. Alkabani, Y., Koushanfar, F.: Consistency-based characterization for IC Trojan detection. In: Proceedings of IEEE/ACM International Conference on Computer-Aided Design, pp. 123–127 (2009)
30. Wolff, F., Papachristou, C., Bhunia, S., Chakraborty, R.S.: Towards Trojan free trusted ICs: problem analysis and detection scheme. In: Proceedings of ACM Design, Automation and Test in Europe Conference, pp. 1362–1365 (2008)
31. Voyiatzis, A.G., Stefanidis, K.G., Kitsos, P.: Efficient triggering of Trojan hardware logic. In: 2016 IEEE 19th International Symposium on Design and Diagnostics of Electronic Circuits & Systems (DDECS), Kosice, pp. 1–6 (2016)
32. Li, M., Davoodi, A., Tehranipoor, M.: A sensor-assisted self-authentication framework for hardware Trojan detection. IEEE Des. Test **30**(5), 74–82 (2013)
33. Narasimhan, S., Wang, X., Du, D., Chakraborty, R.S., Bhunia, S.: TeSR: A robust temporal self-referencing approach for hardware trojan detection. In: Proceedings of IEEE International Symposium on Hardware-Oriented Security and Trust, pp. 71–74 (2011)

34. Hicks, M., Finnicum, M., King, S.T., Martin, M., Smith, J.M.: Overcoming an untrusted computing base: detecting and removing malicious hardware automatically. In: IEEE Symposium on Security and Privacy, pp. 64–77 (2010)
35. Zhang, J., Yuan, F., Wei, L., Sun, Z., Xu, Q.: VeriTrust: verification for hardware trust. In: ACM/EDAC/IEEE Design Automation Conference (DAC), pp. 61:1–61:8 (2013)
36. Waksman, A., Suozzo, M., Sethumadhavan, S.: FANCI: identification of stealthy malicious logic using Boolean functional analysis. In: Proceedings of the 2013 ACM SIGSAC Conference on Computer & Communications Security (CCS), pp. 697–708 (2013)
37. Çakir, B., Malik, S.: Hardware Trojan detection for gate-level ICs using signal correlation based clustering. In: Proceedings of the 2015 Design, Automation & Test in Europe Conference & Exhibition (DATE), pp. 471–476 (2015)
38. Oya, M., Shi, Y., Yanagisawa, M., Togawa, N.: A score-based classification method for identifying hardware-Trojans at gate-level Netlists. In: Proceedings of the 2015 Design, Automation & Test in Europe Conference & Exhibition (DATE), pp. 465–470 (2015)
39. Salmani, H., Tehranipoor, M., Karri, R.: On design vulnerability analysis and trust benchmark development. In: IEEE International Conference on Computer Design (ICCD) (2013)
40. Xiao, K., Tehranipoor, M.: BISA: Built-in self-authentication for preventing hardware Trojan insertion. In: Proceedings of IEEE International Symposium on Hardware-Oriented Security and Trust (HOST), pp. 45–50 (2013)
41. Tehranipoor, M., Salmani, H., Zhang, X.: Integrated Circuit Authentication Hardware Trojans and Counterfeit Detection. Springer (2014)
42. Salmani, H., Tehranipoor, M., Plusquellic, J.: A novel technique for improving hardware Trojan detection and reducing trojan activation time. In: IEEE Trans. Very Large Scale Integr. (VLSI) Syst. **20**(1), 112–125 (2012)
43. Salmani, H., Tehranipoor, M.: Layout-aware switching activity localization to enhance hardware trojan detection. IEEE Trans. Inf. Forensics Secur. **7**(1), 76–87 (2012)
44. Wang, X., Zheng, Y., Basak, A., Bhunia, S.: IIPS: infrastructure IP for secure SoC design. IEEE Trans. Comput. **64**(8), 2226–2238 (2015)
45. Jayashankara Shridevi, R., Rajamanikkam, C., Chakraborty, K., Roy, S.: Catching the Flu: emerging threats from a third party power management unit. In: 2016 53nd ACM/EDAC/IEEE Design Automation Conference (DAC), Austin, TX, pp. 1–6 (2016)
46. Frey, J., Yu, Q.: A hardened network-on-chip design using runtime hardware Trojan mitigation methods. Integration (VLSI J.) (2016)
47. Kulkarni, A., Pino, Y., French, M., Mohsenin, T.: 2016. Real-time anomaly detection framework for many-core router through machine-learning techniques. J. Emerg. Technol. Comput. Syst. **13**(1) Article 10 (June 2016), 22 pp. doi:http://dx.doi.org/10.1145/2827699
48. Mokhoff, N., Wallace, R.: Outsourcing trend proves: complex by design. EE Times. http://www.eetimes.com/document.asp?doc_id=1152570 (2005)

Chapter 11
In-place Logic Obfuscation for Emerging Nonvolatile FPGAs

Yi-Chung Chen, Yandan Wang, Wei Zhang, Yiran Chen and Hai (Helen) Li

11.1 Introduction

Nowadays, embedded systems are widely adopted in handheld devices, automobile control, aircraft autopilot, medical instrumentation, and many other applications. Thus, hardware security becomes very important to prevent piracy on design or leakage of sensitive data. Compared to *application-specific integrated circuit* (ASIC) chips, systems built in *field-programmable gate arrays* (FPGAs) do not need expose design details to foundry or untrusted outsourcing and hence innately have a higher security level [12]. However, attackers aiming at pirating FPGA configurations, including *intellectual property* (IP) owned by system designers and sensitive data owned by users, cannot be fully prevented. More specific, state-of-the-art FPGAs face the following security issues [3, 13]:

- Conventional FPGAs built with SRAMs need to load logic configuration from external nonvolatile media during system initialization [2, 32]. Probing attack on external connections is a common way to steal FPGA design.

Y.-C. Chen · Y. Wang · Y. Chen (✉) · H. (Helen) Li
University of Pittsburgh, Pittsburgh, PA, USA
e-mail: yic63@pitt.edu

Y. Wang
e-mail: yaw46@pitt.edu

Y. Chen
e-mail: yic52@pitt.edu

H. (Helen) Li
e-mail: hal66@pitt.edu

W. Zhang
Hong Kong University of Science and Technology, Hong Kong, China
e-mail: wei.zhang@ust.hk

S. Bhunia et al. (eds.), *Fundamentals of IP and SoC Security*,
DOI 10.1007/978-3-319-50057-7_11

- *Nonvolatile memory* (NVM) FPGAs utilize antifuse [21] or Flash memory [22] to maintain configuration data on-chip. Physical attack, e.g., probing after reverse engineering [4], is the major security threat.
- Partial run-time reconfiguration emerges as an important security issue too. For example, Xilinx products realize run-time reconfiguration via *internal configuration access port* (ICAP), which is less secure due to port vulnerability [3, 13]. Updating configurations remotely could go through public and insecure network, which requires more than one authentication schemes to enhance security level.

Many novel FPGA architectures were proposed by utilizing the emerging NVM technologies, such as *phase change memory* (PCM), *spin-transfer torque RAM* (STT-RAM), and *resistive RAM* (RRAM) [5, 20, 27]. On the one hand, the use of NVM technologies promises fast operations as conventional *SRAM-based FPGAs* (SRAM-FPGA), increases configuration capacity, and lowers system power consumption significantly [18]. On the other hand, the nonvolatile storage of logic configuration in these architectures raises a big concern in design security: powerful attackers could access configuration memory which indeed contains the entire design without any further protection. Note that the situation does not exist in SRAM-FPGA in which data cannot be retained during powering off. Certainly, user can erase the data in configuration memory after usage and initialize it from external or in-package memory when needed. The security concern in such an operation mode then becomes similar to that of SRAM–FPGA at communication port. In summary, logic and storage components made of NVMs are more vulnerable to physical attacks, making IP protection, and data security even more challenging.

This work targets at the security issue in NVM-based FPGAs. Particularly, a hardware security scheme is proposed for *RRAM-based FPGA* (RRAM-FPGA), in which RRAM devices are used to construct *look-up tables* (LUTs) for logic functions as well as *block RAMs* (BRAMs) for configuration and temporary data storage [7]. The design demonstrates a high density of logic integration and well supports partial run-time reconfiguration. The hardware security scheme in the work protects RRAM-FPGA in three aspects:

1. *An obfuscated configuration* is loaded to BRAMs, combining a *Chip DNA* for logic function identification. The FPGA system operates the designed functionality only when all the pieces of the logic configuration are correctly selected from BRAMs and assembled in a proper sequence.
2. When a higher security level is needed, the system enters *the blank mode* by erasing the contents on nonvolatile logic and routing elements. Even attackers obtain the obfuscated configuration on BRAMs through physical attacks, the design cannot be revealed or reproduced without *Chip DNA*.
3. We combine the communication ports of initialization and run-time reconfiguration in RRAM-FPGA. The bitstream loading scheme is enhanced by *the encrypted addressing*, which enables partial random configuration loading and secret key updating to resist bitstream piracy and protocol-based denial-of-service attack.

The three key components together offer a high level protection on the hardware and data communication of RRAM-FPGA. Our evaluations show that at acceptable system loading and execution performance, the proposed scheme can resist *level 3 attackers* [1, 3]. Meanwhile, the communication port protected by the encrypted addressing demonstrates a much lower probability of protocol-based denial-of-service attack compared to the modern FPGAs with AES encryption.

The rest of the paper is organized as follows. Section 11.2 gives a brief introduction on hardware security in FPGAs and the preliminary of RRAM-FPGA design. Section 11.3 describes the threat models in RRAM-FPGA and the corresponding solutions in this work. Section 11.4 presents the design details of the proposed security scheme. We present the security evaluation and system performance analysis in Sect. 11.5. At the end, Sect. 11.6 concludes the paper.

11.2 Background

11.2.1 Hardware Security in FPGAs and Related Work

Most of the commercial FPGAs use SRAM-based *look-up tables* (LUTs) to realize logic functions [2, 32]. An external memory is needed to store design configuration and initialize system during powering up. The connection between FPGA and its external memory, therefore, is the weakest point in data protection. Attackers could probe the signal at the connection to discover the bitstream and even reform the system into denial-of-service [13]. Encryption technologies, such as AES, are widely adopted to protect bitstreams in modern FPGAs [34]. *Physical unclonable function* (PUF) is another popular solution in preventing attacks of bitstream reverse engineering [17].

Nonvolatile FPGAs equipped with antifuse [19] or Flash memory [24] do not require external memory, and therefore have higher security level than SRAM-FPGAs. However, the on-chip logic configuration could be pirated through probing attack after reverse engineering [13]. For example, Lattice products [19] use a *in-system programmable scheme*, which integrates NVMs of bitstream storage and SRAMs of function logics into one package. Distributed security bits are placed in the silicon as security fuses in loading configuration data. The technique aims at non-invade attack and provides a *moderately high* (MODH) security level. Microsemi, previously known as Actel, supplies FPGAs with functional memory in Flash fabric, which is innately MODH device [24]. A large variety of cryptography services, e.g., AES-128, SHA-256, and PUF, are offered [23]. Moreover, anti-tamper protection scheme is provided to further protect design from physical attacks. It includes a physical containment to detect physical attacks and a system level protective loop to detect the disturbs of protective mesh [24]. Once tamper is over the limitation, penalty such as erasure of the entire design would be kicked in.

As attractive features of FPGA, the remote, and run-time reconfigurability can significantly enhance the design flexibility. However, the security protection of such systems is more challenging. For example, Xilinx products support run-time configuration through *internal configuration access port* (ICAP) [34]. These ports provide logic reconstruction only at the specific locations and are more vulnerable to attacks. Remote reconfiguration transmits bitstreams through networks, which could be accessed by anonymous users. To prevent piracy or protocol-based denial-of-service attack during bitstream transmission, multiple authentications are needed [3].

11.2.2 High Density RRAM-FPGA

FPGAs built with various emerging NVMs have been proposed previously [5, 7, 20, 27]. In this work, the FPGA architecture built with RRAM technology is taken as the example case for its extremely high density, fast execution speed, and better support on run-time reconfiguration [7].

As illustrated in Fig. 11.1, the smallest *reconfiguration unit* (RU) in RRAM-FPGA includes not only the logic and routing elements but also a *block RAM* (BRAM). The BRAM stores temporary data and logic configuration to enhance functionality flexibility, and execution performance [7]. In this architecture, the logic configuration is divided into two steps.

- *Step 1*: Through bitstream loading, design configuration is broadcast to and stored in BRAMs.
- *Step 2*: Each RU distributes the configuration in BRAM to the corresponding logic and routing elements through special tracks.

By leveraging RRAM technology, RRAM-FPGA significantly reduces leakage power consumption and increases logic integration density [6]. However, the system and data protection becomes more challenging.

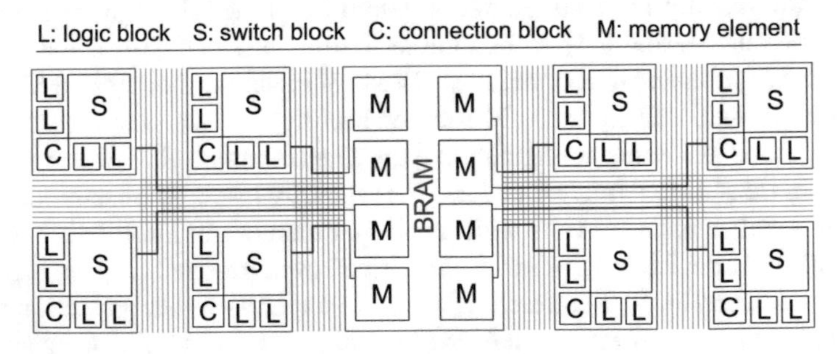

Fig. 11.1 The *reconfiguration unit* (RU) in RRAM-FPGA

11.2.3 Security Advantages and Concerns of Nanoscale Device

Nanoscale memory devices such as PCM, STT-RAM, and RRAM, demonstrate superior advantages on security primitives as compared with complementary metal oxide semiconductor (CMOS)-based memory devices in modern *silicon on-chip* (SOC) designs. Nanoscale memory devices have native characteristics of ultra low power consumption, fast accessing time, and stronger robustness, which can be applied to address security issues, such as piracy, counterfeiting, and side channel attacks. Emerging solutions of hardware security including *physical unclonable functions* (PUF), *public physical unclonable functions* (PPUFs), *nonvolatile memories* (NVMs), *memristor-based true random number generator* (MTRNG), unique signatures, tamper detection circuits, and cryptographic architectures, have received intense study in recent years. The following is an introduction of PUF, PPUFs, NVMs, and MTRNG in modern SOC designs.

1. PUF is a popular solution to address hardware security issue, which provides a hardware specific unique signature or identification. The principle for PUF is based on the intrinsic process variations within integrated circuits. Emerging nanoscale devices such as memristor behaves process variations characteristic, which becomes a promising device for hardware security solutions [29].
2. PPUFs are advanced technologies based on PUFs which overcome the limitations of PUFs of being a secret key technology. Gap-based, matching-based, and digital PPUFs are main PPUFs families, which behave the advantages of energy efficiency, high throughput, low latency, a small footprint, flexibility in the creation of new classes of security protocols, and permanent integration with sensing and computing systems to enable trustable flow of information [28].
3. Emerging NVMs, such as resistive STT-RAM and RRAM, demonstrating some intrinsic randomness in their physical stochastic mechanisms are also explored as one solution to hardware security. These intrinsic randomness, such as resistance variation, random telegraph noise, and probabilistic switching behaviors, can be utilized as entropy source [15].
4. MTRNG leverages the stochastic property when switching a device between its binary states, which significantly reduces the design cost, offering high operating speed, and low power consumption. MTRNG plays a crucial role in system protection and many other security applications [31].

Though nanoscale memory devices have advantages in SOC designs, innate nonvolatility of the devices incurs a concern of data leak in applications of memory system since data lasts longer time as compared to conventional memory devices, such as SRAM and DRAM. For a high density RRAM-FPGA, the concern becomes even serious since FPGAs rely on on-chip memory system to configure logic functions. In the following sections, we will introduce a simple and effective solution to overcome the security concerns of nanoscale memory devices in FPGAs. The technique also helps to address conventional FPGA security flaw, such as spoofing and replay attacks [8, 10].

11.3 Threat Model of RRAM-FPGA and Our Proposed Hardware Solutions

Figure 11.2 summarizes three major security threats in RRAM-FPGA and the corresponding hardware solutions proposed in this work. The design and implementation details of the proposed security scheme shall be described in Sect. 11.4.

Threat 1: *Pirating configurations in BRAMs*. As aforementioned, RRAM-FPGA first loads a design into distributed BRAMs [7]. It is unlikely to encrypt the logic configuration at this step because BRAMs are also be used as data memory in system operation. Moreover, introducing an encryption scheme to each BRAM can severely increase design area and complexity. Thus, the physical attack is a major threat when distributing a RRAM-FPGA with preloaded IP: attackers may obtain the BRAM content through probe attack and then duplicate it on other FPGAs. As illustrated in Fig. 11.2a, we propose to leverage the extremely high storage density of RRAM technology and place *obfuscated* copies of configurations. A *Chip DNA* is used to enable the logic function. As such, even an unauthorized attacker has obtained the data in BRAMs, without the *Chip DNA*, he/she is not able to identify the correct logic combination or discover the system functionality.

Threat 2: *Physical attack on logic components*. After FPGA initialization, attackers can obtain the exact design information by probing the memory units used for

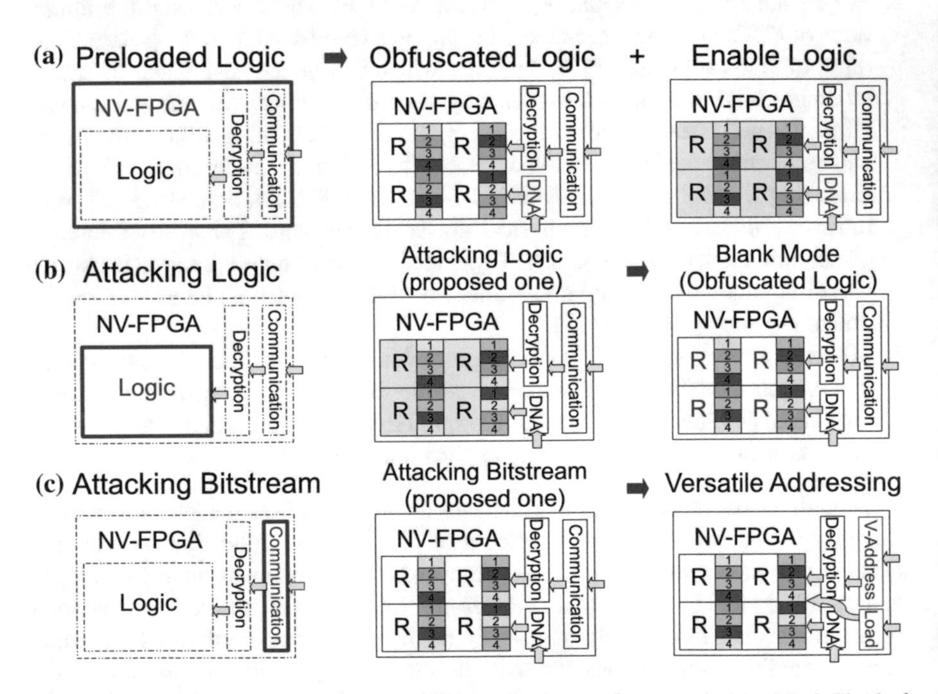

Fig. 11.2 Three major threats in RRAM-FPGA: **a** Pirating configuration in BRAMs. **b** Physical attack on logic components. **c** Attacking bitstream at communication port

logic and routing operations. Therefore, it is even more important to protect the data in logic and routing components. For the designs with a higher security level, we introduce *a blank mode* in which the configuration data on logic and routing elements will be erased at the end of normal operations or as power failure occurs, as shown in Fig. 11.2b. Note that the design is still maintained in BRAMs and can be re-initialized with *Chip DNA*.

Threat 3: *Attacking bitstream at communication port*. The remote and partial runtime reconfigurations are naturally supported in RRAM-FPGA. Hence, bitstream protection is important to prevent piracy or protocol-based denial-of-service attack. Bitstream piracy targets at revealing logic function by predicting configurations from load sequence or comparing bitstreams from different customers. Protocol-based denial-of-service attack, including random pattern injection and replay attacking, can also hurt system integrity. Here, we propose a *encrypted addressing scheme*. As illustrated in Fig. 11.2c, it blends a configuration bitstream by mixing up loading sequence and adding redundant pieces. Meanwhile, it can support the change of secret key for encryption and decryption to prevent replay attacking in remote reconfiguration.

Attacker model: *Powerful attacker*. This work assumes powerful attackers who have equipment to perform physical attacks and possess cutting edge supercomputers for brute force attacks. They have the capability to fetch data stored in RRAM through reverse engineering. They also have knowledge of attacks at communication port through network or eavesdropping. Such attackers are *level 3 attackers* based on IBM's report [1]. In the proposed secure RRAM-FPGA, three elements—the preloaded logic configuration with obfuscation, the *Chip DNA*, and the control of communication network—are needed to acquire logic configuration data and activate the chip functionality. In the work, we assume powerful attackers can obtain either two but not all of the three elements. Attackers shall be able to reveal the logic design based on the logic configuration and the *Chip DNA*. However, probing attack to obtain the data from all the RRAM cells is extremely time and cost consuming. Attackers with the logic configuration and the control of communication port will still require the *Chip DNA* to activate the chip functionality. Otherwise, it is highly impossible to acquire the real design from the obfuscated data. By taking the control of communication port and possessing *Chip DNA*, attackers may perform denial-of-service attack. However, there is no way to acquire logic function.

11.4 Security Protection of RRAM-FPGA

Figure 11.3 gives an overview of the proposed secure RRAM-FPGA architecture, which utilizes the obfuscated configurations for logic protection, the blank mode to resist physical attack, and the encrypted addressing to guard from the bitstream attacks. The details of these features are explained in this section.

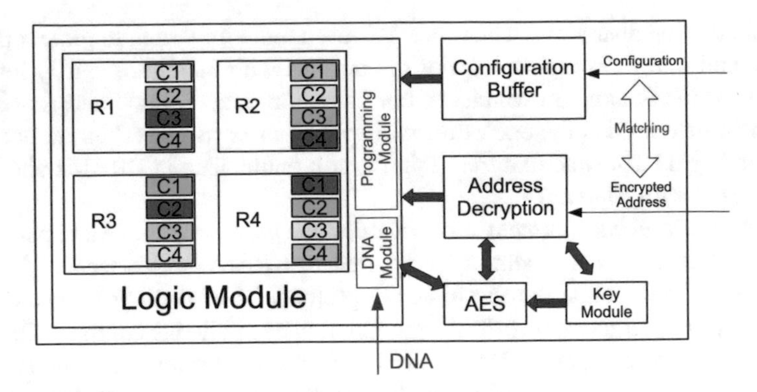

Fig. 11.3 An overview of the proposed secure RRAM-FPGA architecture

11.4.1 Obfuscated Configurations and Chip DNA

By blending the real design with some redundant data, the obfuscated configurations are generated and then loaded into BRAMs. Each BRAM contains only one required configuration. All the remaining copies result in denial-of-service. The FPGA system operates normally only when the required logic configurations are properly selected and assembled. The indices of correct copies in all the RUs form a configuration indication, which is defined as *Chip DNA*.

Here, let's use a FPGA design with four RUs ($R1 \sim R4$) in Fig. 11.3 to demonstrate the configuration obfuscation. We assume a RU contains up to four different design configurations ($C1 \sim C4$). The different colors in the figure represent the logics belonging to different designs. The FPGA could operate in the following conditions:

1. *Blank configuration*. Even the configuration data has been loaded in BRAMs, the logic and routing elements cannot be set up without *Chip DNA*. In another words, the FPGA is not functional yet.
2. *Normal configuration*. The FPGA operates normally under a given design configuration (i.e., Yellow function) after applying a proper DNA sequence (i.e., $C1$-$C2$-$C4$-$C3$) to RUs (i.e., $R1$-$R2$-$R3$-$R4$).
3. *Faulty configuration* occurs if an incorrect DNA is applied. For instance, the DNA sequence $C1$-$C2$-$C4$-$C4$ triggers unmatched logic functions from different RUs: $R4$ is set up for Cyan design, which is not compatible to the Yellow functions of the other RUs. Thus, the FPGA cannot work properly.
4. *Multi-boot* is naturally supported in the proposed design. For example, beside the aforementioned Yellow function, the sample FPGA can also realize Magenta function when applying another DNA sequence $C3$-$C4$-$C2$-$C1$. To further enhance the security level, we can generate the obfuscated data by packing logics with similar routing connections within the same RU or intentionally inject cer-

Table 11.1 Area comparison of FPGAs with different BRAM configurations

SRAM-FPGA			RRAM-FPGA		
No BRAM	w/BRAM 1 copy	w/BRAM 4 copies	No BRAM	w/BRAM 1 copy	w/BRAM 4 copies
1	1.185	1.742	0.373	0.55	0.66

tain redundant routing information. In this way, guessing the logic pattern through connection relation would become even harder.

During system initialization, the *Chip DNA* is loaded through the DNA module (refer Fig. 11.3). The DNA sequence shall be encrypted by AES to resist wiretap attack. After loading into the FPGA, the DNA sequence will be partitioned and distributed to corresponding RUs. In hardware implementation, we can allocate an index to each RU, or share the same piece of DNA among a group of RUs to reduce the length of DNA.

The introduction of obfuscated copies requires larger BRAMs and potentially increases the size of RUs, which is a major concern of the proposed scheme. Fortunately, RRAM technology offers very compact data storage ($\sim 40\times$ of SRAM) and can be integrated in 3D monolithic stacking structure [14], making high density BRAM possible. Based on the design parameters in [7], we analyzed and compared the area cost of BRAM and RRAM-FPGA. The results are summarized in Table 11.1. Here, the area values of these designs are all normalized to that of the baseline SRAM-FPGA which does not utilize BRAMs. Note that the area increment in RRAM-FPGA has a nonlinear relationship with BRAM capacity: adding a BRAM with one or four copies of configuration logic induces 47.5 % or 76.9 % area overhead to the design without BRAM. This is because the 3D structure is adopted in constructing BRAMs [7]. Overall, a RRAM-FPGA with four copies of configuration logic occupies 34 % less area compared to the baseline SRAM-FPGA.

The capacity of logic confiscation shall be users' choice. For economical usage, users can create obfuscation logic with BRAM of two copies to save area cost. In the work, we suggest the BRAM capacity of four copies of configuration data. As such, the system can protect high sensitive designs by utilizing four copies of logic obfuscation. Some less sensitive applications, instead, can take only half of the capacity for logic obfuscation and the rest memory to store temporary data, which improves system performance by reducing IO accesses.

11.4.2 Blank Mode Operation

In a general purpose application built with nonvolatile FPGA, the configuration data is kept in logic and routing elements. However, the nonvolatile configuration data is in danger of physical attack, e.g., probing attack, potentially resulting in security concerns. Here, *blank mode* operation is proposed for the RRAM-FPGA to prevent physical attacks. Whenever the system completes normal operations or detects power failure, it automatically erases the content on logic and routing elements with assist of a *power-off erasing scheme*. At the end of normal operations, an instruction shall be issued to enable the blank configuration before the system is powered off. In the case of power failure, a backup power is necessary to erase at last partial configuration data. Since the programming energy of RRAM is relative small, on-board soldered battery [19] or super-capacitor [16] shall be sufficient enough. The blank configuration in Sect. 11.4.1 indeed is a form of *blank mode*. Even attackers obtain the content in BRAMs, recovering the design functionality is difficult because the correct configuration copies are hidden by the dummy copies.

Note the blank mode operation cannot be supported in the conventional and many emerging NVM FPGAs, in which the logic configuration is stored only in the logic and routing elements. In contrast, our design has a copy of obfuscated design in BRAMs. Therefore, the design can always be recovered as far as the end-user keeps the proper *Chip DNA*.

11.4.3 Encrypted Addressing

Conventional SRAM-FPGAs usually have two sets of configuration ports: (1) the port used for system initialization and (2) the *internal configuration access port* (ICAP) for partial run-time reconfiguration. Benefiting from the nonvolatile storage and logic elements, RRAM-FPGA only requires one-time programming for design loading. Hence, a dedicated initialization port is not necessary. A bitstream can be encrypted with AES and loaded in sequence though one communication port.

Here, we propose an alternative solution that provides better loading speed and leverages the data protection by obfuscated logic. The scheme utilizes the address-based bitstream [33] and imports the RU addresses and the corresponding logic configurations separately through the two ports. The short address stream needs to be protected through AES [26]. Simple operation modes such as ECB [11] shall be sufficient since address is a non-repeated data in the sequence. The long configuration data is already under the protection of obfuscated logic so it can keep the same format as conventional bitstream. The small size of unencrypted configuration data promises fast loading and reduced stall time which is critical in many partial reconfiguration applications [9]. If higher level of security is required, AES can be applied to encrypt the configuration data. More complicated operation modes might be necessary because there is a chance of repeated configuration in sequence. In the work, we set the address length as 128 bits to be compatible with AES standard.

To guarantee a piece of configuration data is shipped to its target RU, the simple *serial loading scheme* can be applied. As shown in Fig. 11.4a, the RU address increases sequentially and is loaded followed by its configuration. Then the configuration goes to the buffer and the address is decrypted, which takes about 20 $\sim$ 40 ns [30]. The related hardware components, including AES, Key module, and ad-

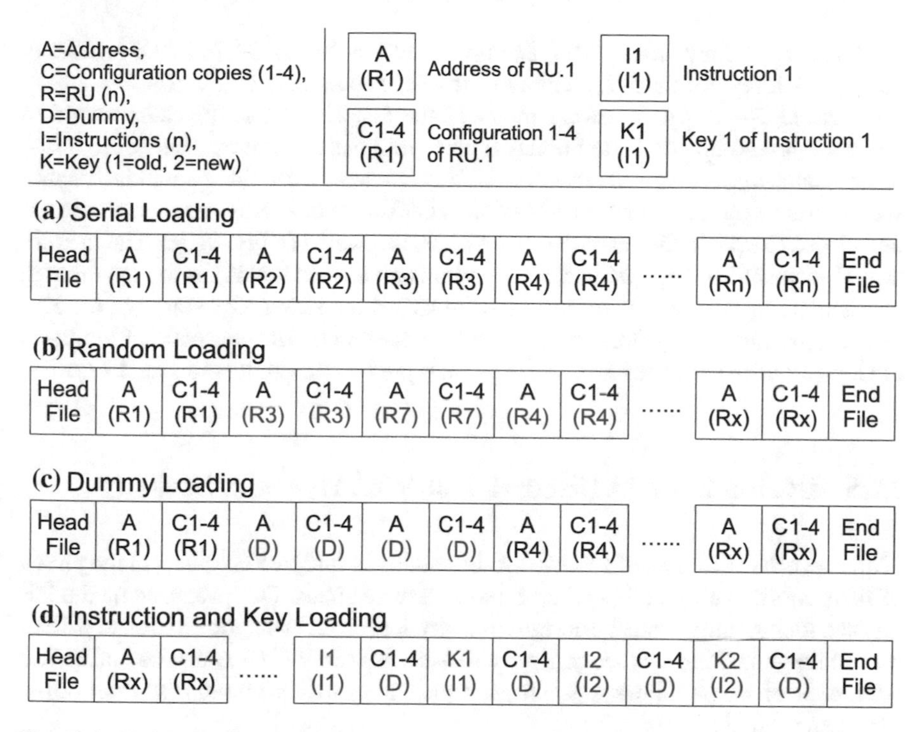

Fig. 11.4 Various loading schemes: **a** serial loading; **b** random loading; **c** dummy loading; and **d** instruction and key loading

dress decryption, are shown in Fig. 11.3. However, such a simple loading scheme inevitably faces security issues, such as bitstream piracy and protocol-based denial-of-service attack, due to the sequential address loading and unprotected configuration. In this work, we propose the *encrypted addressing* to protect the communication ports aiming at different conditions.

Scenario 1—Bitstream piracy: Attackers could reveal configurations from loading sequence by comparing two or multiple bitstreams belonging to different customers. *Our Solution—Random loading scheme* can be easily realized in the address-based design by hashing the loading sequence as shown in Fig. 11.4b. As a result, the address sequences of different bitstreams are not comparable. The address hashing shall be done by designers/IP providers at software level. It doesn't introduce extra hardware cost to RRAM-FPGA system.

Scenario 2—Protocol-based denial-of-service attack: While proceeding partial reconfiguration, attackers may record the encrypted addresses and inject fake configuration at these addresses. They even do not have to find out decrypted address to make system malfunction. *Our Solution—Dummy loading:* We can intentionally inserts dummy data during loading process, as illustrated in Fig. 11.4c. These dummy data eventually will be thrown away once the system detects that the dummy address corresponds to invalid RU location. Note that a large pool of invalid addresses are available for dummy loading. For example, a FPGA having 32,768 RUs need only 15 bits to address all the RUs, while length of coded address is 128 bits.

Scenario 3—Replay attacking indicates the situation when attackers record (a piece of) the old bitstream and replay it after designer configures a new design into the system. *Our Solution – Instruction and key loading:* We define a small instruction set by leveraging a few unused bits of the address code. A Key-renew operation is introduced to update the secret Key of AES as illustrated in Fig. 11.4d. The instruction **I1** requests the key verification operation and the old key **K1** is loaded through configuration port for comparison. If **K1** matches the secret Key stored in the Key module, the system approves the key updating operation. The instruction **I2** initiates the loading of the new key **K2**, which is encrypted based on the old secret key.

11.5 Evaluation and Discussion of RRAM-FPGA Security

Major security concerns of FPGAs can be cataloged into two aspects: (1) the piracy of logic configuration and (2) the denial-of-service attack. Our proposed methodology can successfully protect a design from physical attack, analytic attack and pattern recognition attack. In this section, we will use a RRAM-FPGA composed of 32,768 RUs, each of which allows four design copies, to evaluate the security level of the proposed scheme.

11.5.1 Protection of Logic Piracy in FPGA

An end-user can design his own logic function or buy IP core from other IP companies or silicon vendors. In RRAM-FPGAs, obfuscated configuration, encrypted address, and *Chip DNA* utilize different protection schemes. Without a complete set of the three parts, it is very difficult for an attacker to pirate the correct logic function.

For powerful attackers who have FPGA board and get all obfuscated configuration data, they still need *Chip DNA* to activate the function. Here, the length of *Chip DNA* determines the complexity of *analytic attack*, which in general is very high. Theoretically, the number of rounds through brute attack is $2^{65,536}$, which is extraordinary large to figure out the correct DNA.

Loading bitstream to FPGA over insecure network is under protocol eavesdropping. Attackers could get unencrypted obfuscated logic and encrypted address. Thus, it is still hard to place a piece of logic function into correct location. Attackers may be able to duplicate the bitstream to other FPGAs. However, different AES key can be applied, indicating that the location mapping of logic function pieces is different. Even AES module is under side channel attack and not secure, we still have the *Chip DNA* to protect the chip function: an incorrect DNA cannot activate device or reveal logic configuration as explained in Sect. 11.4.1.

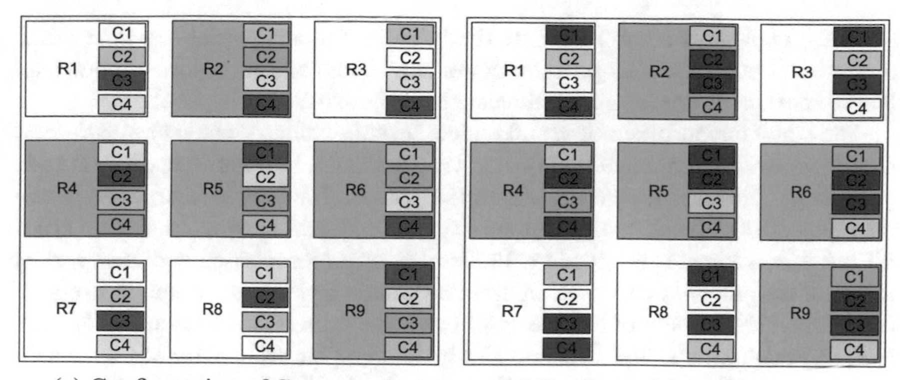

(a) Configuration of Customer 1 (b) Configuration of Customer 2

Fig. 11.5 Different logics activated by different DNAs

Here is an extensive application of *Chip DNA*. For IP companies/silicon providers who distribute IP core bitstream to many customers, the best way to protect IP cores is to generate unique configuration bitstream file and *Chip DNAs* for each customer. Such strategy can be easily realized by the proposed RRAM-FPGA. An effective method is changing the index of each RU and the corresponding *Chip DNA*. For example, Fig. 11.5 shows that an identical design can be coded into two different bitstreams and distributed to two customers. Accordingly, they need different *Chip DNAs* to enable the logic.

As aforementioned, potentially attackers could compare the obfuscated bitstream from the two customers and guess the genuine configuration, that is *the pattern recognition attack*. For instance, attackers may match the code pieces of R2 configurations in two bitstreams and distinguish the Green piece, the only common code existing in both designs. Two approaches can resist this attack. The designer can either insert the same dummy configurations in every RU so that attacker cannot distinguish the real one based on its appearance. Or, as presented in Sect. 11.4.3, the random loading scheme supported in the encrypted addressing can blend the bitstream sequence so that comparing and matching two bitstreams becomes extremely difficult.

11.5.2 Protocol-Based Denial-of-Service Attack

The run-time and remote reconfigurations have become more and more popular on FPGA-based embedded systems, e.g., space system and set top boxes [3]. Modern FPGAs use ICAP for partial run-time reconfiguration, which is more vulnerable compared to initialization port. Remote configuration usually transmit data through existing network, which could be insecure. Multiple authentications are needed to guarantee the authentication process not pirated by the attackers. The proposed

RRAM-FPGA does not differentiate the initialization and partial run-time reconfiguration. Both operations go through the same communication port, are protected by the same hardware scheme, and hence have the same security level.

The major threats in run-time and remote reconfiguration include *protocol-based denial-of-service attack* and *replay attacks*. Among the 128 bits of an address code, only 15 bits are valid for the RRAM-FPGA with 32,768 RUs. Hence, when attackers inject false or blank configurations to damage system integration, the chance to hit the valid addresses is very low. Probability of protocol-based denial-of-service attack of the proposed encrypted addressing when varying address length from 128 bits to 256 bits. The size of RRAM-FPGA changes from 16,364 RUs to 65,536 RUs. For the FPGA with 32,768 RUs and 128-bit address, the probability of a successful protocol-based denial-of-service attack is as low as $2^{-116.3}$. Large FPGA design with more RUs has a higher attack probability because more bits in address are valid. For such a system, properly extending the length of address should be sufficient. The replay attack could be successful if attackers record the information during transferring and resend it through the same network to attack the FPGA system. The secret key updating of the encrypted addressing loading scheme can prevent this type of attack. When attackers replay the old configurations, the decrypted address based on the new secret key will be mapped to an invalid address. So nothing will be loaded into RUs. In contrast, the system designers with the new secret key are still able to change ciphertext of the address and DNA and load function for the next remote configuration.

11.5.3 Miscellaneous Discussion

The obfuscated configuration of the secure RRAM-FPGA results in bigger bitstream file and longer loading time. The size of configuration bitstream is determined by not only the design function but also the resources on routing elements. As an example, we investigated the length of bitstreams for the RRAM-FPGA with different *Look-Up Table* (LUT) designs. Figure 11.6 shows the bitstream size for FPGAs with 4-input, 6-input, or 8-input LUTs, assuming all the FPGAs have 32,768 RUs. After including the cost at auxiliary components, we can see the bitstream length increases fast with the LUT input number. The system based on 8-input LUT with four obfuscated copies requires an even larger bitstream file than the designs based on 4-input or 6-input LUTs with eight copies. And the bitstream size grows linearly as the number of obfuscated configuration copies increases.

To evaluate the overhead on loading time, we tested the loading operation of 20 big MCNC benchmarks [25] assuming using USB 2.0 at 480 Mbps. The results are given in Fig. 11.6c. The FPGA designs with 4-input LUTs and 6-input LUTs were examined. The red line shows the ratio of the loading time (LUT 6/LUT 4). Because the design with four-input LUTs needs less configuration bits and hence spends less loading time. The ratio, though, is strongly related to the benchmarks. Though the ratio is strongly related to the benchmarks, all the benchmarks can complete their loading within 0.02 s.

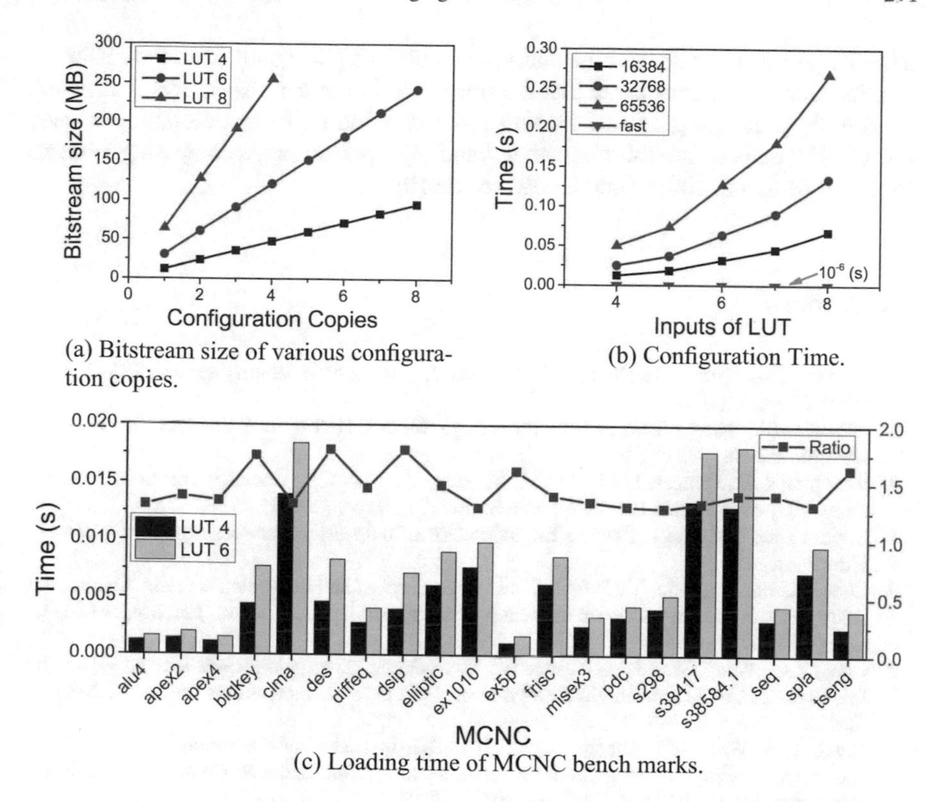

(a) Bitstream size of various configuration copies.

(b) Configuration Time.

(c) Loading time of MCNC bench marks.

Fig. 11.6 Performance analysis

The time to distribute the genuine configurations from BRAMs to logic and routing elements is irrelevant to the number of obfuscated copies. Figure 11.6b shows the configuration time of different RRAM-FPGAs with various capacity. Because the programming mechanism is designed to serially program memory cells for less area cost, the design with 8-input LUTs needs longer configuration time. And larger FPGAs with more RUs takes longer time in configuration if DNA index is loaded to RU in sequence. However, if we assume the Chip DNA is prefetched and stored, and functions in all RUs are distributed in parallel (“`fast`”), the configuration time can be shorten to 10^{-6} to 10^{-5} s.

11.6 Conclusion

We proposed a secure configuration scheme for a RRAM-FPGAs, in which the high density of emerging resistive memory technology is leveraged to build logic and storage elements. The proposed architecture integrates three techniques to enhance system security—the obfuscated configuration for logic protection, a blank mode to resist physical attack, and the encrypted addressing to guard from the bitstream attacks. Our simulations show that the proposed design can resist level 3 attackers by slightly prolonging system loading and execution performance. The proposed RRAM-FPGA is suitable for embedded and high performance computing systems for better design flexibility and hardware security.

References

1. Abraham, D., Dolan, G., Double, G., Stevens, J.: Transaction security system. IBM Syst. J. **30**(2), 206–229 (1991)
2. Altera: Logic Array Blocks and Adaptive Logic Modules in Stratix V Devices (2011). http://www.altera.com
3. Badrignans, B., Danger, J., Fischer, V., Gogniat, G., Torres, L.: Security Trends for FPGAS: From Secured to Secure Reconfigurable Systems. Springer (2011)
4. Bottom Line Technlogy: Reverse Engineering/Re-Engineering Services (2011). http://www.bltinc.com
5. Chen, Y., Zhao, J., Xie, Y.: 3D-NonFAR: three-dimensional non-volatile FPGA architecture using phase change memory. In: International Symposium on Low Power Electronics and Design (ISLPED), pp. 55–60 (2010)
6. Chen, Y.C., Wang, W., Li, H., Zhang, W.: Non-volatile 3D stacking RRAM-based FPGA. In: International Conference on Field Programmable Logic and Applications (FPL), pp. 367–372 (2012a)
7. Chen, Y.C., Wang, W., Zhang, W., Li, H.: uBRAM-based run-time reconfigurable FPGA and corresponding reconfiguration methodology. In: International Conference on Field-Programmable Technology (FPT), pp. 80–86 (2012b)
8. Devic, F., Torres, L., Badrignans, B.: Secure protocol implementation for remote bitstream update preventing replay attacks on FPGA. In: IEEE International Conference on Field Programmable Logic and Applications (FPL), pp. 179–182 (2010)
9. Dimou, K., Wang, M., Yang, Y., Kazmi, M., Larmo, A., Pettersson, J., Muller, W., Timner, Y.: Handover within 3gpp lte: design principles and performance. In: 2009 IEEE Vehicular Technology Conference Fall (VTC), pp. 1–5 (2009)
10. Drimer, S., Kuhn, M.G.: A protocol for secure remote updates of FPGA configurations. In: International Workshop on Applied Reconfigurable Computing, pp. 50–61 (2009)
11. Dworkin, M.: Recommendation for Block Cipher Modes of Operation. Technical report, DTIC Document (2001)
12. Huffmire, T., Brotherton, B., Sherwood, T., Kastner, R., Levin, T., Nguyen, T., Irvine, C.: Managing security in FPGA-based embedded systems. IEEE Des. Test Comput. **25**(6), 590–598 (2008)
13. Huffmire, T., Irvine, C., Nguyen, T., Levin, T., Kastner, R., Sherwood, T.: Handbook of FPGA Design Security. Springer (2010)
14. ITRS: International Technology Roadmap for Semiconductors 2011 Edition (2011). http://www.itrs.net/
15. Karam, R., Liu, R., Chen, P.Y., Yu, S., Bhunia, S.: Security primitive design with nanoscale devices: a case study with resistive RAM. In: ACM Great Lakes Symposium on VLSI (GLVLSI), pp. 299–304 (2016)
16. Knoth, S.: Supercaps Can Be a Good Choice Over Batteries for Backup Applications (2012). http://www.eetimes.com/document.asp?doc_id=1280982
17. Kumar, S.S., Guajardo, J., Maes, R., Schrijen, G.J., Tuyls, P.: The butterfly PUF protecting IP on every FPGA. In: IEEE International Workshop on Hardware-Oriented Security and Trust (HOST), pp. 67–70 (2008)

18. Kuon, I., Tessier, R., Rose, J.: Fpga architecture: survey and challenges. Found. Trends Electron. Des. Autom. **2**(2), 135–253 (2008)
19. Lattice: FPGA Design Security Issues: Using the ispXPGA Family of FPGAs to Achieve High Design Security (2003). http://www.latticesemi.com
20. Liauw, Y., Zhang, Z., Kim, W., Gamal, A., Wong, S.: Nonvolatile 3D-FPGA with monolithically stacked RRAM-based configuration memory. In: IEEE International Solid-State Circuits Conference Digest of Technical Papers (ISSCC), pp. 406–408 (2012)
21. Microsemi: Axcelerator Family FPGAs (2012a). http://www.actel.com
22. Microsemi: IGLOO Low Power Flash FPGAs (2012b). http://www.actel.com
23. Microsemi: Introduction to the SmartFusion2 and IGLOO2 Security Model (2013a). http://www.microsemi.com
24. Microsemi: Overview of Data Security Using Microsemi FPGAs and SoC FPGAs (2013b). http://www.microsemi.com
25. Minkovich, K.: MCNC benchmark (2007). http://cadlab.cs.ucla.edu/~kirill/
26. Nechvatal, J., Barker, E., Bassham, L., Burr, W., Dworkin, M.: Report on the Development of the Advanced Encryption Standard (AES). Technical report, DTIC Document (2000)
27. Paul, S., Mukhopadhyay, S., Bhunia, S.: A circuit and architecture codesign approach for a hybrid CMOS-STTRAM nonvolatile FPGA. IEEE Trans. Nanotechnol. (TNANO) **10**(3), 385–394 (2011)
28. Potkonjak, M., Goudar, V.: Public physical unclonable functions. Proc. IEEE **102**(8), 1142–1156 (2014)
29. Rose, G.S., Rajendran, J., McDonald, N., Karri, R., Potkonjak, M., Wysocki, B.: Hardware security strategies exploiting nanoelectronic circuits. In: IEEE Asia and South Pacific Design Automation Conference (ASP-DAC), pp. 368–372 (2013)
30. Suh, G., Clarke, D., Gasend, B., Van Dijk, M., Devadas, S.: Efficient memory integrity verification and encryption for secure processors. In: Annual IEEE/ACM International Symposium on Microarchitecture (MICRO), pp. 339–350 (2003)
31. Wang, Y., Wen, W., Li, H., Hu, M.: A novel true random number generator design leveraging emerging memristor technology. In: ACM Great Lakes Symposium on VLSI (GLVLSI), pp. 271–276 (2015)
32. Xilinx: 7 Series FPGAs Overview (2011a). http://www.xilinx.com
33. Xilinx: Partial Reconfiguration of Xilinx FPGAs Using ISE Design Suite (2011b). http://www.xilinx.com
34. Xilinx: Virtex-5 FPGA Configuration User Guide (2012). http://www.xilinx.com

Chapter 12
Security Standards for Embedded Devices and Systems

Venkateswar Kowkutla and Srivaths Ravi

12.1 Introduction

Embedded system-on-chips (SoCs) are widely used in a variety of applications today. These range from general purpose and compute to end equipment specific such as mobile, industrial, networking, payment card systems, and automotive. A majority of these embedded SoCs capture, store, manipulate, and/or access sensitive/critical code, and data. Not surprisingly, these embedded SoCs or the systems deploying them are prone to a wide range of security attacks. It has been documented in numerous surveys that the cost of not having adequate security mechanisms in the face of these threats can be significantly high. Classical IT security threats have been estimated to have costed the U.S companies on an average $12.7 million in 2014, up by nearly 10 % year on year [1]. In 2012, security incidents due to mobile computing have been pegged as costing nearly half a million US dollars just at the enterprise level. This cost is only increasing with the proliferation of mobile devices [2]. The growing networked nature of any embedded device in the Internet of Things (IoT) era exposes new security issues on a daily basis—be it, smart home hacks [3], insulin pump hijacks [4], smartwatch security flaws [5], remote car commandeering [6], etc.

Given the above factors, security has now become a critical consideration in the design cycle of every embedded SoC or system to the point that *every embedded SoC needs to provide some level of security* [7]. It is only the “level of security” that is debated based on the critical nature of the applications and the associated

V. Kowkutla (✉)
Texas Instruments, Mail Station E4000 12500 TI Blvd, Dallas, TX 75243, USA
e-mail: venkat@ti.com

S. Ravi
Texas Instruments India, 66/3 Bagmane Tech Park, C V Raman Nagar, Bangalore 560093, India
e-mail: srivaths.ravi@ti.com

S. Bhunia et al. (eds.), *Fundamentals of IP and SoC Security*,
DOI 10.1007/978-3-319-50057-7_12

"cost-benefit-risk" tradeoffs. In many end equipments or applications, there are well-defined standards that can be used to guide this process. Let us pick a couple of examples.

- **Point of sale (PoS) systems:** PoS systems deal with accessing, processing, and storing sensitive data pertaining directly or indirectly to monetary payments. Hence, these systems require high level of security and are heavily governed with multiple security standard committees with involvement of payment card institutions like Visa, MasterCard, American Express, and Discover. The two main security standard committees are Payment Card Industry Data Security Standard (PCI DSS) [8] and Europay Mastercard and Visa (EMV) [9]. These standards define high-level specifications to ensure that end system is in compliance with the desired security goals. SoCs targeted for these applications must adhere to these specifications and support required features to allow end customer pass PCI/EMV certification. In this chapter, we look at the security standards tied to this application in depth.
- **Automotive systems:** With the advent and proliferation of innovative in-vehicle infotainment (IVI) technology and development of next-generation advanced driver assistance systems (ADAS), automakers are relying heavily on embedded SoCs to deliver these advanced safety and convenience features. Automotive safety applications based on vehicle-to-vehicle and vehicle-to-infrastructure communication have been identified as a means for decreasing the number of fatal traffic accidents in the future. Examples of such applications are local danger warnings and electronic emergency brakes. While these functionalities herald a new era of traffic safety, new security requirements need to be considered in order to prevent attacks on these systems. This has become urgent with the rapid increase in the number of electronic components and software content in a car in recent years.

A "Smart Car" may be equipped with up to 200 embedded ECUs (electronic control units) for a variety of functions using several million lines of code. The ECUs are connected via various vehicular buses (e.g., CAN, MOST, MLB, ETHERNET), forming a complex distributed system. The increasingly interconnected nature of these electronic components, significant software usage along with the availability of vehicular communication interfaces immediately exposes the automotive surface to a wide range of security threats. In [10], the authors describe two different methods of gaining indirect access to the car's bluetooth unit. Software vulnerability in the bluetooth interface code running in telematics unit is then used to execute arbitrary code.

Hence, functional safety ("the state in which a vehicle function does not cause any intolerable endangering states") and security ("protection against malicious attacks") are today critical requirements for automotive SoCs. Security features must include not just physical access and protection of confidential information, but also protect critical safety subsystems [28]. Safety requirements are governed by regulatory standards such as ISO 26262 [11] and IEC 61508 [12]. While there is no uniform automotive security standard in place yet, security has been largely defined by the general purpose standard Common Criteria [13], which been largely concerned with the protection of assets against malicious attackers. This standard will also be reviewed later in this chapter.

In the rest of this chapter, we describe the two layers of a hierarchical security model that the embedded system designer typically uses to achieve his or her application specific security needs.

- Foundation level security targeted at basic security services such as privacy, authentication, and integrity is achieved predominantly using the powerful mathematical functions called cryptographic algorithms. Cryptographic algorithms are the subject of excellent books in data and network security [14, 15]. In Sect. 2, we attempt to give a brief insight into these building blocks and their coverage through standards.
- Security protocols like TLS or SSL [16], IPSec [17], etc, have been the classical overlay atop foundational cryptographic algorithms to achieve computer communication and network security. In the embedded world, both general purpose and end equipment or application level security standards are now becoming *niche* overlays that encompass not just foundational cryptographic algorithms but also a "basket of cryptographic and non-cryptographic solutions" to achieve system security goals. One example of such a security goal can be "Zero-ize secure memory in a payment terminal chip" upon detection of an attack. In Sects. 3 and 4, we look at examples of security standards that are actively used by chip and system vendors in the embedded space.

It would also be worth noting that both these layers are effectively implemented using a combination of hardware and software techniques in embedded SoCs. Examples of various applicable solutions themselves are not covered in this chapter, and can be found in [7] and other chapters in this book.

12.2 Foundational Cryptographic Algorithms and Associated Standards

The cryptographic toolkit [18] is a repository for recommended or approved cryptographic algorithms from the National Institute of Standards and Technology (NIST). Table 12.1 lists the various algorithms that are approved by NIST for various cryptographic operations at this time.

The following subsections take a further look into commonly used block ciphers—AES and Triple DEA or Triple DES, and the secure hash standard (SHS).

Table 12.1 NIST cryptographic toolkit and recommended/approved algorithms

Category	Recommended or Approved algorithms
Block ciphers	AES, Triple DES/DEA, and Skipjack
Digital Signatures	DSA, RSA, and ECDSA
Secure hashing	SHA-1, SHA-224, SHA-256, SHA-384, SHA-512, SHA-512/224, and SHA-512/256
Message authentication	Message Authentication Code (MAC or DAC) and Keyed-Hash Message Authentication Code (HMAC)
Random number generation	Approved deterministic random bit generators (DRBGs) or pseudorandom number generators

12.2.1 *Advanced Encryption Standard (AES)*

FIPS-197 [19] is the specification for the Advanced Encryption Standard (AES), which is a symmetric block cipher that can be used to encrypt and decrypt data. The AES algorithm (Rijndael was selected as NIST AES standard in Nov. 2001) encrypts (decrypts) data in blocks of 128 bits. Encryption (decryption) can be performed using cryptographic keys of 128, 192, or 256 bits.

A representative block diagram of AES is shown in Fig. 12.1, where encryption and decryption of a block of data happens through an iterative application of four basic transformations. Each iteration, called a round, works on a block of data generated from the previous iteration and a round key which is derived from the AES key through a key expansion cycle. The four basic transformations in an

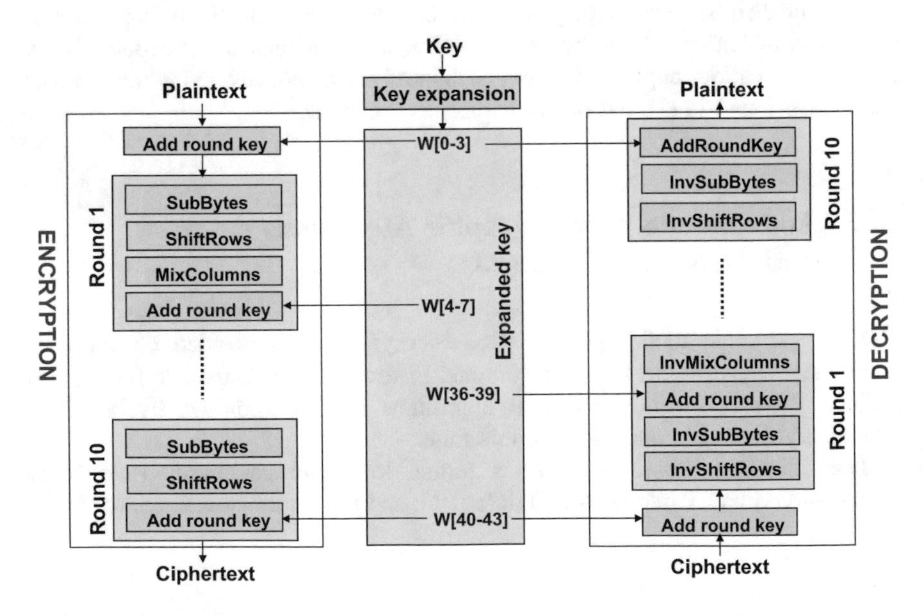

Fig. 12.1 Representative block diagram of AES Encryption and Decryption Operations

encryption round are SubBytes (substitution using a lookup table), ShiftRows ("macro" permutation), MixColumns (fixed polynomial multiplication of columns in $GF(2^8)$, and Add RoundKey (bitwise xor). Details of the AES operation can be found in an accessible form at [20]. Decryption is similar to encryption with inverse transformations now pulled into each round.

AES is known to be cryptanalytically strong at this time with no known theoretical attacks that can break it with reasonable time or storage. However, information leakage in implementations of AES has been exploited through a range of side-channel attacks including timing and fault analysis attacks to infer the cryptographic key.

12.2.2 Triple DES (3DES)/Triple DEA

FIPS 46-3 [21] is the standard specification of the Data Encryption Algorithm (DEA) or Data Encryption Standard (DES). DES/DEA has now become outdated due to the inherent weakness of a 56-bit key to brute force attacks with evolution of computational power and has been withdrawn as a standard in 2005 [22]. Please note that the 56-bit key with additional 8 error detection bits form the 64-bit key that one commonly associates with DEA.

Triple DES or Triple DEA at its simplest level involves applying the DEA algorithm three times. When the keys of each application of DEA are independent, the key strength is equivalent to 168 key bits. Though DES is no longer a NIST approved standard, Triple DEA with DEA as the core engine is an approved standard as specified by NIST Special Publication (SP) 800-67 [23].

With the origins of the AES influenced by the DEA algorithm, the structure of the DEA algorithm is in many ways similar. The DEA algorithm is an iterative application of mathematical transformations on an input data stream which is divided into blocks. The block size is 64 bits and each iteration in a DEA algorithm is called a round. There are 16 rounds and each round processes the 64-bit block in two 32-bit halves. One half is processed by a transformation function called F-function, while the other half is xored with the result of the F-function. The halves are then swapped before the next round. The F-function in each round is a sequential application of four transformations—expansion, key mixing, substitution and permutation. Decryption is again a sequence of inverse transformations with the same key.

12.2.3 Secure Hash Standard (SHS)

FIPS 180-4 [24] specifies the approved algorithms to compute the condensed digest or hash of a message. A cryptographic hash can conceptually be thought of as similar to the checksum representation for a given piece of data, but it is mathematically more secure. The security comes from two angles—for a given secure hash algorithm, it is computationally infeasible to (a) find a message that corresponds to a given message digest, or (b) to find two different messages that produce the same message digest.

The FIPS 180-4 standard specifies SHA-1, SHA-224, SHA-256, SHA-384, SHA-512, SHA-512/224, and SHA-512/256 as the approved secure hash algorithms. For these algorithms, the message digests range in length from 160 to 512 bits. As per NIST's 2012 policy on hash functions [25], recently discovered weaknesses in SHA-1 have limited its usage to certain applications. The other approved SHA algorithms today are collectively called SHA-2. There is a plan to have the next-generation hash algorithms (KECCAK) standardized as SHA-3 algorithms in the future. Secure hash algorithms are also used in combination with other cryptographic algorithms, such as digital signature algorithms and keyed-hash message authentication codes, or in the generation of random numbers (bits).

12.3 Generic Security Standards

In this section, we will present a brief overview of the generic security standards which are also employed in the embedded space.

12.3.1 Federal Information Processing Standard (FIPS) 140

Federal Information Processing Standards (FIPS) 140 is a series of security standards published by the National Institute of Standards and Technology (NIST) that specifies security requirements for cryptographic module—loosely, we can define a cryptographic module as any system with cryptographic components used to secure sensitive information. The latest version is FIPS 140-2 [27], while there is a draft version of FIPS 140-3 still under the review/approval process. The security requirements specified in FIPS 140-2 span 11 specific areas related to secure design and implementation of cryptographic modules as indicated in Table 12.2.

A given cryptographic module is assessed in each area and receives a security level rating: from Level 1 (lowest) to Level 4 (highest) depending on the extent to which requirements are met. An overall rating is then issued to the cryptographic module which depends on minimum of the independent ratings received and fulfillment of other requirements specified in the standard. The four security levels provided by this standard are compared below in Table 12.3 against various dimensions such as levels of security, physical security, security against environmental conditions exploits, and the underlying operating system on which the firmware of the cryptographic module runs.

Table 12.2 A listing of cryptographic design and implementation aspects of a cryptographic module covered by FIPS 140-2

Category	Description
Cryptographic module specification	Specification of a cryptographic module that may have hardware and software components
Cryptographic module Ports and Interfaces	All physical ports and logical interface that define entry/exit points for a cryptographic module
Roles, services, and authentication	Definition and separation of roles accorded to operators of the cryptographic module, along with the corresponding services and access authentication
Finite state model	Definition of all operational and error states of a cryptographic module, along with the state transition specification
Physical security	Protection of sensitive or critical information in the cryptographic module from physical access
Operational environment	Management of the software and hardware components required to operate the cryptographic module
Cryptographic key management	Key management spanning the entire lifecycle of cryptographic keys from key generation, distribution, storage to destruction
Electromagnetic interference/compatibility	Proof of conformance of cryptographic module to electromagnetic interference and compatibility
Self-test	Self-tests (power up/conditional) to ensure that the cryptographic module is functioning properly
Design assurance	Use of various best practices by the vendor of a cryptographic module during module development
Mitigation of other attacks	Mitigation against emerging attacks that could include (but not restricted to) power analysis, timing analysis, fault induction, electromagnetic analysis

Table 12.3 Sample comparison of FIPS 140-2 security levels along various dimensions

FIPS 140-2 security level	Level of security	Physical security	Security against environmental conditions exploits	Operating system
Security level 1	Lowest	Not required	Not required	Unevaluated operating system
Security level 2	↓	Tamper evidence	Not required	Evaluated at the CC evaluation assurance level EAL2 (or higher)
Security level 3		Tamper Detection and Response (High Probability)	Not required	Evaluated at the CC evaluation assurance level EAL3 (or higher)
Security level 4	Highest	Tamper Detection and Response (Highest Probability). Zeroize security parameters	Required	Evaluated at the CC evaluation assurance level EAL4 (or higher)

If we review any dimension in FIPS 140-2 in detail, we will see the levels of security requirements increase from one level to another. Picking physical security as an example, we can see that

- Physical security mechanisms are not required for Security Level 1.
- Security Level 2 enhances the physical security aspects of a Security Level 1. Per FIPS 140-2, it includes the requirements for tamper-evidence such as, use of tamper-evident coatings/seals or for pick-resistant locks on removable module covers or doors to protect against unauthorized physical access.
- Security Level 3 provides tamper resistant physical security—preventing the intruder from gaining access to critical security parameters held within the cryptographic module. This level has a high probability of detecting and responding to physical tampering of the cryptographic module. Physical security mechanisms include tamper detection/response circuitry that erases all critical security parameters when the removable covers/doors of the cryptographic module are opened.
- Security Level 4 requires physical security mechanisms to detect and respond to all unauthorized accesses. This level offers high probability of detecting all physical tamper attacks resulting in the immediate eraser of critical security parameters. This level of security is appropriate for applications deployed in physically unprotected environments needing high security.

This is true for other security dimensions as well (in some cases, they could be same as the previous level). Additional observations on each security level are listed below for further illustration.

- In Security Level 1, security requirements for cryptographic modules are very basic—e.g., use one approved algorithm or security function. This level is appropriate for low-cost applications where the security requirements are very low and does not need physical/network security. From an operating system perspective, cryptographic software and firmware can be executed on a general purpose computing system using an unevaluated operating system. An example of a Security Level 1 is a personal computer (PC) encryption board.
- Security Level 2 requires role-based authentication to authorize an operator to assume a specific role to perform corresponding set of tasks. Cryptographic software and firmware can be executed on a general purpose computing system using an operating system that has been evaluated at the Common Criteria (CC) evaluation assurance level EAL2 (or higher). We will be reviewing CC in the next section.
- Security Level 3 requires identity-based authentication mechanisms to authenticate the identity of an operator and verifies that the identified operator is authorized to assume a specific role to perform corresponding set tasks. Security Level 3 requires encryption on critical secure parameters while writing into or reading from the cryptographic modules. These writing/reading ports must be physically separated from ports on other interfaces. The underlying operating

system should have been evaluated at the Common Criteria (CC) evaluation assurance level EAL3 (or higher).

- Security Level 4 provides a complete envelope of protection around the cryptographic module. This level provides protection against all types of tamper attacks including security vulnerabilities due to environmental conditions or fluctuations outside of the module's normal operating ranges for voltage and temperature.

Overall, these security levels provide a cost-effective way to rate a cryptographic module covering a wide range of products and application environments in which it may be deployed. Users of cryptographic modules can make their selection based on the individual area ratings and overall rating depending on the environment in which the cryptographic module will be implemented.

12.3.2 Common Criteria (CC)

Common Criteria (CC) [13] is an internationally recognized standard used by consumers, government agencies, and other organizations to assess the security and assurance of technology products. With various security criteria existing within the international community, it was developed in an effort to resolve the differences (conceptual or technical) and enable standardization right from specification to implementation to evaluation.

Common Criteria approaches security in a holistic way—Highest security can be achieved only when security is considered during all phases of the product life cycle including specification, development, evaluation, and operation. It has two key components which are defined below.

- Protection Profile (PP): A "Protection Profile (PP)" defines an implementation-independent set of security requirements and objectives. A PP is intended to be reusable and to define requirements which are known to be useful and effective in meeting the identified objectives.
- Evaluation Assurance Level (EAL): An "Evaluation Assurance Level (EAL)" defines how thoroughly the product is tested. Evaluation Assurance Levels are scaled from 1–7, with one being the lowest level of security assurance and seven being the highest level of security assurance. Consumers, Developers and Evaluators can then specify the security requirements in terms of standard security Protection Profile (PP) and independently select the Evaluation Assurance Levels (EAL) from the defined set of assurance levels. As seen in last section, FIPS 140-2 also uses the definitions of EALs in Common Criteria to evaluate specific areas of a system. The seven EALs are defined in Table 12.4.

With the above definitions, the security-aware product development cycle envisioned by CC is outlined in Fig. 12.2. In specification and development phase, CC-defined protection profiles are first used to create a standardized set of security

Table 12.4 Evaluation assurance levels

Evaluation assurance levels	Type of testing	Intent and description
EAL1	Functionally tested	EAL1 is the lowest assurance level intended to detect obvious security errors. This level of assurance is used to support the contention that personal information is protected and is not exposed. This will not detect subtle security weaknesses
EAL2	Structurally tested	EAL2 requires developers to deliver minimal amount of design information and test results. Intent is to keep additional development and test cost low on the developer side. This level of assurance is useful for a low to moderate level of security
EAL3	Methodically tested and checked	EAL3 requires developers to follow certain level of security standards during development process. Again, substantial changes to existing development practices are not intended. This level of assurance is useful if a moderate to high level of security is desired
EAL4	Methodically designed, tested and reviewed	EAL4 requires developers to apply certain security standards based on specialized commercial practices. This level is believed to be the highest assurance level that is economically feasible to retrofit to an existing product line
EAL5	Semi-formally designed and tested	Applicable when specialized security techniques is used to protect high value assets
EAL6	Semi-formally verified design and tested	Applicable when high level of specialized security techniques is used in a rigorous development to protect high value assets
EAL7	Formally verified design and tested	Applicable when highest level of specialized security techniques is used for high cost/risk scenarios

requirements which meet the needs of products. The product security specification named as "Security Targets" in the chart is used to define the security objectives, attacks under consideration, specification of the security functions, and the assurance measures. This then forms the input to the evaluation phase where the expected result is a confirmation that Security Targets meet the desired assurance levels. The results from the evaluation phase help both the product developer and end system user. Finally, when the system is in operation, it is highly possible that new vulnerabilities may surface or the deployment environment assumptions change. Reports are then made to the developer for changes back to the specification and development, and the product cycle is repeated.

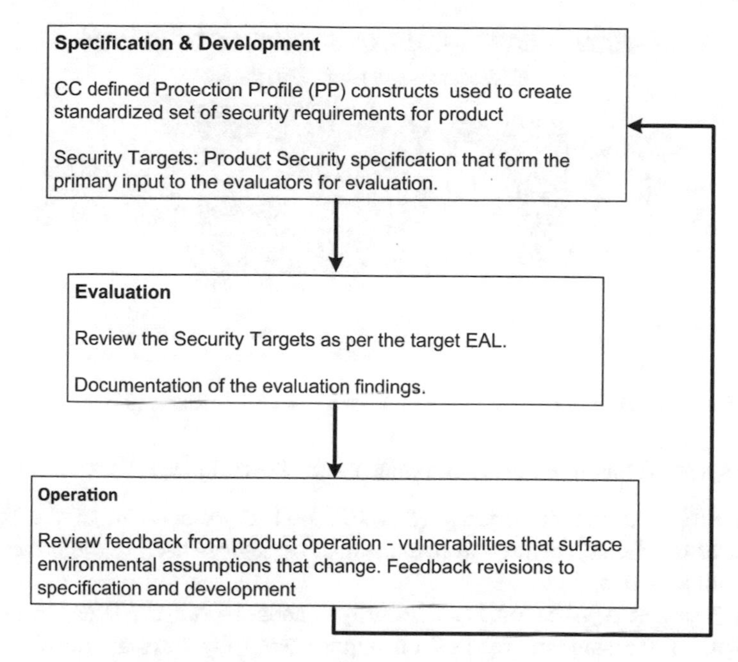

Fig. 12.2 A representative product development cycle with CC

12.4 Application/Market Specific Security Standards

In this section, we will present a brief overview of various standards as dictated by their respective markets.

12.4.1 Payment Card Industry (PCI) Standard

The Payment Card Industry (PCI) council is responsible for developing and managing the security standards for payment card transactions. The PCI security standards define the technical and operational requirements mandated by the Payment Card Industry Security Standards Council (PCI SSC) for two primary objectives (a) security of credit, debit, and cash card transactions and (b) protection of cardholders against misuse of their personal information. These standards cover end-to-end security requirements starting from the point of entry of card data into a system, to how the data is processed, through secure payment applications. Compliance with the PCI set of standards is enforced by the Council's five founding global payment brands—American Express, Discover Financial Services, JCB International, MasterCard Worldwide and Visa Inc.

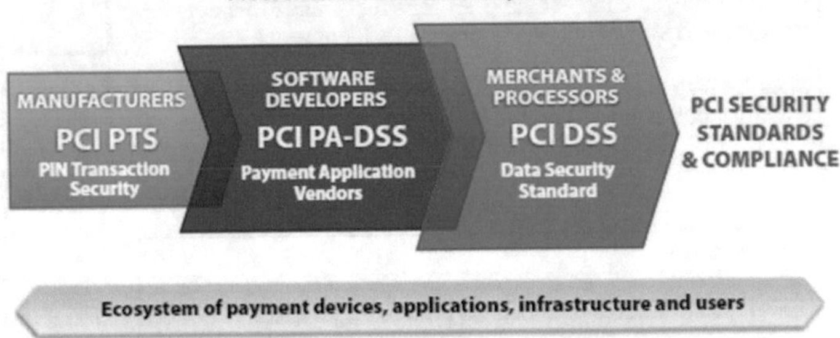

Fig. 12.3 An overview of the payment card industry security standards [26]

The standard has three main components as shown in Fig. 12.3.

- PCI PIN Transaction Security (PCI PTS)—This component of the standard focuses on characteristics and management of devices used in the protection of cardholder PINs
- PCI Payment Application Data Security Standard (PCI PA-DSS)—This component of the standard focuses on requirements for software developers and integrators of payment applications. Payment applications store, process or transmit cardholder data as part of authorization or settlement, hence form a key component of the critical data chain
- PCI Data Security Standard (PCI DSS)—This component of the standard focuses on requirements related to secure handling of card holder data

The following subsections provide a high-level summary of physical and logic security requirements for the secure hardware as mandated by the PCI standard [8]:

Physical Security Requirements

The device must have robust tamper detection, protection, and response mechanisms against various physical and side-channel attacks.

- Direct physical attacks: Examples include drilling, lasers, chemical solvents, opening covers, splitting the casing (seams), and using ventilation openings.
- Indirect physical attacks: Examples include fluctuations in environmental conditions and/or operational conditions—device may be subjected maliciously to temperatures or operating voltages outside the stated operating ranges. Care should be taken so that the attacks do not put the device in vulnerable state. Indirect attacks also include attacks that monitor information leakage through side-channels such as electromagnetic radiation, power rail noise or sound.
- Multilevel security: Tamper protection of the system should strongly consider building redundancy into the security mechanisms. This will ensure safeguarding for the scenario that failure of a single security mechanism compromises device security.

- Secure memory and key protection: Critical secure parameters or data must be stored in protected area(s) of the device and must be protected from modification. There should be no feasible way to determine any PIN-security-related cryptographic key resident in the device or make any modifications to secure information.
- Secure interfaces: Data entering or leaving the secure device must always be encrypted. One strong requirement is that there should be no feasible way to determine any entered and internally transmitted PIN digit by any means.
- Tamper Reaction/Response: The device must implement tamper response mechanisms. Upon tamper detection, one of the key responses must be to erase any sensitive data that may be stored in the device and immediately make the device inoperable.

Logical Security Requirements

The logical security requirements are predominantly focused on the software and system aspects of the device.

- Anomalous input handling: The device's functionality should not be compromised by code or data anomalies. Examples of anomalies include unexpected command or initialization sequences, invalid commands or data which could result in the device outputting the unencrypted PIN or other sensitive data.
- Firmware updates: Any firmware and software changes must be inspected and reviewed using a documented and auditable process. The firmware/software must be certified as being free from hidden and unauthorized or undocumented functions. Any firmware updates must be cryptographically authenticated and if the authentication fails then the firmware update must be rejected and deleted.
- Application Isolation: The system must ensure that it is not possible for one application to interfere with or tamper with another application or the OS of the device. Application isolation should ensure that there is separation between multiple applications. Further, the operating system must be configured securely and must contain only the necessary components for the intended operation.
- Transaction PIN protection: The device must not display the entered PIN digits. PINs are encrypted within the device immediately after PIN entry is complete and has been signified.
- Transaction memory residue protection: The device must automatically clear its internal buffers after completing a transaction or when the device has timed out waiting for the response.
- Periodic Self Testing: Secure device must perform periodic self-test, including integrity and authenticity tests to check whether the device is in a compromised state. In the event of a test failure, the device must activate the tamper protection mechanism.

12.4.2 Europay Mastercard Visa (EMV) Standard

The Europay Mastercard Visa (EMV) Standard is a security standard for payment cards and terminals with embedded smart cards (also known as chip card). Chip cards store data on an embedded secure microchip rather than magnetic stripes. The embedded microchip provides enhanced transaction security features and other application capabilities not possible with traditional magnetic stripe cards. The transaction protection benefits offered by EMV standard cover the following.

- Protection against counterfeit fraud: The standard employs stronger authentication methods and unique transaction elements such as advanced encryption to verify that the card is genuine. This is applicable to both online and offline transactions.
- Embedded card risk analysis capabilities: The standard defines the conditions under which the issuer will permit the transaction to be conducted offline and the conditions that force transactions online for authorization if offline limits have been exceeded.
- Card/data tamper protection: Offers protection against tampering of card and POS data during online transaction processing by attaching a dynamic cryptogram to each authorization and clearing transaction.
- Cardholder verification: Robust cardholder verification methods to protect against lost and stolen card fraud

Thus, the levels of protection available against chip card account data thefts and counterfeit fraud are significantly enhanced.

EMV Co manages, maintains and enhances the EMV Integrated Circuit Card Specifications for chip-based payment cards and acceptance devices, including point of sale (POS) terminals and ATMs. The EMV standard is currently owned by American Express, JCB, MasterCard, and Visa. EMV specifications ensure interoperability between chip-based payment cards and terminals. Specifications encompass both contact and contactless cards. Contact cards are cards which must be physically inserted into a card reader for transactions; Contactless cards are cards capable of transmitting data wirelessly over short distances using radio-frequency communication technology. The EMV Chip Specifications are based on existing International Organization for Standardization (ISO) standards as follows:

- ISO/IEC 7816: Identification Cards: Integrated Circuit(s) Cards
- ISO/IEC 14443: Identification Cards: Contactless Integrated Circuit(s) Cards—Proximity Cards

SoCs designs targeting chip-based payment cards and acceptance devices must meet EMV security standards as defined in the EMV Integrated Circuit Card Specifications [9]. EMV also establishes and administers testing and approval processes to evaluate compliance with the EMV Specifications.

12.4.3 Interplay of EMV and PCI Standards

EMV and PCI Standards together offer an enhanced security for payment transaction data in the following ways. EMV chip provides an additional level of authentication at the point of sale that increases the security of a payment transaction and reduces chances of fraud. Once the card is entered into the merchant's system, the cardholder's confidential information is transmitted and stored on their network in a clear (unencrypted) exposing it for variety of frauds. This is where PCI Standards come in. On top of EMV chip at the POS, they offer protections for the POS device itself and provide layers of additional security controls. It covers end-to-end security requirements starting from the point of entry of card data into a system, to how the data is processed, through secure payment applications.

12.5 Security Standard Compliance Assessment

Systems or chips that need to comply with various standards need to go through a rigorous compliance assessment with the standards consortiums or third-party labs that are approved to evaluate them.

For example, the PCI Security Standards Council is responsible for managing the security standards, while compliance with the PCI Security Standards is enforced by the payment card brands. The PCI Security Standards Council operates a number of programs to train, test, and certify organizations and individuals who can then assess and validate adherence to PCI Security Standards. PCI Security Standards Council maintains an up to date list of Qualified Security Assessors—internal and external organizations who have been qualified by the Council to assess compliance to PCI Standards.

Similarly, the EMV Standards Committee has setup a Security Evaluation Process based on a complete set of published EMV Security Standard documents (specifications, requirements, and security guidelines). These documents are made available to product providers and security evaluation laboratories for the development and security evaluation of their products. Security Evaluation Process is intended to provide organizations with valuable and practical information relating to the general security performance characteristics and the suitability of use for smart card related products, Platforms and ICs.

EMVCo uses independent security evaluation laboratories to perform security evaluations. An up to date list of these organizations is maintained on its website [9]. The EMV security guidelines support product providers in developing and testing their products, and test laboratories in performing security evaluations.

Security Evaluation Process evaluates the security features of IC, Platform, and Integrated Chip Card products. Evaluation includes:

- Integrated circuit hardware with its dedicated software, Operating System, Real Time Environment
- Firmware and software routines required to access the security functions of the IC
- Payment application software running on the platform

Once the product passes the evaluation assessment, a certificate will be released indicating the level of security compliance of the product with respect to the Security Standards requirements.

12.6 Summary

Security standards such as FIPS and Common Criteria have so far been the bulwark of general purpose systems—software and hardware—that need to offer security services. In this chapter, we first looked at the need for embedded security, fundamental cryptographic algorithms and associated standards, as well as general purpose security standards. We then also saw how market specific end-to-end security concerns have led to the evolution of application specific standards. A specific application we looked at in detail is payment transactions. We saw how security standards such as PCI and EMV are building on the general security principles and defining a custom set of security requirements for the payment ecosystem. We hope the overview provided in this chapter and the references for further reading will give a good launchpad for any embedded SoC/system developer to understand their security requirements better and define the next steps in addressing them.

References

1. HP, Understand the cost of cyber security crime. http://www8.hp.com/us/en/software-solutions/ponemon-cyber-security-report/index.html
2. Symantec, 2012 State of Mobility Survey. http://www.symantec.com/about/news/release/article.jsp?prid=20120221_02
3. How to hack and crack the connected home. http://www.bbc.com/news/technology-27373328
4. Black Hat hacker details lethal wireless attack on insulin pumps. http://www.extremetech.com/extreme/92054-black-hat-hacker-details-wireless-attack-on-insulin-pumps
5. Smartwatches have security flaws says HP. http://www.bbc.com/news/technology-33642728
6. Greenberg, A.: Hackers Remotely Kill a Jeep on the Highway—With Me in It. http://www.wired.com/2015/07/hackers-remotely-kill-jeep-highway/
7. Kocher, P., Lee, R., McGraw, G., Raghunathan, A., Ravi, S.: Security as a new dimension in embedded system design. In: Proceedings of the Design Automation Conference (DAC), 2004, pp. 753–760, July 2004

8. Payment Card Industry (PCI) PIN Transaction Security (PTS) Point of Interaction (POI), Modular Security Requirements Version 4.0, June 2013. https://www.pcisecuritystandards.org/documents/PCI_PTS_POI_SRs_v4_Final.pdf
9. EMV Requirements Security Requirements—http://www.emvco.com
10. Checkoway, S., et al.: Comprehensive experimental analyses of automotive attack surfaces. In: Proceedings of the Usenix Security Symposium 2011
11. ISO 26262-1:2011 "Road vehicles—Functional safety". www.iso.org
12. International Electrotechnical Commission (IEC)—Functional Safety and IEC 61508. http://www.iec.ch/functionalsafety/
13. Common Criteria for Information Technology Security Evaluation Part 1: Introduction and general model, Sept 2012, Version 3.1, Revision 4, CCMB-2012-09-001
14. Stallings, W.: Cryptography and Network Security: Principles and Practice, 6th edn. Pearson (2013)
15. Schneier, B.: Applied Cryptography: Protocols, Algorithms and Source Code in C. Wiley (2015)
16. Transport Layer Security (TLS), https://datatracker.ietf.org/wg/tls/documents/
17. IP Security Maintenance and Extensions (ipsecme). http://datatracker.ietf.org/wg/ipsecme/documents/
18. NIST, Cryptographic Toolkit. http://csrc.nist.gov/groups/ST/toolkit/index.html
19. Federal Information Processing Standards Publication 197 Nov 26, 2001 "Specification for the Advanced Encryption Standard (AES)"
20. Advanced Encryption Standard, https://en.wikipedia.org/wiki/Advanced_Encryption_Standard
21. Federal Information Processing Standards Publication 46-3, 1999 Oct 25 "Data Encryption Standard (DES)"
22. NIST Withdraws Outdated Data Encryption Standard. http://www.nist.gov/itl/fips/060205_des.cfm
23. NIST Special Publication 800-67, Recommendation for the Triple Data Encryption Algorithm (TDEA) Block Cipher. http://csrc.nist.gov/publications/nistpubs/800-67-Rev1/SP-800-67-Rev1.pdf (2012)
24. Federal Information Processing Standards Publication 180-4, March 2012, Secure Hash Standard (SHS). http://csrc.nist.gov/publications/fips/fips180-4/fips-180-4.pdf
25. NIST'S Policy on Hash Functions. http://csrc.nist.gov/groups/ST/hash/policy.html
26. PCI Quick Reference Guide. https://www.pcisecuritystandards.org/pdfs/pci_ssc_quick_guide.pdf
27. Federal Information Processing Standards Publication 140-2, May 25, 2001 "Standard for Security Requirements for Cryptographic Modules"
28. Safety & security architecture for automotive ICs, Yash SainiArun Jain, 25 Sept 2013

Chapter 13
SoC Security: Summary and Future Directions

Swarup Bhunia, Sandip Ray and Susmita Sur-Kolay

Secure and trustworthy operation of SoCs used in diverse applications is a critical need for modern computing systems. SoCs serve as the trust anchors for the software stack, and hence any security/trust issue at SoC level violates the fundamental concept of hardware trust anchor and can essentially prevent secure operation of the whole system. The security and trust issues in SoC are varied and span different stages of its life cycle—from design to fabrication to deployment. These issues depend on the business model of the chip manufacturers and the original equipment manufacturers (OEMs) as well as the target application space for a SoC product. However, the horizontal business model that emerged with the current trend of globalization, which favors a long globally distributed manufacturing and production process, brings large number of untrusted parties in the SoC life cycle. Hence, the threat of hardware IP piracy, reverse engineering, tampering, and counterfeiting are becoming ever more significant for SoCs. The book has presented an overview of the SoC threat space and attack vectors throughout its life cycle.

Secure high-assurance SoC design will require appropriate design and test methodologies and CAD tools. The design process starts with defining the security requirements, then developing a security architecture, implementing targeted security solutions in the architecture, and verifying the security properties before fabrication. The prevalent practice of IP-based SoC design involves an important

---

S. Bhunia (✉)
University of Florida, 216 Larsen Hall, 968 Center Dr., Gainesville, FL 32611, USA
e-mail: swarup@ece.ufl.edu

S. Ray
NXP, Austin, USA

S. Sur-Kolay
Indian Statistical Institute, Kolkata, India

S. Bhunia et al. (eds.), *Fundamentals of IP and SoC Security*,
DOI 10.1007/978-3-319-50057-7_13

new challenge. These IPs are often obtained from untrusted third-party vendors, which creates trust concerns—in particular, the IPs may potentially come with malicious implants or Trojans. Similarly, third-party design tools used in the SoC integration process may be untrusted and can cause malicious design changes. Designing trusted SoCs with untrusted components has emerged as a major challenge with modern SoC design flow. Secure SoC design methodologies need to be accompanied with appropriate post-silicon validation approaches, which aim at ensuring that the fabricated SoC works in a system in secure trustworthy fashion and is impervious to known attack vectors.

Design and validation of SoCs require addressing two broad classes of security and trust issues: (1) protection against the hardware security threats, which include all forms of side-channel attacks (power analysis, timing, EM, and fault injection), hardware IP piracy, hardware Trojan attacks, micro-probing attacks, and scan based attacks; and (2) protection of multitude of security assets on chip against malicious access by software running on the SoC. With respect to the first class of protection, it is worth noting that existing IP level solutions do not scale well at SoC level. For example, an IP level Trojan detection solution may not work at SoC level. It would almost certainly miss a Trojan instance in the interconnect fabric. Similarly, analysis of and protection against side-channel attacks for crypto IPs in isolation may not be as precise or effective when integrated into an SoC. Hence, it requires consideration of other IP blocks and interaction among them for more accurate security analysis and for developing countermeasures that provide right level of protection for an SoC at optimal overhead. Furthermore, even though IP level solutions are effective to protect individual IPs, they need to be integrated at the SoC level to enable holistic protection of the SoC. It would require a centralized infrastructure in SoC to control the protection mechanisms inside individual IPs.

Modern SoC designs contain a number of critical security assets, including keys, firmware, cryptographic modules, fuses, and private user data, which are sprinkled across multiple IPs often cross-cutting hardware and software boundaries. Consequently, developing an SoC design critically requires: (1) specifying, implementing, and validating security policies to govern the access and interaction of these assets during field operation, (2) developing on-chip security architectures to enable effective validation of SoC resiliency against security attacks and vulnerabilities; and (3) comprehensive validation of these security policies both before and after fabrication to ensure they are protected against undesired leakage or manipulation.

Several chapters of this book have been dedicated to examine the state of the research and practice in the important areas of design and validation of SoC security and to emphasize the cooperation and trade-offs between the two areas. We have also presented an overview of the industrial practices in security assurance and validation of modern SoC designs. The goal has been to provide an understanding of the current state of the practice, and to describe the different pieces of a highly complex ecosystem that must interact and cooperate to ensure security and trustworthiness of our computing devices.

Recent years have seen an explosion of deeply embedded, smart, and highly connected computing devices in diverse form factors. In particular, wearable and implant technologies, cyber physical systems (CPS), and Internet of Things (IoT) have made significant forays into nearly all aspects of our life. With advances in technology, design of new and advanced sensors, pervasive connectivity, and the trend in business towards cloud-driven data-centric solutions, the future is projected to see an even higher proliferation of systems comprising of such devices that coordinate through cloud to solve complex, distributed tasks. Commensurate with computing capability, the applications have also scaled in complexity by several factors, e.g., from smart phones to smart cities. The evolving paradigm of computing systems and their wide-spread deployment in diverse fields would impose increasingly strong demands on security and trust of SoCs used in these systems. The problem is accentuated by the features of smartness and ubiquitous connectivity of these systems, which give rise to new opportunities for attacks. Hence, SoC security designers and validators would face ever-growing challenges in meeting the security demands of future systems. It would require major research activities and innovations in architecture, design methodologies, security analysis, and pre-/post-silicon validation. One related future research direction that we believe will gain strong momentum and would complement design and validation of high-assurance SoC will be development of software, which can ensure trusted system operation in the presence of untrusted hardware—in particular, untrusted SoC.

Given the broad spectrum of potential vulnerabilities and corresponding mitigation strategies, the subject of SoC security today is highly fragmented. Different research groups focus on different aspects of the problem, often without a good understanding of the trade-offs and synergies involved in applying the different approaches on the same artifact. For example, there has been little work on integrating techniques for supply-chain security with architectural initiatives for design-level security implementation. Consequently, security research in different communities run the danger of reinventing the "wheel" already employed in another context, or creating a solution that conflicts with the fundamental requirements of another one. Hence, effective collaboration between researchers working in various fields of SoC design and validation is urgently needed.

Although we have covered a broad spectrum of activities on SoC security in this book and tried to provide a comprehensive overview of SoC security and trust challenges and solutions, we have only depicted a small but important part of the SoC design and validation process. There are more complexities involved in the overall process, including trade-offs with power management, physical design, testing, as well as complex supply-chain issues, which we have touched peripherally. The readers interested in deeper exploration are encouraged to explore into some of the references, which include challenges and surveys of specific components, and use the discussions in this book as a glue for connecting the different pieces. The editors strongly hope that the book will provide adequate background and stimulate interest of the readership in exploring more relevant knowledge in this field and in pursuing research towards innovations of critical needs.

Acknowledgements The authors sincerely acknowledge support from number of collaborators, students as well as funding sources. The work is funded in part by National Science Foundation (NSF) grants 1603475, 1563924, 1623310, 1603483, and 1662976, as well as research funding supports from Cisco, Raytheon, Draper and Semiconductor Research Corporation (SRC).

Printed in the United States
By Bookmasters